The Poincaré Half-Plane

─────────── **Jones and Bartlett Books in Mathematics** ───────────

Baum, R.J., *Philosophy and Mathematics*
ISBN 0-86720-514-2

Beltrami, E., *Mathematical Models in the Social and Biological Sciences*
ISBN 0-86720-292-0

Eves, H., *Fundamentals of Modern Elementary Geometry*
ISBN 0-86720-247-5

Jones, F., *Lebesgue Integration on Euclidean Space*
ISBN 0-86720-203-3

Loomis, L.H., and Sternberg, S., *Advanced Calculus*
ISBN 0-86720-122-2

Protter, M.H., and Protter, P.E., *Calculus, Fourth Edition*
ISBN 0-86720-093-6

Redheffer, R., *Differential Equations: Theory and Applications*
ISBN 0-86720-200-9

Redheffer, R., *Introduction to Differential Equations*
ISBN 0-86720-289-0

Ruskai, M.B., et al., *Wavelets and Their Applications*
ISBN 0-86720-225-4

Stahl, S., *The Poincaré Half-Plane: A Gateway to Modern Geometry*
ISBN 0-86720-298-X

The Poincaré Half-Plane
A Gateway to Modern Geometry

Saul Stahl

Department of Mathematics
University of Kansas
Lawrence, Kansas

Jones and Bartlett Publishers
Boston   *London*

Editorial, Sales, and Customer Service Offices

Jones and Bartlett Publishers
One Exeter Plaza
Boston, MA 02116
617-859-3900
1-800-832-0034

Jones and Bartlett Publishers International
7 Melrose Terrace
London W6 7RL
England

Copyright © 1993 by Jones and Bartlett Publishers, Inc.

All rights reserved. No part of the material protected by this copyright notice may be reproduced or utilized in any form, electronic or mechanical, including photocopying, recording, or by any information storage and retrieval system, without written permission from the copyright owner.

Library of Congress Cataloging-in-Publication Data

Stahl, Saul.
  The Poincaré half-plane; a gateway to modern geometry / Saul Stahl.
    p. cm.
  Includes bibliographical references and index.
  ISBN 0-86720-298-X
  1. Geometry, Non-Euclidean.   I. Title.
QA685.S79   1993
516.9–dc20   92-35850
                                                              CIP

The excerpts from Euclid's *The Elements* are reprinted with the kind permission of Dover Publications, Inc. Hilbert's axioms are reprinted from *The Foundations of Geometry* by David Hilbert (E. J. Townsend, translator), by permission of Open Court Publishing Company, La Salle, Illinois. The translation is copyrighted by The Open Court Publishing Co. 1902.

Printed in the United States of America
97 96 95   10 9 8 7 6 5 4 3

To Susan

Table of Contents

Preface xi

Chapter Dependencies xiii

CHAPTER 1. EUCLIDEAN GEOMETRY 1

1. A brief history of Euclidean geometry 1
2. Excerpts from Euclid's *Elements* 3
3. Hilbert's axiomatization of Euclidean geometry (optional) 22
4. Variations on Euclid's fifth 27
5. Exercises 31

CHAPTER 2. EUCLIDEAN RIGID MOTIONS 35

1. Introduction 35
2. Rigid motions 36
3. Translations, rotations, and reflections 37
4. Glide-reflections 43
5. The main theorems 46
6. Rigid motions and absolute geometry 48
7. Exercises 49

CHAPTER 3. INVERSIONS 51

1. An interesting non-rigid transformation 51
2. An application of inversions (optional) 59
3. Exercises 61

CHAPTER 4. THE HYPERBOLIC PLANE 63

1. The hyperbolic distance 63
2. Hyperbolic straight lines 66
3. Hyperbolic angles 70
4. Hyperbolic rigid motions 71
5. Riemannian geometry (optional) 73
6. Exercises 76

CHAPTER 5. EUCLIDEAN VERSUS HYPERBOLIC GEOMETRY 79

1. Euclid's postulates revisited 79
2. Absolute geometry 88
3. Hyperbolic geometry 88

4. Hyperbolic rigid motions 88
5. Exercises 90

CHAPTER 6. THE ANGLES OF THE HYPERBOLIC TRIANGLE 93

1. Introduction 93
2. The standard position of a triangle 93
3. The sum of the angles of the hyperbolic triangle 95
4. A new congruence theorem 103
5. Regular tesselations (optional) 104
6. Exercises 107

CHAPTER 7. HYPERBOLIC AREA 109

1. The general definition of area 109
2. The area of the hyperbolic triangle 114
3. Exercises 116

CHAPTER 8. THE TRIGONOMETRY OF THE HYPERBOLIC TRIANGLE 119

1. The trigonometry of hyperbolic line segments 119
2. Hyperbolic right triangles 122
3. The general hyperbolic triangle 125
4. Exercises 128

CHAPTER 9. COMPLEX NUMBERS AND RIGID MOTIONS 131

1. Complex numbers and the Euclidean rigid motions 131
2. Hyperbolic rigid motions 136
3. Euclidean flow diagrams 143
4. Hyperbolic flow diagrams — rotations 145
5. Hyperbolic flow diagrams — translations 145
6. Hyperbolic flow diagrams — the general case 151
7. Hyperbolic rigid motions — construction 155
8. Exercises 158

CHAPTER 10. ABSOLUTE GEOMETRY AND THE ANGLES OF THE TRIANGLE 161

1. The sum of the angles of the triangle 161
2. Exercises 166

CHAPTER 11. SPHERICAL TRIGONOMETRY AND ELLIPTIC GEOMETRY 167

1. Introduction 167

2. Geodesics on the sphere 168
3. Spherical trigonometry 172
4. Spherical areas 175
5. A digression into geodesy (optional) 177
6. Elliptic geometry 179
7. Exercises 180

CHAPTER 12. DIFFERENTIAL GEOMETRY AND GAUSSIAN CURVATURE 183

1. Differential geometry 183
2. A review of lengths and areas on surfaces 190
3. Gauss's formula for the curvature at a point 196
4. Riemannian geometry revisited 198
5. Exercises 205

CHAPTER 13. THE CROSS RATIO AND THE UNIT DISK MODEL 207

1. Introduction 207
2. Conformal transformations 207
3. The cross ratio 209
4. The unit disk model and its flow diagrams 213
5. Explicit rigid motions of the unit disk model 221
6. The Riemann metric of the unit disk model 225
7. Regular tesselations of the unit disk model 228
8. Exercises 230

CHAPTER 14. THE BELTRAMI–KLEIN MODEL 233

1. Introduction 233
2. The Beltrami–Klein Model 234
3. Exercises 245

CHAPTER 15. A BRIEF HISTORY OF NON-EUCLIDEAN GEOMETRY 247

1. History 247
2. Exercises 254

CHAPTER 16. SPHERES AND HOROSPHERES 257

1. Introduction 257
2. Hyperbolic space and its rigid motions 258
3. Hyperbolic geodesics 262

4. The stereographic projection of spheres 266
  5. The geometry of spheres and horospheres 271
  6. Exercises 275

APPPENDIX 277

  Proofs of some of Euclid's propositions 277

BIBLIOGRAPHY 293

INDEX 295

Preface

My high school mathematics teacher, Mr. Yossef Mashee'ach, once told us that our textbook's geometry was not the only geometry in existence. In fact, he said, there is another geometry, called non-Euclidean geometry, which is far richer than the geometry of Euclid. Since I found high school geometry so challenging and fascinating to begin with, I looked forward to learning about this esoteric geometry. Later, in college, I looked into some books whose titles probably contained the words "non-Euclidean geometry" and was greatly disappointed. The investigations of Saccheri quadrangles seemed tedious and the theorems themselves excited no interest. Compare, for example, the vagueness of the statement that the sum of the angles of the hyperbolic triangle is less than π with the sheer elegance and tautness of the statement that the sum of the angles of a Euclidean triangle is π. I just could not see what it was that motivated my teacher's comparison. Moreover, the subtleties of synthetic non-Euclidean strained my abilities, and, since the theorems seemed so drab, I lost my interest in this topic.

Subsequently, while still in college, I took a projective geometry course that included a discussion of projective metrics. The latter helped give non-Euclidean geometry some reality in my mind, but the approach was too abstract for my taste. Also, while the new theorems were different from those of Euclidean geometry, they did not seem nearly as elegant. Sometime during my undergraduate career I read W. W. Sawyer's *Prelude to Mathematics,* in which he mentions the unit disk model and the fact that its geodesics are the arcs of circles orthogonal to the bounding circle. This had just the right flavor. The concreteness and simplicity of the model made the non-uniqueness of parallelism believable. The fact that the geodesics are circular arcs gave me the same aesthetic pleasure that I had derived from Euclidean geometry and which I could not find anywhere else in my undergraduate curriculum.

I came across the same model several times during my graduate career, but, due to a variety of distractions I failed to pursue my interest in it. These distractions included the writing of a dissertation which dealt with the construction of certain maps on closed surfaces, a subject whose visual appeal was nearly as strong as that of high school geometry. Subsequently, while searching for the historical roots of my mathematical specialty, I came across the Poincaré upper half-plane and the Fuchsian groups. Only then did I come to understand and share my teacher's enthusiasm for the richness of non-Euclidean geometry. For in Poincaré's work I saw a Riemann metric with interesting geometrical properties, that gave me a new understanding of the Gauss-Bonnet Theorem. I saw a geometry wherein complex numbers, Moebius transformations, and the cross ratio played very concrete roles. I saw groups acting on a topological space and yielding familiar quotient structures. In fact,

I saw an undergraduate geometry course that would motivate many of the ideas that are central to graduate mathematics. This book constitutes my attempt to create this course.

The Poincaré upper half-plane is the hinge on which this book turns. Each of the chapters has a very direct connection to this model. Such a well focused approach to so diverse a discipline as geometry is bound to have some disadvantages. The failure to include a rigorous development of Euclidean geometry and the omission of any discussion of logic and proof is likely to deter many instructors of Modern Geometry courses from using this as their text. My decision to exclude these topics is based on the belief that students who have forgotten their high school geometry, or who never understood it to begin with, will benefit more from a course that builds on that topic than they will from a course that explicitly reviews it. Many exercises were included in this text that will force such students to dust off their high school geometry texts and to reexamine them in a new light. I have also offered a compromise in the form of a first chapter that provides the framework for a review of Euclidean geometry, and an appendix that supplies some of the missing details. As for explicit instruction on the nature of proof, I feel that this time would be better spent elsewhere. It is doubtful that the students who have gone through the high school mathematics curriculum without understanding proofs will benefit from yet another pass over this ephemeral topic. Certainly they could learn much interesting mathematics in this book. It is my hope that this concrete and calculational approach to geometry will provide a foundation on which to base the abstractions they will encounter later. I do not mean to say that I have ignored proofs here, but I have tried to deemphasize them somewhat.

Chapters 1–10 contain more than enough material for a one semester junior level course, especially if the class takes time to work on the several series of exercises that compare and contrast Euclidean geometry with its alternates. The exposition of these chapters is fairly leisurely. High school algebra, geometry, and trigonometry are repeatedly used in the development, as are the topics of integration, polar coordinates, and the multivariate chain rule.

Chapters 11–16 call for more sophistication and background on the part of the reader. The pace is somewhat faster and familiarity with three dimensional vector geometry is assumed. Nevertheless, the required background material is contained in any of the tomes used as texts in the standard preengineering calculus courses.

It gives me pleasure to express my thanks to Frank Kujawski and other students for asking me the right questions: to Marshall Cohen, David Lerner, and Albert Sheu for teaching from early drafts of this book; to Stanley Lombardo and Tom Tuozzo for help with classical matters; to Ray Carry, William Goldman, David Henderson, Gerald Meike, Philip Yasskin, as well as other anonymous reviewers for their encouragement and/or constructive criticisms; and to Carl Hesler and all the helpful people at Jones and Bartlett Publishers, Inc.

Chapter Dependencies

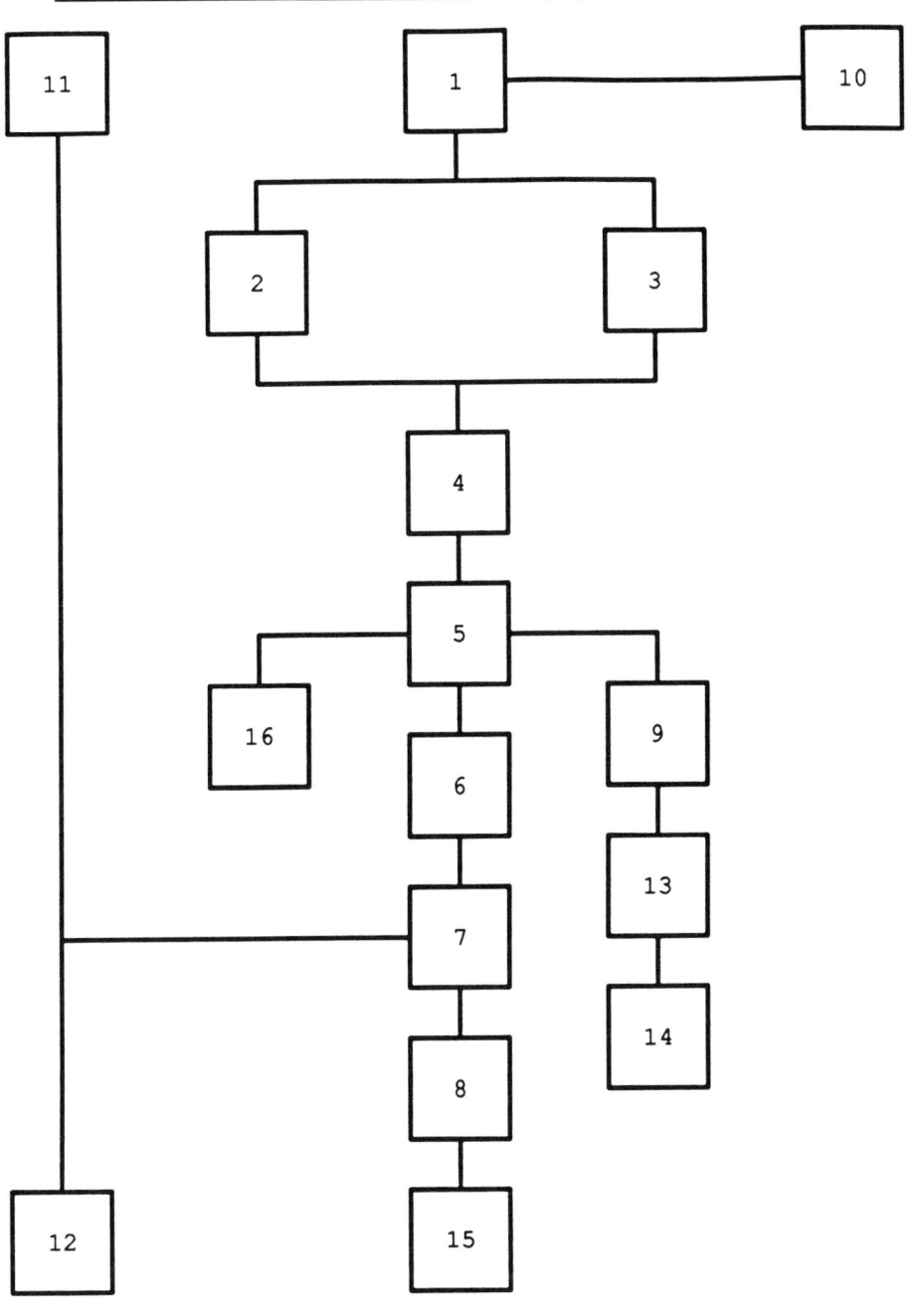

xiii

1
Euclidean Geometry

1.1 A BRIEF HISTORY OF EUCLIDEAN GEOMETRY

The subject matter of this book is non-Euclidean geometry. To define such an alternate geometry one needs to know what Euclidean geometry is. Unfortunately, the definition of this well known geometry is a more difficult task than might at first be suspected. In fact, it took mathematicians thousands of years to produce the various currently accepted descriptions, which are all based on the relatively recent work of David Hilbert (1862–1943). Needless to say, future generations may deem this century's versions to be lacking and opt for yet other characterizations. In view of these considerations the readers may not be surprised to learn that the author has chosen to evade the issue by simply providing them with a short history of the subject and an incomplete description of its contents.

Geometry in the sense of mensuration of figures was spontaneously developed by many cultures and dates to several millenia B.C. The science of geometry as we know it, namely, a collection of abstract statements regarding ideals figures, the verification of whose validity requires only pure reason, was created by the Greeks. Historians agree that the origin of geometry can be further narrowed down to the person of Thales of Miletos. Neither the exact dates of his birth and death, nor the precise nature of his accomplishments have been ascertained. However, since he is reputed to have predicted the eclipse of 585 B.C., it is safe to assume that he lived during the sixth century B.C. Geniuses like Thales were probably born and forgotten in many other countries and centuries. It was fortunate for him that he was born into a culture that found his contributions useful and cared enough to preserve them for the future by incoporating them into its educational system. Since the Greece of that day was evolving both the democratic form of government and the idea of a trial by a jury of peers, it was imperative to every citizen that he

2 THE POINCARÉ HALF-PLANE

should be able to argue both cogently and persuasively. During the fifth and sixth centuries B.C. the task of educating their fellow Hellenes in the skills required by this new political system was assumed by the sophists. These were itinerant pedagogues who made their fortunes by training their pupils mostly in the subject of rhetoric. However, since they found both the abstraction and the logical dynamics of geometry to be useful tools in this training, Thales's creation found a niche in their curriculum and was preserved for posterity.

It is very likely that Democritus (ca. 410 B.C.), who is better known for being the first person to promulgate an atomic theory of nature, already knew more geometry than is taught in today's high schools. It is more than likely, however, that his demonstrations were not still not fully rigorous.

Eudoxus (408–355?B.C.) was one of the first to formally organize the theorems of geometry into a structure that begins with axioms and goes on to derive theorems in a systematic manner. His book has not survived and our information about his accomplishments is second hand. There is no doubt, however, but that he was a first rate mathematician with an excellent understanding of the subtleties of his subject matter. On some points, his mathematical successors did not catch up with him until the late nineteenth century.

Euclid wrote his famous book *The Elements* circa 300 B.C. This book codified some of the state of the art of geometry at the time, although much was omitted too. It actually consists of thirteen books whose contents are

Book I–IV The Geometry of Triangles and Circles

Book V–VI Theory of Geometric Proportions

Book VII–IX Theory of Numbers

Book X Theory of Irrational Numbers

Book XI–XIII Solid Geometry.

Over the centuries, Euclid's opus was translated into numerous languages and it essentially defined the content of geometry for many cultures. In the sequel, we shall adopt the same convention. Namely, by Euclidean geometry we shall mean geometry as it is developed in the books of Euclid. This has the advantage of providing us with a well defined and well known body of knowledge to which we can refer when the need arises. Moreover, it will now be possible for us to define the term *non-Euclidean* by simply negating one of Euclid's axioms. The disadvantage to identifying Euclidean geometry with Euclid's geometry is that this is incorrect. Euclid's axioms are in fact insufficient, and his proofs are less than perfect. Much work was done during the nineteenth century to attain the perfection that Euclid sought. These efforts were culminated by Hilbert in his book *The Foundations of Geometry* wherein he put forth a complete set of axioms and showed how the body of knowledge commonly known as *Euclidean geometry* could be derived from them. This work attracted a fair amount of attention during the first part of this century, and many other mathematicians formulated alternate axiomatizations of Euclidean geometry. All these systems share the pedagogical defects that they contain numerous axioms and the derivation of even "obvious" facts from these axioms is onerous. Consequently the interest in providing a foolproof logical framework for Euclidean geometry has somewhat waned. (E. Moise's excellent book *Elementary Geometry from an Advanced Standpoint* is a notable exception to this broad generalization.) It is because of these defects that the author decided to return to Euclid for a description of Euclidean geometry, despite his flaws. The next section of this chapter contains a complete list of Euclid's axioms, to which he refers as definitions, postulates, and common notions, Propositions 1–4 of Book I together with their proof, and the statements of most of the other propositions of the same book. Several propositions from later books of Euclid are also included because of their significance to the subsequent chapters of this book. The reader is referred to Heath's book for a complete translation of, and extensive commentary on, Euclid's Elements. The optional Section 3 contains a brief description of Hilbert's axiomatization of Euclidean geometry; it can be omitted without causing the reader any difficulties in the sequel. The chapter concludes with a discussion of Euclid's Parallel Postulate and several of its equivalent formulations.

1.2 EXCERPTS FROM EUCLID'S *ELEMENTS*

FROM BOOK I

DEFINITIONS

1. A point is that which has no part.

Since one cannot define something by simply listing all the properties it does *not* have, this cannot be taken as a genuine definition. Euclid must have understood that every definition must rely on previously defined terms and so it is logically impossible to define all of one's terms. Instead, this should be viewed as an attempt on Euclid's part to tell his readers that his point is something like a dot one marks with a pencil, but at the same time it is an idea rather than a physical entity. Definitions 2 and 5 below should be understood in the same way.

2. A line is breadthless length.

Euclid's *line* is our *curve*.

3. The extremities of a line are points.

Since both points and lines have already been defined, and the word *extremity* is subsequently used to denote the boundary of any figure whatsoever, it is hard to see exactly what is being defined here. It is possible that Euclid felt that having defined both points and lines he now needed to clarify the relationship between them.

It is implicit in this definition that every line has endpoints. Consequently, Euclid's line is in fact a finite arc, and his straight lines, to be defined next, are in fact line segments. Much has been made of the Greeks' insistence on the finiteness of the objects of their investigations. Nevertheless, Euclid is quite pragmatic about this issue. When the need arises, as it does in the statement of Proposition 12 below, Euclid is not above bending his own rules and mentioning infinite straight lines.

4. A *straight line* is a line which lies evenly with the points on itself.

This too is a somewhat obscure sentence. It is possible that the points to which this definition refers are the extremities of the previous one. Thus, this definition should be understood as saying that of all the (curved) lines joining a pair of points the straight line is the one which consists of all the points that lie directly in between the two given points.

5. A surface is that which has length and breadth only.

6. The extremities of a surface are lines.

It is implicit in this definition that every surface is necessarily of finite extent.

7. A *plane surface* is a surface which lies evenly with the straight lines on itself.

Possibly this means that the extremities of a plane surface are straight lines and that the surface lies evenly with them.

8. A *plane angle* is the inclination to one another of two lines in a plane which meet one another and do not lie in a straight line.

9. And when the lines containing the angle are straight, the angle is called *rectilineal*.

It is clear from the above two definitions that Euclid has some interest in curvilinear angles, i.e., angles formed by curves rather than straight lines. Nevertheless, he subsequently refers to such angles only once, in Proposition 16 of Book III, where he says that the angle between the tangent to a circle and its circumference is smaller than any rectilineal angle. It has been posited that geometricians of the time were dallying with such angles without coming to any serious conclusions, and that Euclid felt the need to acknowledge their efforts.

10. When a straight line set up on a straight line makes the adjacent angles equal to one another, each of the equal angles is *right*, and the straight line standing on the other is called a *perpendicular* to that on which it stands.

This innocuous looking definition suffers from two difficulties. It is not at all clear what equality of angles means to Euclid. Does he have equality in measure in mind here, or is he referring to the stronger notion of congruence? Moreover, the issue of the existence of such right angles, whatever Euclid may understand by equality is neatly sidestepped. The latter issue is settled by Proposition 11. As for the issue of equality, the author believes that Euclid is referring to equality of measure rather than congruence. A more detailed discussion of the use of equality in *The Elements* is given in the paragraphs dealing with the Common Notions below.

11. An *obtuse* angle is an angle greater than a right angle.

12. An *acute* angle is an angle less than a right angle.

13. A *boundary* is that which is an extremity of anything.

14. A *figure* is that which is contained by any boundary or boundaries.

15. A *circle* is a plane figure contained by one line such that all the straight lines falling upon it from one point among those lying within the figure are equal to one another.

16. And the point is called the *centre* of the circle.

17. A *diameter* of the circle is any straight line drawn through the centre and terminated in both directions by the circumference of the circle, and such a straight line also bisects the circle.

18. A semicircle is the figure contained by the diameter and the circumference cut off by it. And the centre of the semicircle is the same as that of the circle.

19. *Rectilineal figures* are those which are contained by straight lines, *trilateral* figures being those contained by three, *quadrilateral* those

contained by four, and *multilateral* those contained by more than four straight lines.

20. Of trilateral figures, an *equilateral triangle* is that which has its three sides equal, an *isosceles triangle* that which has two of its sides alone equal, and a *scalene triangle* that which has its three sides unequal.

21. Further, of trilateral figures, a *right-angled triangle* is that which has a right angle, an *obtuse-angled triangle* that which has an obtuse angle, and an *acute-angled triangle* that which has its three angles acute.

22. Of quadrilateral figures, a *square* is that which is both equilateral and right-angled; an *oblong* that which is right-angled but not equilateral; a *rhombus* that which is equilateral but not right-angled; and a *rhomboid* that which has its opposite sides and angles equal to one another but is neither equilateral nor right angled. And let quadrilaterals other than these be called *trapezia*.

23. *Parallel* straight lines are straight lines which, being in the same plane and being produced indefinitely in both directions, do not meet one another in either direction.

POSTULATES

Euclid's book is an attempt to systematize the many geometrical theorems produced by his predecessors. By a geometrical theorem we understand here a valid statement regarding the interplay of points, straight lines, angles, and equality. Each such theorem is backed by an argument that deduces its validity from some definitions, and possibly other theorems. In order to eliminate the element of cyclical reasoning from such a system, it is necessary to order the theorems in such a manner that each makes use of only previously listed theorems (and definitions, of course). It must have been the cause of much frustration to the early Greek mathematicians to discover that such an ordering turns out to be impossible if one wishes to include such interesting results as the Theorem of Pythagoras and the fact that the sum of the angles of every triangle is two right angles. It turned out that some theorems must always be accepted on faith, without justification. These are called *postulates*. Just which theorems should be listed as postulates is a question that must be resolved on subjective grounds. It is very likely that Euclid's choice was the culmination of hundreds of years of lively discussion amongst Greek mathematicians. Later generations modified his choices, and one well known system is dicussed in some detail in the next section.

It is important to note that postulates are identical with theorems as far as the nature of their respective contents are concerned. It is only in their justification that they differ. The former are accepted without justification, whereas each of the latter must be accompanied by a proof that relies on previous theorems, postulates, or definitions.

This view of *The Elements* as a well grounded and logically consistent ordering of theorems is to be understood as an ideal. It is a well acknowledged fact there are many flaws in Euclid's organization of geometry. He made repeated use of both undefined terms and unstated postulates. Some of these errors will be pointed out in the sequel. Nevertheless, because of its vision and because of its logical strength, Euclid's opus is justly regarded as one of the supreme achievements of Greek Civilization in particular and of the human mind in general. We now go on to list Euclid's choice of postulates.

Let the following be postulated:

1. *To draw a straight line from any point to any point.*

As this postulate stands it simply says that every pair of distinct points can be joined by a straight line. However, in view of some of the arguments given in the proof of Proposition 4, it would seem that Euclid understood this statement to include the additional assumption that any two points can be joined by at most *one* straight line.

If the given points are A and B, then Euclid denotes the straight line joining them by AB, a notation that runs counter to modern conventions which denote this object by $\overline{AB}$. In Proposition 12 the same symbol AB is also used to denote the infinite straight line joining A and B. Euclid's notation (and ambiguities) will be adopted even in the subsequent chapters in the belief that the context will invariably clarify what is intended. This has the major pedagogic advantage of simplicity and the disadvantage of logical inaccuracy; the author believes that the former outweighs the latter.

2. *To produce a finite straight line continuously in a straight line.*

This is a clever way of stating the assumption that the plane extends infinitely far in all directions. Whenever possible Greek mathematicians avoided the explicit mention of infinity, probably because such formulations led to too many logical complications.

3. *To describe a circle with any centre and distance.*

The statement of Proposition 2 below makes it clear that this postulate is to be interpreted in a very narrow sense. Namely, given a point A and a line segment AB there exists a circle with centre A and radius AB. This assumption is sometimes rephrased by modern mathematicians by saying that their Greek counterparts used *collapsible* compasses whose legs lost their angle whenever the compass was lifted off the paper.

4. *That all right angles are equal to one another.*

The right angle is Euclid's unit for measuring all rectilineal angles and so he needs to know that all right angles are indeed equal. Since he has failed to provide for the rigid motions that would have made it possible for him to prove the congruence (and hence also equality) of all right angles, some such postulate is necessary. For a further discussion of equality the reader is referred to the paragraph introducing the Common Notions below.

5. *That, if a straight line falling on two straight lines make the interior angles on the same side less than two right angles [in sum], the two straight lines, if produced indefinitely, meet on that side on which are the angles less than the two right angles.*

Many of Euclid's readers and successors felt that this postulate was unnecessary and that its validity could be demonstrated on the basis of the others. However, their repeated attempts, over two millennia, to substantiate this feeling invariably failed. A brief history of these attempts is provided in Chapter 15. One may very well consider the rest of this book as an explanation of their lack of success. For that reason this postulate will not be further discussed here except to note that some of its better known variants appear in the last section of this chapter.

COMMON NOTIONS

The common notions listed below are also postulates, but they differ from the above postulates in that they are mainly concerned with the concepts of equality and inequality. Unfortunately, there is a fair amount of uncertainty as to what Euclid understood by these terms. In view of his Proposition 35 of Book I (parallelograms which are on the same base and in the same parallels are equal to one another) it is clear that at least part of the time Euclid was referring to equality of area. We propose here that Euclid was actually consistent in his employment of this term and always used it to denote *equality in size (measure)*. In other words, when Euclid says that two parallelograms are equal he means that they have the same area, when he says that two line segments are equal he means that they have the same length, and when he says that two angles are equal he means something of the same nature. Thus, we suggest that Euclid has an underlying and unstated assumption that all geometric objects have an aspect of numerical size whose properties are set forth in the Common Notions below, so named because they describe the properties that are shared by length, area, volume, and angular measure. This explanation accounts for Euclid's failure to provide any other definition of the notions of area and volume notwithstanding his many propositions about these very concepts. The Common Notions, we contend, constitute Euclid's simultaneous definition of length, area, volume, and angular measure. The reader who is familiar with modern measure theory will notice that if this interpretation of the Common Notions is valid then there is a very striking resemblance between them and definition of a Haar measure. An informal discussion of this measure is to be found in the first section of Chapter 7.

This interpretation of Euclid's Common Notions is hereby adopted as a notational device throughout the remainder of this book. The symbol "=" is used to denote equality in size or measure. Thus

$$\angle ABC = \angle DEF$$

means that the two said angles have the same measure. Of course, they are also congruent, but that is usually irrelevant to the argument. Similary,

$$AB = CD$$

means that the straight line segments joining A to B and C to D have equal lengths.

Let us now examine Euclid's definition of equality.

1. *Things which are equal to the same thing are also equal to one another.*

2. *If equals be added to equals, the wholes are equal.*

3. *If equals be subtracted from equals, the remainders are equal.*

The modern reader may be puzzled by the need for the third postulate, since it seems to be already subsumed by the previous one. The Greeks, however, did not recognize the existence of negative numbers, and so Euclid found it necessary to include both Common Notions 2 and 3 in his list.

4. *Things which coincide with one another are equal to one another.*

In view of the proof of Proposition 4 this should be understood as saying that things which can be made to coincide with one another are equal to one another in measure. In other words, congruent figures have equal areas or lengths, as the case may be. This can be viewed as Euclid's first mention of congruence in the sense of a rigid transformation. Euclid's ambivalence about the use of such transformations is made obvious by the contradictory attitudes he displays in the proofs of Propositions 2 and 4. In the first of these he avoids simply moving a line segment from one location to another at the cost of producing an elaborate proof for an intuitively obvious fact. On the other hand, the proof of Proposition 4 starts with an application of one triangle to another. In other words, one triangle is lifted and placed on top of the the other – a clear relinquishment of standards on the part of Euclid. It is generally conceded that Euclid's treatment of transformation and congruence, or rather the lack thereof, constitutes one of the more serious flaws in *The Elements*.

5. *The whole is greater than the part.*

Observe that if we assumed that all geometrical figures have size 0 then the first four common notions would still hold relative to this trivial notion of content. This last common notion excludes the possibility of such a degenerate notion of size since it clearly implies that some objects have non-zero size. It also turns out to be very convenient in many reductio ad absurdum arguments.

PROPOSITIONS

Proposition 1. *On a given finite straight line to construct an equilateral triangle.*

Let AB be the given finite straight line. Thus it is required to construct an equilateral triangle on the straight line AB (Fig. 1.1).

With centre A *and distance* AB let the circle BCD be described; again, with centre B and distance BA let the circle ACE be described; and from the point C in which the circles cut one another, to the points A, B let the straight lines CA, CB be joined.

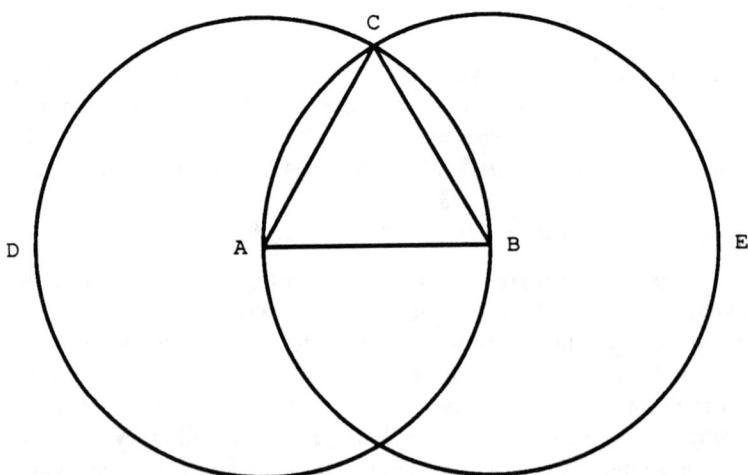

FIGURE 1.1

Now since the point A is the center of the circle CDB, AC is equal to AB.
Again, since the point B is the centre of the circle CAE, BC is equal to BA.
But CA was also proved equal to AB; therefore each of the straight lines CA, CB is equal to AB.

And things which are equal to the same thing are also equal to one another; therefore CA is also equal to CB. Therefore the three straight lines CA, AB, BC are equal to one another. Therefore the triangle ABC is equilateral; and it has been constructed on the given straight line AB.

(Being) what it was required to do.

This proposition's proof demonstrates both some of Euclid's strengths and some of his weaknesses. On the positive side we find him a careful reasoner and expositor who is unwilling to accept as obvious the existence of a triangle which most people take for granted. Unfortunately, he is not careful enough. Specifically, he implicitly accepts that the two auxiliary circles drawn in this proof necessarily intersect. Now, as physical objects, these figures clearly must intersect, but as abstract entities, whose properties must be reducible to Euclid's definitions, postulates, and common notions, this claim calls for verification. This is not a minor point. The fact is that Euclid failed to provide a framework within which the interiors and exteriors of configurations can be discussed and this is one of the major defects of his logical edifice.

Proposition 2. *To place at a given point (as an extremity) a straight line equal to a given straight line.*

Let A be the given point and BC the given straight line (Fig. 1.2). Thus it is required to place at the point A (as an extremity) a straight line equal to the given straight line BC.

From the point A to the point B let the straight line AB be joined; and on it let the equilateral triangle DAB be constructed. Let the straight lines AE, BF be produced in a straight line with DA, DB; with centre B and distance BC let the circle CGH be described; and again, with centre D and distance DG let the circle GKL be described.

Then since the point B is the centre of the circle CGH, BC is equal to BG.
Again, since the point D is the centre of the circle GKL, DL is equal to DG.
And in these DA is equal to DB; therefore the remainder AL is equal to the remainder BG. But BC was also proved equal to BG; therefore each of the straight lines AL, BC is equal to BG. And things which are equal to the same thing are also equal to one another; therefore AL is also equal BC.

Therefore at the given point A the straight line AL is placed equal to the given straight line BC.

(Being) what it was required to do.

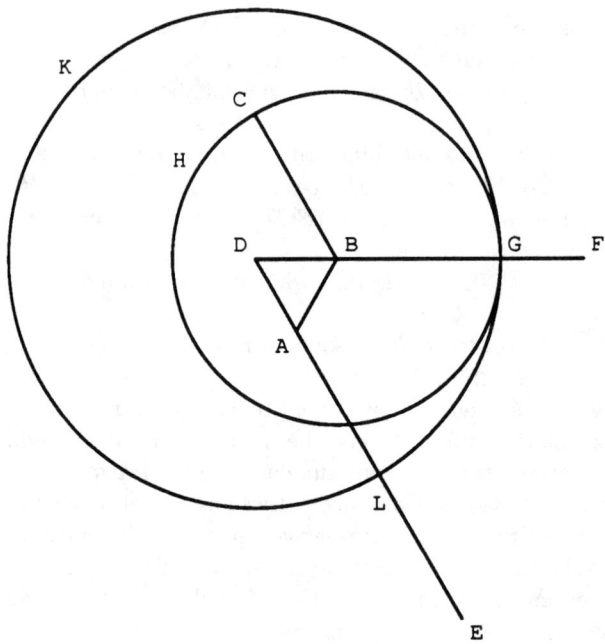

FIGURE 1.2

At first glance it might seem that Proposition 2 belabors the obvious. In this author's humble opinion, though, this proposition and its proof are nothing short of marvellous. As was mentioned above, this proposition indicates that Euclid had collapsible compasses in mind when he wrote his books. The subtlety of this notion still strains the understanding of many of today's students. Moreover, Euclid's willingness to handicap himself by assuming as little as he possibly could get away with demonstrates a faith in the power of human reasoning that was lost with the collapse of the Greek civilization and did not resurface until the Renaissance. Euclid could have assumed a rigid compass in Postulate 4, an assumption that would have taken not many more words to state and would have obviated the need for the nontrivial proof Proposition 2. The fact that he chose not to do this indicates that he enjoyed flexing his mental muscles just for the joy of using them.

Proposition 3. *Given two unequal straight lines, to cut off from the greater a straight line equal to the less.*

Let AB, C be the two given unequal straight lines, and let AB be the greater of them (Fig. 1.3).

Thus it is required to cut off from AB the greater a straight line equal to C the less.

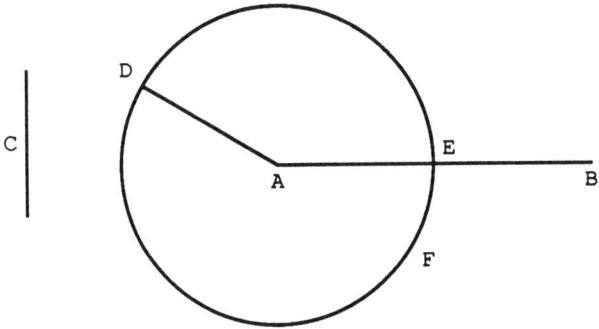

FIGURE 1.3

At the point A let AD be placed equal to the straight line C; and with the center A and distance AD let the circle DEF be described.

Now since the point A is the centre of the circle DEF, AE is equal to AD. But C is also equal to AD.

Therefore each of the straight lines AE, C is equal to AD; so that AE is also equal to C.

Therefore, given the two straight lines AB, C from AB the greater AE has been cut off equal to C the less.

(Being) what it was required to do.

Proposition 4. *If two triangles have the two sides equal to two sides respectively, and have the angles contained by the equal straight lines equal, they will also have the base equal to the base, the triangle will be equal to the triangle, and the remaining angles will be equal to the remaining angles respectively, namely those which the equal sides subtend.*

Let ABC, DEF be two triangles having the two sides AB, AC equal to the two sides DE, DF respectively, namely, AB to DE and AC to DF, and the angle BAC equal to the angle EDF (Fig. 1.4).

I say that the base BC is also equal to the base EF, the triangle ABC will also be equal to the triangle DEF, and the remaining angles will be equal to the remaining angles respectively, namely those which equal sides subtend, that is, the angle ABC to the angle DEF, and the angle ACB to the angle DFE.

For if the triangle ABC be applied to the triangle DEF, and if the point A be placed on the point D and the straight line AB on DE, then the point B will also coincide with E, because AB is equal to DE.

Again, AB coinciding with DE, the straight line AC will also coincide with DF, because the angle BAC is equal to the angle EDF.

But B also coincided with E; hence the base BC will coincide with the base EF. [For if, when B coincides with E and C with F, the base BC does not

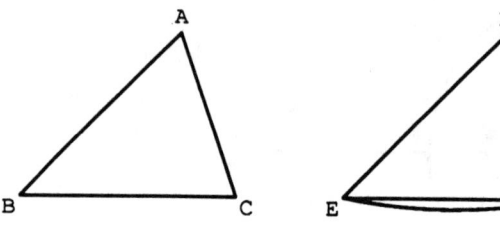

FIGURE 1.4

coincide with the base *EF*, two straight lines will enclose a space: which is impossible. Therefore the base *BC* will coincide with *EF*] and will be equal to it.

Thus the whole triangle *ABC* will coincide with the whole triangle *DEF* and will be equal to it. And the remaining angles will also coincide with the remaining angles and will be equal to them, the angle *ABC* to the angle *DEF* and the angle *ACB* to the angle *DFE*.

Therefore etc.

(Being) what it was required to prove.

The reader will recognize Proposition 4 as the one that is more commonly referred to as the side-angle-side (SAS) congruence theorem. As was pointed out earlier, its proof suffers from the flaw that it makes use of a notion of *application* for which no definitions or axioms are to be found. The statement of the proposition is also awkward, a flaw that recurs in some later propositions as well.

This concludes the detailed discussion of the first few pages of Euclid's *Elements*. The author hopes that despite the exposed flaws in the logic, despite the tediousness of some of the arguments, and despite the verbosity of some of the statements of the propositions, the reader will agree that Euclid is well deserving of the acclaim he has received over the centuries, both for what he accomplished and for what he conceived of, though failed to carry out. We now go on to state Propositions 5–28 of Book I as well as a few others. In some cases Euclid's proof is included; the proofs of the others can be found in the appendix. This serves the dual purpose of better acquainting the reader with the scope of Euclid's work and also of providing us with necessary references.

Propositions 5–7, while of some interest in their own right, are essentially lemmas for Proposition 8, which is the well known SSS congruence theorem.

Proposition 5.   *In isosceles triangles the angles at the base are equal to one another; and, if the equal straight lines be produced further, the angles under the base will be equal to one another.*

Proposition 6. *If in a triangle two angles be equal to one another, the sides which subtend the equal angles will also be equal to one another.*

Proposition 7. *Given two straight lines constructed on a straight line (from its extremities) and meeting in a point, there cannot be constructed on the same straight line (from its extremities) and on the same side of it, two other straight lines meeting in another point and equal to the former two respectively, namely, each to that which has the same extremity.*

Proposition 8. *If two triangles have two sides equal to two sides respectively, and have also the base equal to the base, they will also have the angles equal which are contained by the equal straight lines.*

Propositions 9–12 assert the feasibility, within Euclid's narrow framework, of some standard constructions: bisection of angles and line segments, and construction of lines perpendicular to a given straight line. It is interesting to note that the statement of Proposition 12 includes a reference to an infinite straight line, despite the clear implication of the definitions that every straight line is necessarily bounded. Euclid was not above bending his own rules

Proposition 9. *To bisect a given rectilinial angle.*

Proposition 10. *To bisect a given finite straight line.*

Proposition 11. *To draw a straight line at right angles to a given straight line from a given point on it.*

Proposition 12. *To a given infinite straight line, from a given point which is not on it, to draw a perpendicular straight line.*

When examining Proposition 13, the readers should bear in mind that according to Definition 8 the sides of any angle cannot lie on a single straight line. In more modern terminology, Euclid's system does not allow for angles whose measure is either $0°$ or $180°$. One must remember that zero was not recognized as a bona fide number until more than a thousand years later. However, it was argued above that Euclid implicitly assumed that all angles (as well as segments, figures, and solids) had some numerical size. So what angular size could he possibly assign to a pair of coincident line segments? None, in fact, and so he must perforce exclude the *zero angle*. Once this angle is excluded, Proposition 13 dictates that the *straight angle* (which measure $180°$) must also be excluded. The reader is warned against attaching too much importance to these niceties. They are issues of style rather than of substance.

The purpose of this proposition is to facilitate the comparison of the sum of two angles of a triangle to two right angles below. Proposition 14 is its predecessor's converse. Proposition 15 is both elementary and fundamental. As we shall see, it serves as a crucial lemma for the Proposition 16.

Proposition 13. *If a straight line set up on a straight line make angles, it will make either two right angles or angles [whose sum is] equal to two right angles.*

Proposition 14. *If with any straight line, and at a point on it, two straight lines not lying on the same side make the adjacent angles to two right angles, the two straight lines will be in a straight line with one another.*

Proposition 15. *If two straight lines cut one another, they make the vertical angles equal to one another.*

The following proposition is overshadowed in many students' recollection by the stronger one which asserts that any exterior angle of a triangle is actually equal to the the sum of the two interior and opposite angles. The proof of this latter version relies on either Euclid's Postulate 5 or on some logically equivalent assumption such as Playfair's Postulate or the fact that the sum of the interior angles of the triangle equals two right angles (see Section 4 of this chapter). Euclid, however, takes great pains to avoid using this postulate for as long as possible, and so he states this partial result here. Partial as it is, it does provide the foundations necessary to prove later (Proposition 27) that parallel lines do indeed exist. For this reason, and in order to refresh the students' memory, Euclid's proof is summarized below.

Proposition 16. *In any triangle, if one of the sides be produced, the exterior angle is greater than either of the interior and opposite angles.*

PROOF: Let ABC be a triangle, and let one side of it BC be produced to D (Fig. 1.5); it will be shown that the exterior $\angle DCA$ is greater than the interior and opposite $\angle BAC$ and $\angle ABC$.

Let AC be bisected at E, and let BE be joined and produced in a straight line to F; let EF be made equal to BE and let FC be joined.

Then $\triangle AEB \cong \triangle CEF$ since $AE = CE$, and $BE = FE$, and $\angle AEB = \angle CEF$.

Consequently $\angle BAC = \angle FCA < \angle DCA$.

A similar argument using a bisection of the side BC results in the conclusion that $\angle DCA > \angle ABC$. q.e.d.

The sum of the angles of a triangle is the theme that unites most of the chapters of this book, and in a more generalized form is one of the central issues of modern geometry. The following proposition constitutes the first variation on this theme.

EUCLIDEAN GEOMETRY 17

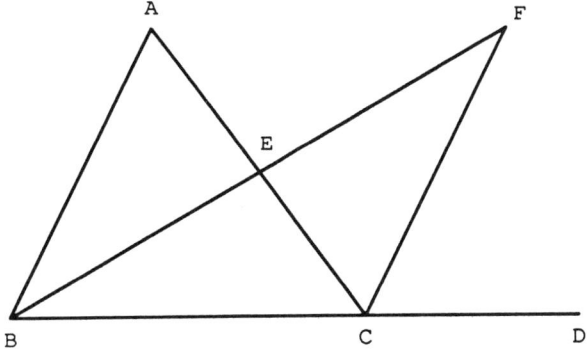

FIGURE 1.5

Proposition 17. *In any triangle two angles taken together in any manner are less than two right angles.*

PROOF: Let ABC be a triangle (Fig. 1.6); it will be shown that $\angle CBA + \angle ACB$ is less than two right angles.

Let BC be produced to D. Then since $\angle DCA$ is an exterior angle of $\triangle ABC$, it is greater than the interior and opposite $\angle CBA$. Consequently $\angle CBA + \angle ACB < \angle DCA + \angle ACB =$ two right angles.   q.e.d.

Most people, when asked to characterize a straight line, will respond by saying that it is the shortest distance between two points. This is quite reasonable and we shall later make use of the same underlying principle when defining straight lines in other geometries. Euclid, however, has a different

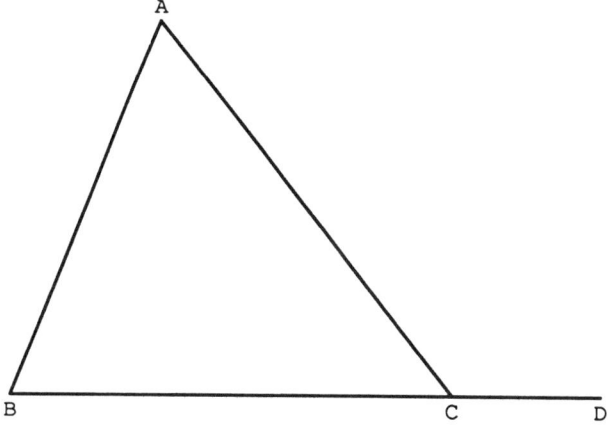

FIGURE 1.6

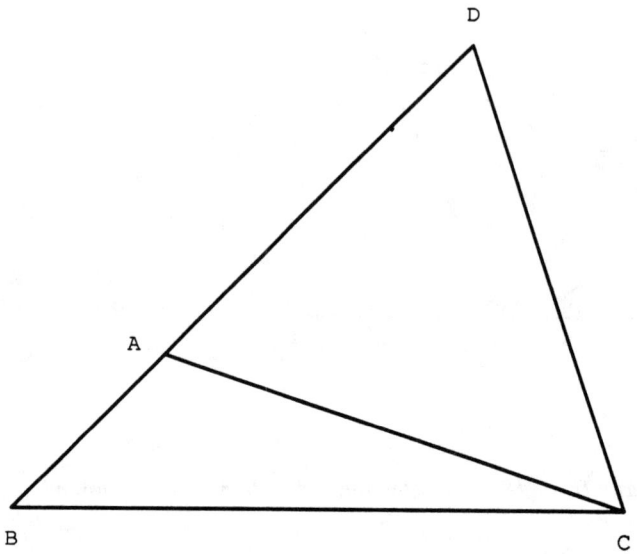

FIGURE 1.7

definition for straight lines and so it behoves him to prove that his straight line does in fact possess the shortest distance property dictated by common sense. That, in essence, is the purpose of the next group of propositions, culminating in Proposition 20.

Proposition 18. *In any triangle the greater side subtends the greater angle.*

Proposition 19. *In any triangle the greater angle is subtended by the greater side.*

Proposition 20. *In any triangle two sides taken together in any manner are greater than the remaining one.*

The reader will recall that the proof of Proposition 4 (SAS) entailed an application of one triangle onto another, a move of questionable validity within Euclid's framework. Euclid now sets out to prove two more congruence theorems, ASA and SAA in Proposition 26. This time he succeeds in avoiding the use of applications. This success, however, comes at the cost of a very complicated proof, necessitating the buildup contained in Propositions 21–25. The second of these preliminary propositions essentially obviates the need for any subsequent applications. It is very tempting to speculate that Euclid's capitulation and employment of applications in the proof of Proposition 4 came only after many vain and strenuous efforts to produce an untainted proof.

Proposition 21. *If on one of the sides of a triangle, from its extremities, there be constructed two straight lines within the triangle, the straight lines so constructed will be less than the remaining two sides of the triangle, but will contain a greater angle.*

Proposition 22. *Out of three straight lines which are equal to three given straight lines, to construct a triangle: thus it is necessary that two of the straight lines, taken in any manner, should be greater than the remaining one.*

Proposition 23. *On a given straight line and at a point on it to construct a rectilinial angle equal to a given rectilineal angle.*

Proposition 24. *If two triangles have the two sides equal to two sides respectively, but have the one of the angles contained by the equal straight lines greater than the other, they will also have the base greater than the base.*

Proposition 25. *If two triangles have the two sides equal to two sides respectively, but have the base greater than the base, they will also have the one of the angles contained by the equal straight lines greater than the other.*

Proposition 26. *If two triangles have the two angles equal to two angles respectively, and one side equal to one side, namely, either the side adjoining the equal angles, or that subtending one of the equal angles, they will also have the remaining sides equal to the remaining sides and the remaining angle to the remaining angle.*

Having proved all the standard congruence theorems Euclid now turns to the topic of parallelism. In the next two propositions he indicates several methods for constructing parallel lines.

Proposition 27. *If a straight line falling on two straight lines make the alternate angles equal to one another, the straight lines will be parallel to one another.*

Proposition 28. *If a straight line falling on two straight lines make the exterior angle equal to the interior and opposite angle on the same side, or the interior angles on the same side equal to two right angles, the straight lines will be parallel to one another.*

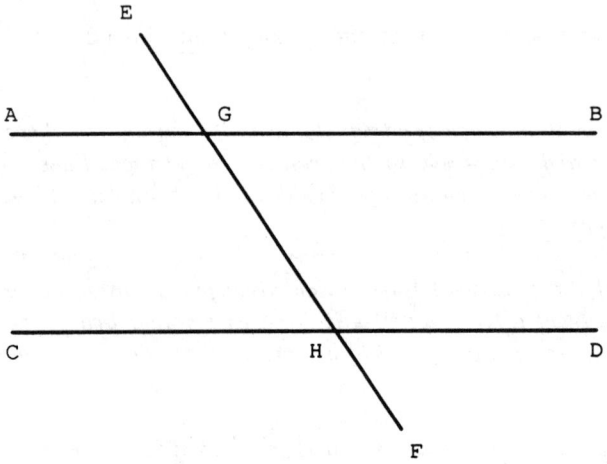

FIGURE 1.8

None of the propositions listed so far make use of Postulate 5, and it has been argued that this is evidence of a conscious reluctance on the part of Euclid to rely on it. Be that as it may, in view of later developments it has become convenient to give a different name to that part of Euclidean geometry which does not depend on this postulate. Janos Bolyai (1802–1860), one of the founders of non-Euclidean geometry, referred to this subset of Euclidean geometry as *absolute geometry*. Nowadays it is also common to refer to it as *neutral geometry*. Euclid's Propositions 1–28 are all theorems of absolute geometry. Theorem 1.1 of this chapter, as well as the contents of Chapter 10 constitute several more examples of propositions of absolute geometry. We go on to list some more propositions of Euclid that do rely on the parallel postulate and to which we will have occasion to refer later.

Proposition 29. *A straight line falling on parallel straight lines makes the alternate angles equal to one another, the exterior angle equal to the interior and opposite angle, and the interior angles on the same side equal to two right angles.*

PROOF: Let the straight line EF fall on the parallel straight lines AB, CD (Fig. 1.8); it will be shown that

$$\angle AGH = \angle GHD, \angle EGB = \angle GHD,$$

$$\angle BGH + \angle GHD = \text{two right angles}.$$

If ∠AGH and ∠GHD are unequal, then one of them is greater. Let ∠AGH be the greater. Hence,

$$\angle GHD + \angle BGH < \angle AGH + \angle BGH = \text{two right angles}.$$

Hence, by Postulate 5, the straight lines AB, CD, if produced indefinitely, will meet; but they do not meet because they are by hypothesis parallel.

Therefore ∠AGH = ∠GHD. The proposition's other assertions now follow easily and the details are omitted.   q.e.d.

The following two propositions are included here because of their immense importance. They constitute the two most important statements of Euclidean geometry, if not of all mathematics. We shall try to substantiate this claim in the subsequent chapters of this book.

Proposition 32. *In any triangle, if one of the sides be produced, the exterior angle is equal to the two interior and opposite angles, and the three interior angles of the triangle are equal to two right angles.*

Proposition 47. *In right-angled triangles the square on the side subtending the right angle is equal to the squares on the sides containing the right angle.*

FROM BOOK III (CIRCLES)

Euclidean circles play a fundamental role in this book's development of non-Euclidean geometry. For this reason it is convenient to restate here some of the standard theorems about circles, their tangents, and their angles.

Proposition 18. *If a straight line touch a circle, and a straight line be joined from the centre to the point of contact, the straight line so joined will be perpendicular to the tangent.*

Proposition 22. *The opposite angles of quadrilaterals in circles are equal to two right angles.*

Proposition 27. *In equal circles angles standing on equal circumferences are equal to one another, whether they stand at the centers or at the circumferences.*

Proposition 31. *In a circle the angle in the semicircle is right.*

Proposition 32. *If a straight line touch a circle, and from the point of contact there be drawn across, in the circle, a straight line cutting the circle, the angles which it makes with the tangent will be equal to the angles in the alternate segments of the circle.*

FROM BOOK VI (GEOMETRICAL PROPORTION)

Any theory of proportions must confront some formidable difficulties when the issue of incommensurability (irrational ratios) is encountered. In order to surmount these Euclid chose a rather surprising approach. He first dealt with incommensurability in the context of areas and proved Proposition 1 below. This enabled him to transfer questions regarding the ratios of certain lengths to the ratios of corresponding areas. This clever technique has been neglected by modern expositions in favor of more direct approaches and is essentially forgotten. Since the details would take us too far afield, only some of the better known propositions about similarity are restated here, without any attempt to sketch out Euclid's proofs.

Proposition 1. *Triangles and parallelograms which are under the same height are to one another as their bases.*

Proposition 2. *If a straight line be drawn parallel to one of the sides of the triangle, it will cut the sides of the triangle proportionally; and, if the sides of the triangle be cut proportionally, the lines joining the points of the section will be parallel to the remaining side of the triangle.*

Proposition 4. *In equiangular triangles the sides about the equal angles are proportional, and those are corresponding sides which subtend the equal angles.*

Proposition 5. *If two triangles have their sides proportional, the triangles will be equiangular and will have those angles equal which the corresponding sides subtend.*

1.3 HILBERT'S AXIOMATIZATION OF EUCLIDEAN GEOMETRY (**OPTIONAL**)

While all the subsequent generations praised Euclid for his great accomplishments and regarded his work as epitomizing pure reason, many mathematicians were fully aware that this work was incomplete. In his proofs Euclid repeatedly used definitions and postulates that had not been made explicit previously. Attempts to correct these deficiencies were continued over the centuries and gathered great momentum in the second half of the nineteenth century, mostly because of the discovery of non-Euclidean geometry during its first half. We present here the axiomatization offered by David Hilbert in 1899 in his book *The Foundations of Geometry*. As mentioned earlier, this system is considered definitive, but it is at the same time also ignored by most pedagogues because of its complexity and subtlety. What follows is a summary of Chapter I of Hibert's book. For a detailed development of Euclidian geometry from an axiom system that is very close to Hilbert's (although by no means identical with it) the reader is referred to E. E. Moise's book.

In contrast with Euclid's development, Hilbert begins by listing a collection of terms for which no definition is offered. This approach has the advantage of both greater generality and logical correctness. Mathematicians have found it useful upon occasion to interpret the abstract idea of point as something other than an infinitely small dot. The most common of these alternate interpretations is that of a point as a set: sometimes a set of other points, and sometimes a set of lines. The absence of a definition that ties down the meaning of the term *point* to any specific visual image should be construed as a freedom to apply this abstract set of axioms to other logical systems besides Euclid's geometry. In any case, Euclid's first seven definitions make it clear that it is impossible to define everything. One must start with some undefined terms.

0. Undefined Quantities.

A class of undefined elements called points, denoted by Latin capitals A, B, C,
...
A class of undefined elements called lines, denoted by small Latin letters a, b, c,
...
A class of undefined elements called planes, denoted by small Greek letters α, β, γ, ...
Undefined relations: Incidence (being incident, lying in); being in between; congruence; being parallel.

I. Axioms of Connection.

Having listed his undefined elements and relations, Hilbert goes on to state some postulates that connect these as yet unrelated elements with each other. While he refers to these as axioms, there is no logical difference between them and postulates.

I.1. Two distinct points A and B always completely determine a straight line a. We write $AB = a$ or $BA = a$.

I.2. Any two distinct points of a straight line completely determine that line; that is, if $AB = a$ and $AC = a$, where $B \neq C$, then we also have $BC = a$.

I.3. Three points A, B, C, not situated in the same straight line always completely determine a plane α. We write $ABC = \alpha$.

I.4. Any three points A, B, C, of a plane a, which do not lie in the same straight line, completely determine that plane.

I.5. If two points A, B of a straight line a lie in a plane α, then every point of a lies in α.

I.6. If two planes α, β have a point A in common, then they have at least a second point B in common.

I.7. On every straight line there exist at least two points, in every plane at least three points not lying in the same straight line, and in space there exist at least four points not lying in a plane.

From these axioms of connection one can easily prove such simple theorems as the fact that two distinct intersecting planes must intersect in a line.

II. Axioms of order.

This group of axioms is meant to remedy Euclid's failure to provide a logical foundation for the relations of *being between* and *being inside*. These axioms are based on the M. Pasch's article *Vorlesungen uber neuere Geometrie* which appeared in 1882.

II.1. If A, B, C are points of a straight line and B lies between A and C, then B lies also between C and A.

II.2. If A and C are points of a straight line, then there exists at least one point B lying between A and C and at least one point D so situated that C lies between A and D.

II.3. Of any three points situated on a straight line, there is always one and only one which lies between the other two.

II.4 Any four points A, B, C, D of a straight line can always be so arranged that B shall lie between A and C and also between A and D, and, furthermore, so that C shall lie between A and D and also between B and D.

These axioms make it possible to define the line segment AB as the set of all points that lie between A and B. These segments, in turn, are used to formulate the following crucial axiom which effectively says that every triangle has both an outside and an inside.

II.5. (*Pasch's Axiom*). *Let A, B, C be three points not lying in the same straight line and let a be a straight line lying in the plane ABC and not passing through any of the points A, B, C. Then if the straight line a passes through a point of the segment AB, it will also pass through either a point of the segment BC or a point of the segment AC.*

Theorem 1.1 of the next section contains an application of Pasch's Axiom.

Hilbert now goes on to show that the above two groups of axioms are sufficient to clarify the meaning of the inside and outside of any simple (non self intersecting) polygon. He also proves that every straight line in a plane divides the remaining points on it into two well defined halves (sides), and that every plane divides the remaining points of space into two well defined halves (sides). Next comes the axiom of parallels. It is noteworthy that while Hilbert refers to it as *Euclid's Axiom*, he actually states the version that has come to be known as Playfair's Postulate.

III. Axiom of parallels.

III.1. *In a plane α there can be drawn through any point A, lying outside of a straight line a, one and only one straight line which does not intersect the line a. This straight line is called the parallel to a through the given point A.*

IV. Axioms of congruence.

Any axiomatization of Euclidean geometry must either incorporate the rigid motions (to be discussed in great detail in the following chapter) or else provide some axioms that define the notion of congruence. Hilbert follows the latter route. However, before these axioms of the undefined relation of congruence are stated, rays and angles must be defined. Fortunately, the axioms of order make it possible to provide all the necessary definitions. Briefly, a ray emanating from a point consists of one of the sides of that point on a line and an angle at that point consists of a pair of rays that emanate from it (and that do not lie in the same straight line). One also needs definitions of the interior of an angle and of the angles of a triangle at this point, but this would take us too far afield and the readers are asked to use their visual intuition instead.

IV.1. *If A, B are two points on a straight line a, and if A' is a point upon the same or another straight line a', then, upon a given side of A' on the straight line a', we can always find one and only one point B' so that the line segment AB (or BA) is congruent to the segment $A'B'$. We indicate this relation by writing*

$$AB \equiv A'B'.$$

Every segment is congruent to itself; that is, we always have

$$AB \equiv AB.$$

IV.2. *If a segment AB is congruent to the segment $A'B'$ and also to the segment $A''B''$, then the segment $A'B'$ is congruent to the segment $A''B''$; that is, if $AB \equiv A'B'$ and $AB \equiv A''B''$, then $A'B' \equiv A''B''$.*

IV.3. *Let AB and BC be two segments of a straight line a which have no points in common aside from the point B, and, furthermore, let $A'B'$ and $B'C'$ be two segments of the same or another straight line a' having, likewise, no point other than B' in common. Then, if $AB \equiv A'B'$ and $BC \equiv B'C'$, we have $AC \equiv A'C'$.*

IV.4 *Let an angle (h,k) be given in the plane α and let a straight line a' be given in the plane α'. Suppose also that, in the plane α', a definite side of the straight line a' be assigned. Denote by h' a ray of the straight line a' emanating from a point O' of this line. Then in the plane α' there is one and only one ray k' such that the angle (h, k), or (k, h), is congruent to the angle (h', k') and at the same time all interior points of the angle (h', k') lie upon the given side of a'. We express this realtion by means of the notation*

$$\angle(h, k) \equiv \angle(h', k').$$

Every angle is congruent to itself; that is,

$$\angle(h, k) \equiv \angle(h, k)$$

or

$$\angle(h, k) \equiv \angle(k, h).$$

IV.5. *If the angle (h, k) is congruent to the angle (h', k') and to the angle (h'', k''), then the angle (h', k') is congruent to the angle (h'', k''); that is to say, if $\angle(h, k) \equiv \angle(h', k')$ and $\angle(h, k) \equiv \angle(h'', k'')$, then $\angle(h', k') \equiv \angle(h'', k'')$.*

IV.6. *If, in the two triangles ABC and $A'B'C'$, the congruences*

$$AB \equiv A'B', AC \equiv A'C', \angle BAC \equiv \angle B'A'C'$$

hold, then the congruences

$$\angle ABC \equiv \angle A'B'C' \text{ and } \angle ACB \equiv \angle A'C'B'$$

also hold.

All the congruence theorems of Euclidean geometry can now be proved. In, addition Hilbert also demonstrates that all right angles are congruent to one another and that the sum of the angles of any triangle is equal to two right angles.

V. Axiom of Continuity.

The need for the last of the axioms was pointed out by Archimedes in his book *On the Quadrature of the Parabola*. He used it to justify the argument that when a certain quantity is halved, and one of the halves is then halved again, and then one of the quarters is halved again, and so on, then the ultimate piece can be made arbitrarily small by repeating the process a sufficiently large number of times. A similar application occurs in the proof of Theorem 1.1 below.

V.1. (*Archimedean Axiom*) *Let A_1 be any point upon a straight line between the arbitrarily chosen points A and B. Take the points $A_1, A_2, A_3, \ldots$ so that A_1 lies between A and A_2, A_2 between A_1 and A_3, A_3 between A_2 and A_4, etc. Moreover, let the segments*

$$AA_1, A_1A_2, A_2A_3, A_3A_4, \ldots$$

be equal to one another. Then among this series of points, there always exists a certain point A_n such that B lies between A and A_n.

Several years after the first publication date Hilbert found it necessary to augment the Axioms of Continuity. As this augmentation concerns some issues that have no application to Euclidean geometry as it is understood by most geometers, it will not be discussed here.

1.4 VARIATIONS ON EUCLID'S FIFTH

It is more than likely that the reader was suprised by the statement of Euclid's Postulate 5. This postulate is generally known as *The Parallel Postulate* and

28 THE POINCARÉ HALF-PLANE

yet Euclid's statement makes no reference to parallelism. In modern textbooks it appears in a different version that was first formulated by Proclus in the 5th century. As is quite frequently the case in mathematics, it is commonly attributed to another, much latter person. This is essentially the same as Postulate III.1 of the previous section.

Playfair's Postulate: *Given a line m and a point P not on m, there is a unique line n that contains P and is parallel to m.*

It is not immediately obvious that Playfair's Postulate is logically equivalent to Euclid's Postulate 5. This equivalence is demonstrated in the following theorem. In fact, it is convenient to simultaneously prove the equivalence of both of these postulates to the statement that the sum of the interior angles of any triangle is equal to two right angles. Many other equivalent versions of Euclid's Postulate 5 were formulated over the centuries, and some of these are listed in Chapter 10. In the proofs below the symbol π will be used to denote two right angles.

The proof of the equivalence of these three formulations relies on some postulates that were not explicitly stated by Euclid. However, they, or some equivalent versions are indeed employed by him in some of his proofs. The reader will find the modern statements of these postulates in the previous section.

Theorem 1.1. *The following statements are equivalent in the context of absolute geometry:*
  a) *Playfair's Postulate,*
  b) *Euclid's Postulate 5,*
  c) *The sum of the angles of every triangle is π.*

PROOF: $b \Rightarrow a$) Assume the validity of Euclid's Postulate 5. Let the line CD be given and let X be a point not on CD (Figure 1.9). It will be shown that there is a unique line through X that is parallel to CD. Let Y be any point on the line CD, and let AB be a line containing X such that the the alternating angles $\angle DYX$ and $\angle AXY$ are equal. By Euclid's Proposition 27, the line AB is necessarily parallel to the line CD. If $A'B'$ is any other line through X then either

$$\angle A'XY < \angle AXY = \angle DYX = \pi - \angle CYX$$

or

$$\angle B'XY < \angle BXY = \angle CYX = \pi - \angle DYX.$$

In other words, either

$$\angle A'XY + \angle CYX < \pi$$

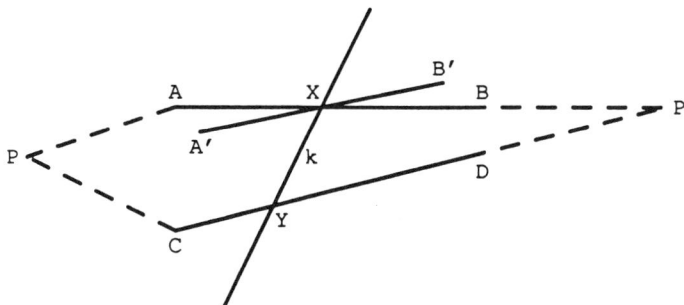

FIGURE 1.9

or

$$\angle B'XY + \angle DYX < \pi.$$

In either case it follows from Euclid's Postulate 5 that the line $A'B'$ intersects the line CD.

$a \Rightarrow c$)   Assume Playfair's Postulate, and let $\triangle ABC$ be given (Figure 1.10). Let DE be the unique line through A parallel to BC. Let $D'A$ be a line such that

$$\angle BAD' = \angle ABC.$$

By Euclid's Proposition 27, $D'A$ is parallel to BC, hence it must be identical

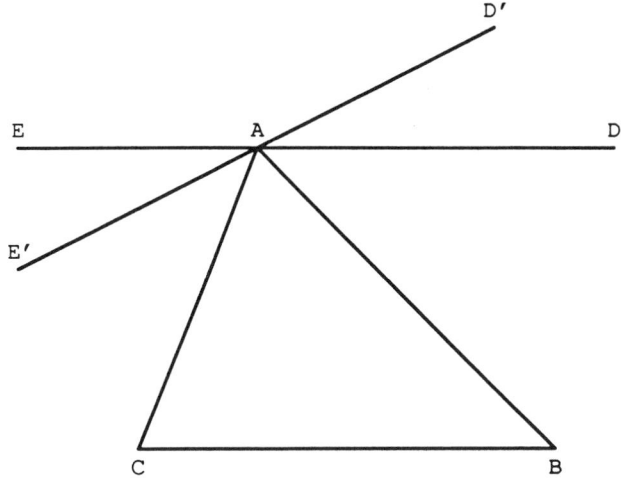

FIGURE 1.10

with DE. Thus

$$\angle BAD = \angle ABC$$

and similarly

$$\angle EAC = \angle BCA.$$

Consequently,

$$\angle ABC + \angle BCA + \angle CAB = \angle BAD + \angle EAC + \angle CAB = \pi.$$

$c \Rightarrow b$) Suppose the sum of the angles of every triangle is π. In Figure 1.11, let the transversal k intersect the two lines AB and CD in the points A and C so that $\alpha = \angle ACD$ and $\beta = \angle CAB$. It will be shown that if

$$\alpha + \beta < \pi$$

then the lines AB and CD intersect.

Define the sequence of points $D_1, D_2, D_3, \ldots$ as follows. Set $D_1 = D$, assuming that D_n has been defined, let D_{n+1} be that point such that D_n is between C and D_{n+1} and $AD_n = D_n D_{n+1}$. Consequently, $\triangle AD_n D_{n+1}$ is

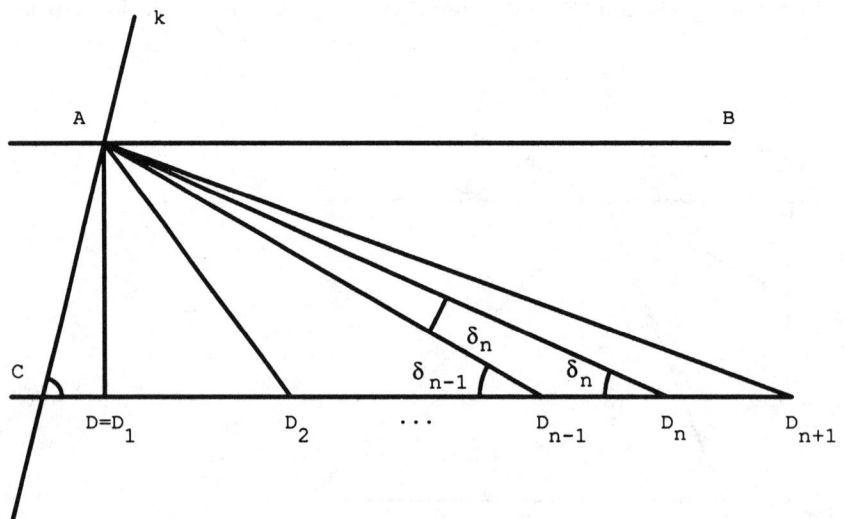

FIGURE 1.11

isosceles, and so, if

$$\delta_n = \angle AD_nC = \angle D_nAD_{n-1},$$

then,

$$\delta_{n-1} = \pi - \angle AD_{n-1}D_n = \pi - (\pi - 2\delta_n) = 2\delta_n.$$

By the Principle of Mathematical Induction,

$$\delta_n = \frac{\delta_1}{2^{n-1}} \quad \text{for } n = 1, 2, 3, \ldots.$$

However, if we set $\beta_n = \angle CAD_n$, then

$$\alpha + \beta_n + \delta_n = \pi,$$

and hence, (by the axiom of Archimedes)

$$\lim_{n \to \infty} \beta_n = \pi - \alpha - \lim_{n \to \infty} \delta_n = \pi - \alpha - 0 > \beta.$$

This means that for some finite n, $\beta_n > \beta$. Consequently, the line AB passes through the interior of triangle ACD_n, and so, by the postulate of Pasch, AB must intersect the line CD. That this intersection occurs on the side of the line k predicted by Postulate 5 follows from the same exterior angle argument that was used in the first case of the proof of this theorem. q.e.d.

1.5 EXERCISES

1. Summarize your high school's geometry course in a five page document.

   Proofs for the propositions cited below which were not included in this chapter appear in Appendix A.

2. Criticize the statements and/or proofs of Propositions 5–9. Rewrite them with more modern terminology and notation.
3. Criticize the statements and/or proofs of Propositions 10–15. Rewrite them with more modern terminology and notation.
4. Criticize the statements and/or proofs of Propositions 16–19. Rewrite them with more modern terminology and notation.
5. Criticize the statements and/or proofs of Propositions 20, 22,23. Rewrite them with more modern terminology and notation.
6. Criticize the statements and/or proofs of Propositions 26–28. Rewrite them with more modern terminology and notation.

   It is strongly recommended that the readers either prove Exercises 7–29 themselves, or else locate their proofs in some high school geometry text. The proofs should rely only on those propositions of Euclid that were quoted in this chapter. Moreover, the readers should decide which of these exercises are valid in absolute geometry and which only in Euclidean geometry.

7. Prove that the locus of all points that are equidistant from two distinct points A and B is the straight line that is perpendicular to the line segment AB and bisects it (the perpendicular bisector of AB).
8. Prove that the perpendicular bisectors of the three sides of any triangle are concurrent (i.e., they all intersect in a common point).
9. Prove that the locus of all the points that are equidistant from the two sides of an angle is that angle's bisector.
10. Prove that the three angle bisectors of any triangle are concurrent.
11. Prove that the sum of the interior angles of any n-sided polygon is $(n-2)\pi$.
12. Prove that the opposite angles of any parallelogram are equal.
13. Prove that if both pairs of opposite angles of a quadrilateral are equal to each other, then the quadrilateral is a parallelogram.
14. Prove that the opposite sides of any parallelogram are equal.
15. Prove that if both pairs of opposite sides of a quadrilateral are equal to each other then the quadrilateral is a parallelogram.
16. Prove that the diagonals of any parallelogram bisect each other.
17. Prove that the locus of all the points that are equidistant from a given line consists of two parallel lines.
18. Prove that the line joining the midpoints of two sides of a triangle is parallel to the third side and equal to half its length.
19. Prove that any two medians cut each other into segments whose lengths have ratio $2:1$.
20. Let P, Q, R be three points on the straight lines BC, CA, and AB, where A, B, and C form a triangle. Prove that the points P, Q, R are collinear if and only if

$$\frac{AR}{RB} \cdot \frac{BP}{PC} \cdot \frac{CQ}{QA} = -1.$$

(Here we follow the convention that if the three points U, V, W are collinear, then the ratio $\frac{UV}{VW}$ is positive or negative according as V is inside or outside the line segment UW. This exercise is known as the Theorem of Menelaus).
21. Let P, Q, R be three points on the straight lines BC, CA, and AB, where A, B, and C form a triangle. Prove that the lines AP, BQ, and CR are concurrent if and only if

$$\frac{AR}{RB} \cdot \frac{BP}{PC} \cdot \frac{CQ}{QA} = 1.$$

(This is the Theorem of Ceva).
22. Prove that the three medians of any triangle are concurrent.
23. Prove that the three altitudes of any triangle are concurrent.
24. Prove that in congruent circles equal central angles determine equal arcs and equal chords.
25. Prove that in congruent circles equal arcs have equal chords and subtend equal central angles.

26. Prove that in congruent circles equal chords cut off equal arcs and subtend equal central angles.
27. Show that the angle at the circumference of a circle is equal to one half the central angle subtended by the same arc.
28. Prove that the locus of all points from which a given line segment subtends the same given angle is the arc of a circle.
29. Prove that the sum of the opposite angles of a quadrilateral that is inscribed in a circle is π.
30. Let T be a point on the circumference of a circle centered at C. Prove that a line m containing T is tangent to the given circle if and only if it is perpendicular to the radius CT.
31. Briefly describe the lives and accomplishments of the following predecessors of Euclid:
    a) Thales
    b) Pythagoras
    c) Democritus
    c) Anaxagoras
    d) Archytas
    e) Eudoxus
    f) Hippocrates of Chios.
32. Briefly describe the lives and accomplishments of the following successors of Euclid:
    a) Archimedes
    b) Apollonius
    c) Ptolemy
    d) Heron
    e) Pappus.
33. Which of the propositions of Euclid's Books III and VI that are listed in this chapter are theorems of absolute geometry?
34. For each of Euclid's postulates and common notions, find its analog, if any, amongst Hilbert's axioms.
35. Draw a diagram illustrating each of Hilbert's axioms.

The following propositions which have been extracted from Hilbert's book are to be proved on the basis of Hilbert's postulates alone.

36. Two straight lines of a plane have either one point or no point in common.
37. Two planes have no point in common or have a straight line in common.
38. A plane and a straight line not lying in it have either no point or one point in common.
39. A straight line and a point not lying in it are contained in one and only one plane.
40. Two intersecting distinct straight lines are contained in one and only one plane.
41. Between any two points of a straight line there is an infinite number of points.
42. Every straight line a, which lies in a plane α, divides the remaining points of this plane into two regions having the following properties:
    i) If A is a point of one region and B is a point of the other, then the segment AB contains a point of the line a;
    ii) If A, A' are points of the same region then the line segment AA' does not intersect the line a.
43. If two straight lines a, b of a plane do not meet a third straight line c of the same plane, then they do not meet each other.
44. If, for the two triangles ABC and $A'B'C'$ the congruences
$$AB \equiv A'B', \quad AC \equiv A'C', \angle A \equiv \angle A'$$
hold, then the two triangles are congruent to each other.
45. If in any two triangles one side and the two adjacent angles are respectively equal, then the two triangles are congruent to each other.

2
Euclidean Rigid Motions

2.1 INTRODUCTION

As we have had occasion to mention before, one of the most serious deficiencies in Euclid's treatment of geometry is his neglect of an explicit consideration of rigid motions, i.e., transformations of the plane that do not distort its configurations. In view of Euclid's goals, this is indeed a serious flaw, as such a transformation is introduced as early as the third paragraph of the proof of Proposition 4. These motions are not mentioned in Hilbert's axiomatization either; instead, they are replaced by several congruence axioms. Other axiom systems, notably that of M. Pieri (1860–1913) do refer directly to them. By 1872 most mathematicians were fully aware that valid alternatives to Euclidean geometry did exist, each with its own collection, or *group*, of rigid motions. One of the problems resulting from this proliferation of geometries was that of classifying them and in that year Felix Klein (1849–1925) set forth his *Erlanger Program* in which he suggested that they be classified by their groups of rigid motions.

Two goals are accomplished in this chapter. First it is shown that every rigid motion of the plane is either a translation, a rotation, a reflection, or a glide-reflection. Second, it is demonstrated that the reflections generate all the other rigid motions. More specifically, every rigid motion can be expressed as the composition of at most three reflections.

This classification of the rigid motions of the Euclidean plane is commonly attributed to M. Chasles (1793-1880). However, in view of Euler's much earlier work on the rigid motions of the Euclidean three dimensional space, it is hard to believe that he was not aware of the simpler two dimensional version.

2.2 RIGID MOTIONS

It is assumed here that the reader is familiar with the concept of a function and with the composition of functions. The composition of the functions f and g will be denoted by $g \circ f$ and is to be read from right to left. Thus

$$(g \circ f)(x) = g(f(x)).$$

Functions of the plane into itself are called *transformations* because of their geometrical nature. If f is such a transformation and P and Q are points, or sets of points, such that $f(P) = Q$, we shall say that f *transforms* or maps P into Q. If P is a point (or a set) such that $f(P) = P$, we shall say that f *fixes* P or, equivalently, that P is a *fixed point* (or set) of f.

We begin by defining rigid motions and describing some of their properties. Attention is focused on the question of what is the minimum amount of information needed to ascertain that two rigid motions are in fact one and the same.

Let $d(P, Q)$ denote the distance between the two points P and Q. A *rigid motion* of the Euclidean plane is a transformation $f(P)$ of the plane into itself such that

$$d(P, Q) = d(P', Q')$$

whenever $P' = f(P)$ and $Q' = f(Q)$.

It is clear that the identity transformation Id which carries every point of the plane onto itself is a rigid motion and that the composition of rigid motions is again a rigid motion.

Proposition 2.1. *Every rigid motion transforms straight lines into straight lines.*

PROOF: Let f be a rigid motion and let m be a given straight line with two points A and B on it. If P is any point of m between A and B, and if $A' = f(A)$, $B' = f(B)$, and $P' = f(P)$, then

$$d(A', P') + d(P', B') = d(A, P) + d(P, B) =$$
$$d(A, B) = d(A', B').$$

Hence P' is on the line $A'B'$ between A' and B'.

Conversely, let P' be any point of $A'B'$ that lies between A' and B', and let P be the unique point of m such that

$$d(A, P) = d(A', P') \text{ and } d(B, P) = d(B', P').$$

Then by the above argument $f(P)$ has the same distances from A' and B' as does P', and so $P' = f(P)$.

A similar proof takes care of the case where P is a point of m that is outside the line segment AB. Hence f transforms every point of the line AB to a point of the line $A'B'$, and every point of the line $A'B'$ is covered by this process. Thus f transforms the line AB to the line $A'B'$. q.e.d.

When attempting to classify a collection of mathematical objects it is useful to have a simple criterion for deciding when two of them are identical. Two functions are of course identical when they have the same value for each point in their domain. In the special case of rigid motions, however, it is not necessary to verify equality at each point of the domain. As we shall soon see, three points suffice. Two functions f and g are said to agree at the point P if $f(P) = g(P)$.

Proposition 2.2. *If two rigid motions agree on two distinct points then they agree everywhere on the line joining those two points.*

PROOF: Exercise 12. □

Theorem 2.3. *If two rigid motions agree at three noncollinear points then they agree everywhere.*

PROOF: Let f and g be two rigid motions that agree at the three noncollinear points A, B, and C. By Proposition 2.2 f and g agree at every point on the lines AB, BC, CA. If P is now any point of the plane then there clearly exists a line through P that intersects the union of the lines AB, BC, and CA in some two distinct points X and Y. Since f and g agree at X and Y, by the above proposition they must also agree at P. q.e.d.

The proof of the following corollary is immediate.

Corollary 2.4. *If a rigid motion fixes three noncollinear points, then it must be the identity.* □

2.3 TRANSLATIONS, ROTATIONS, AND REFLECTIONS

Next we go on to describe some specific rigid motions as well as their inverses. It is shown that every rotation and every translation is the composition of two reflections and this fact is used to explicitly describe the compositions of any two rotations and/or translations.

A *translation* τ of the Euclidean plane is a rigid motion such that the line segments PP' and QQ' have the same length and direction whenever $P' = \tau(P)$ and $Q' = \tau(Q)$. In that case we will also write

$$\tau = \tau_{PP'} = \tau_{QQ'}.$$

We agree to consider the identity as a trivial translation. The following propostion shows that there are many nontrivial translations.

Proposition 2.5. *For any two points A and B there is a translation τ_{AB} that carries A onto B.*

PROOF: If $A = B$, then the identity is the required translation. Suppose now that A and B are distinct. For any point P let P' be a point such that the line segment PP' has the same direction and length as AB. In other words, if P is not on the line AB, then $ABP'P$ is a parallelogram. It follows from the transitivity of equality and parallelism that the function

$$P' = f(P)$$

is the required translation. q.e.d.

Since rigid motions are functions, it natural to ask how they interact. By this we mean that given any two rigid motions one should be able to clearly specify their composition. Such will be the content of many of the following propositions and we begin with an easy one now.

Proposition 2.6. *If A, B, C are any three points, then $\tau_{BC} \circ \tau_{AB} = \tau_{AC}$.*

PROOF: Exercise 13. □

The inverse of the function f is a function g such that $f \circ g = g \circ f = \text{Id}$. In general, a function may or may not have an inverse. It will eventually become clear that every rigid motion does indeed have an inverse. If f does have an inverse g, we write $g = f^{-1}$. At this point we note that translations do possess inverses.

Proposition 2.7. *The inverse of the translation τ_{AB} is the translation τ_{BA}.*

□

An *oriented angle* is an angle together with an orientation as either clockwise or counterclockwise. All positive angles are assumed to have a counterclockwise orientation and all negative angles are assumed to have a clockwise orientation. Let C be a given point, and let α be a given oriented angle. The *rotation* $R_{C,\alpha}$ is the function that associates to any point P the unique point P' such that

$$CP = CP' \text{ and } \angle PCP' = \alpha.$$

Note that for any point C the rotation $R_{C,0}$ is the identity map and for any angle α, $R_{C,-\alpha}$ is the inverse of $R_{C,\alpha}$.

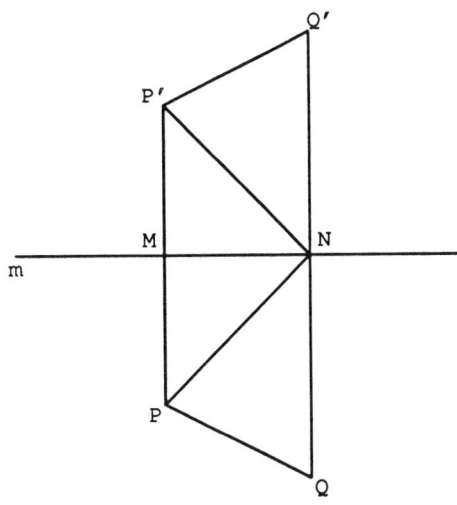

FIGURE 2.1

Proposition 2.8. *The rotation $R_{C,\alpha}$ is a rigid motion.*

PROOF: Exercise 14. □

Given a straight line m, the *reflection* ρ_m is the transformation that fixes every point of m and that associates to each point P not on m the unique point $P' = \rho_m(P)$ such that m is the perpendicular bisector of the line segment PP'. Note that each reflection is its own inverse. Any (non identity) function that possesses this property of being its own inverse is called an *involution*. This involutory nature of reflections will be seen to play a crucial role in our task of classifying the rigid motions. In the next chapter some more involutions, which are not reflections, will be introduced. They serve a similar important purpose in the context of non-Euclidean geometry.

Proposition 2.9. *Every reflection is a rigid motion.*

PROOF: Let m be a given line, and let P and Q be any two points (Fig. 2.1). If $P' = \rho_m(P)$ and $Q' = \rho_m(Q)$, then, by definition, the line m bisects the two line segments PP' and QQ' at, say, M and N respectively, and is perpendicular to both. It follows by the SAS congruence theorem first that $\triangle P'MN \cong \triangle PMN$ and next that $\triangle P'NQ' \cong \triangle PNQ$. Hence, $P'Q' = PQ$.

q.e.d.

Again we ask what the composition of two reflections is. The reader will recall that the composition of two translations is a translation. Such is not the case for reflections. The nature of the answer depends on whether the axes

40 THE POINCARÉ HALF-PLANE

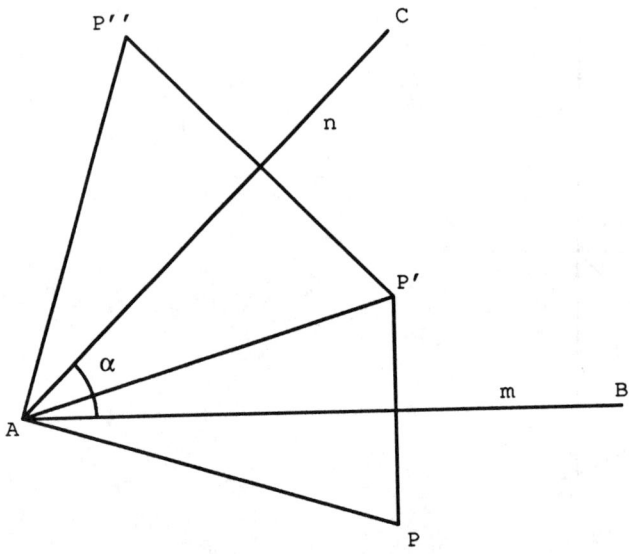

FIGURE 2.2

of the two composed reflecions intersect or not, and the following two propositions deal with these two cases separately.

Proposition 2.10. *Let m and n be two straight lines that intersect at a point A, and let α be the counterclockwise angle from m to n at A. Then*

$$\rho_n \circ \rho_m = R_{A,2\alpha}.$$

PROOF: Let P be a point near line m but outside the oriented angle from m to n. Let $P' = \rho_m(P)$ and $P'' = \rho_n(P')$ as in Figure 2.2. Then m bisects $\angle PAP'$ and n bisects $\angle P'AP''$. Consequently,

$$\angle PAP'' = 2\angle BAP' + 2\angle P'AC = 2\alpha.$$

Thus $\rho_n \circ \rho_m$ and $R_{A,2\alpha}$ agree on the point P. Since it is easy to find three such noncollinear points P, it follows from Theorem 2.3 that $\rho_n \circ \rho_m$ and $R_{A,2\alpha}$ are identical. q.e.d.

Proposition 2.11. *Let m and n be two parallel straight lines. Let AB be a line segment that first intersects m and then n, that is perpendicular to both m and n, and whose length is twice the distance between m and n. Then*

$$\rho_n \circ \rho_m = \tau_{AB}.$$

PROOF: Exercise 15. □

It is clear that the composition of two rotations that have the same center is

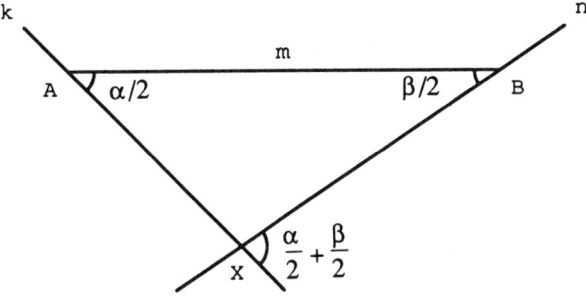

FIGURE 2.3

also a rotation. It is not nearly as obvious that the same holds for almost all pairs of rotations.

Proposition 2.12. *Let A and B be two points, and let α and β be two oriented angles. Then the compostion $R_{B,\beta} \circ R_{A,\alpha}$ is*
  a) *a translation if $\alpha + \beta$ is a mutliple of 2π,*
  b) *a rotation if $\alpha + \beta$ is not a multiple of 2π.*

PROOF: If A and B are the same point, this is obvious. Otherwise, let m be the straight line joining A and B (Figure 2.3). Let k be the line through A such that the oriented angle from k to m is $\alpha/2$, and let n be the line through B such that the oriented angle from m to n is $\beta/2$. Then, by the above two propositions,

$$R_{B,\beta} \circ R_{A,\alpha} = (\rho_n \circ \rho_m) \circ (\rho_m \circ \rho_k) = \rho_n \circ (\rho_m \circ \rho_m) \circ \rho_k = \rho_n \circ \rho_k$$

which is either a translation or a rotation depending on whether the lines k and n are parallel or not. However, the lines k and n are parallel if and only if $\alpha + \beta$ is a multiple of 2π. Hence the proposition is proved.   q.e.d.

Example 2.13. Let A, B be any two points (Figure 2.4). The proof of Proposition 2.12 constitutes a recipe for finding any composition such as, say, $R_{B,\pi/2} \circ R_{A,\pi/3}$. One first joins A and B with a line m. Next draw a line k through A such that the counterclockwise angle from k to m is $\pi/6$ and a line n through B such that the counterclockwise angle from m to n is $\pi/4$. If X is the intersection of the lines k and n, then

$$R_{B,\pi/2} \circ R_{A,\pi/3} = \rho_n \circ \rho_k = R_{X,2(\pi/4+\pi/6)} = R_{X,5\pi/6}.$$

There is an alternate solution to this problem which uses the statement of the above proposition rather than its proof. We know that the given composition is in fact a rotation $R_{X,\pi/2+\pi/3} = R_{X,5\pi/6}$ whose center X is to be determined.

42 THE POINCARÉ HALF-PLANE

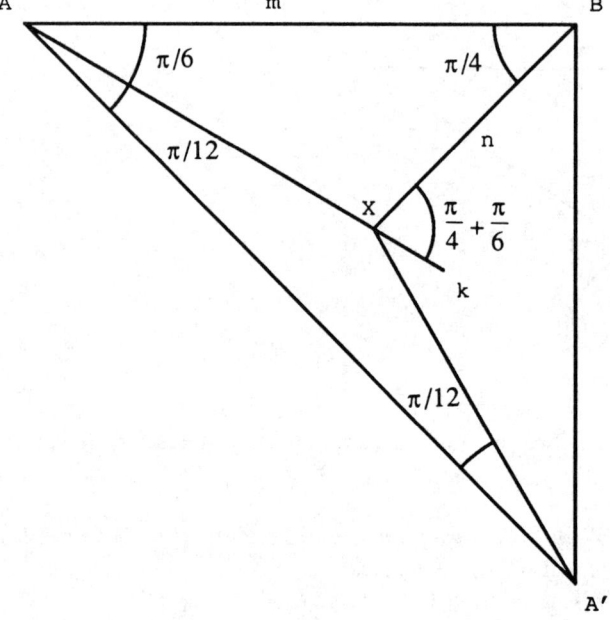

FIGURE 2.4

However,
$$R_{X,5\pi/6}(A) = R_{B,\pi/2} \circ R_{A,\pi/3}(A) = R_{B,\pi/2}(A) = A',$$
and so X is the vertex of an isosceles triangle $AA'X$ in which each of the base angles equals
$$\frac{\pi - \frac{5\pi}{6}}{2} = \frac{\pi}{12}.$$

So far we know how any two translations interact with each other and how two rotations interact with each other. What happens when a translation and a rotation are composed?

Proposition 2.14. *Let R be a rotation which is not the identity and let τ be a translation. Then both $R \circ \tau$ and $\tau \circ R$ are rotations that have the same angle as R.*

PROOF: Let $R = R_{A,\alpha}$, $A' = \tau^{-1}(A)$, and suppose P is the midpoint of the line segment AA' (Figure 2.5). Let k and m be the lines through P and A, respectively, that are perpendicular to AA', and let n be the line through A

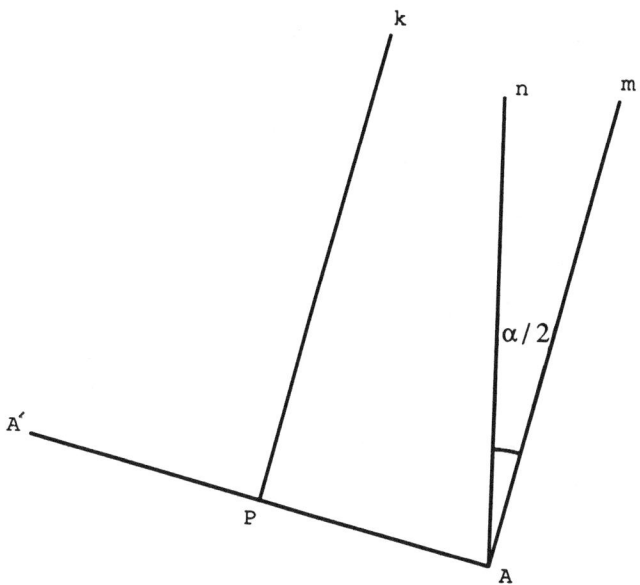

FIGURE 2.5

such that the oriented angle from m to n is equal to $\alpha/2$. Then

$$R \circ \tau = (\rho_n \circ \rho_m) \circ (\rho_m \circ \rho_k) = \rho_n \circ (\rho_m \circ \rho_m) \circ \rho_k$$
$$= \rho_n \circ \rho_k$$

which is a rotation since α is not zero. Since the counterclockwise angle from k to n also equals $\alpha/2$, the rotation $R \circ \tau$ also has angle α.

The proof that $\tau \circ R$ is also a rotation is left to the reader.   q.e.d.

Example 2.15. The proof of Proposition 2.14 also constitutes a recipe for constructing a composition. Thus, if $R = R_{A,\pi/3}$ and $\tau = \tau_{BC}$ (Figure 2.6) we begin by setting $A' = \tau^{-1}(A)$ and we let P be the midpoint of AA'. Through the points P and A respectively, draw lines k and m that are both perpendicular to AA'. Through A draw a line n making a counterclockwise angle of $\pi/6$ from m and let it intersect k at X. Then $R \circ \tau = R_{X,\pi/3}$.

2.4 GLIDE-REFLECTIONS

Having demonstrated how to compose translations and rotations with themselves, we now turn to examine their compositions with reflections. It is convenient to begin with a very special case.

44 THE POINCARÉ HALF-PLANE

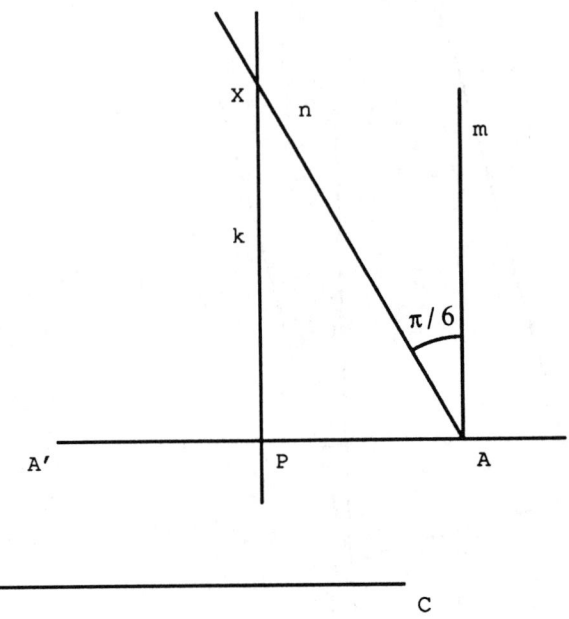

FIGURE 2.6

Lemma 2.16. *Let the line segment AB be perpendicular to the line m. Then*

$$\rho_m \circ \tau_{AB} = \rho_n$$

where n is a line parallel to m.

PROOF: Exercise 17. □

The composition of a reflection with a rotation that is centered at a point on the axis is also a reflection (Exercise 2). However, in general, the composition of a rotation (or a translation) with a reflection is a new kind of rigid motion.

Let A and B be two distinct points. The composition $\rho_{AB} \circ \tau_{AB}$ is called a *glide-reflection* and is denoted by γ_{AB}. It is easily verified that γ_{AB} also equals $\tau_{AB} \circ \rho_{AB}$ and that the inverse of γ_{AB} is γ_{BA}. In order to simplify the statements of some of the subsequent theorems we agree to consider reflections too as glide-reflections. The line AB is called the axis of γ_{AB}, and it is clear that for any point P not on AB, the segment joining P to its image $\gamma_{AB}(P)$ is bisected by the axis. We agree to consider reflections as glide-reflections and note that they are therefore the only glide-reflections with fixed points.

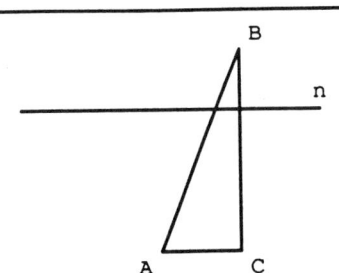

FIGURE 2.7

Proposition 2.17. *Let τ be any translation and ρ any reflection. Then $\rho \circ \tau$ and $\tau \circ \rho$ are both glide-reflections.*

PROOF: Let $\tau = \tau_{AB}$ and $\rho = \rho_m$.

If the lines AB and m are identical, then clearly $\rho \circ \tau = \tau \circ \rho = \gamma_{AB}$.
If the lines AB and m are distinct and parallel, then there exist point A', B' on m such that $\tau = \tau_{AB} = \tau_{A'B'}$. Consequently, by the previous argument,

$$\rho \circ \tau = \tau \circ \rho = \rho \circ \tau_{A'B'} = \tau_{A'B'} \circ \rho = \gamma_{A'B'}.$$

If $AB \perp m$ then we are done by Lemma 2.16. Otherwise, let C be a point such that $AC \parallel m$, and $BC \perp m$ (Fig. 2.7). By Proposition 2.6

$$\rho \circ \tau = \rho_m \circ (\tau_{CB} \circ \tau_{AC}) = (\rho_m \circ \tau_{CB}) \circ \tau_{AC}.$$

Since $BC \perp m$ it follows again from Lemma 2.16 that there is a line n parallel to both m and AC such that $\rho_m \circ \tau_{CB} = \rho_n$ and hence

$$\rho \circ \tau = \rho_n \circ \tau_{AC},$$

which, since $AC \parallel n$, is indeed a glide-reflection.

The proof that $\tau \circ \rho$ is also a glide-reflection is left to the reader. q.e.d.

The composition of translations with reflections called for the introduction of glide-reflections. No new rigid motions are needed to describe the composition of rotations with glide-reflections.

Proposition 2.18. *Let R be any rotation and ρ any reflection. Then $R \circ \rho$ and $\rho \circ R$ are both glide-reflections.*

PROOF: Let $R = R_{A,\alpha}$ and $\rho = \rho_k$. Let m be the line through A parallel to k and let n be that line through A whose oriented angle from m is $\alpha/2$ (Fig. 2.8). Then

$$R \circ \rho = (\rho_n \circ \rho_m) \circ \rho_k = \rho_n \circ (\rho_m \circ \rho_k).$$

46 THE POINCARÉ HALF-PLANE

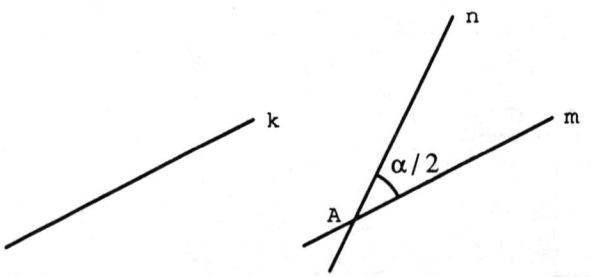

FIGURE 2.8

Since $\rho_m \circ \rho_k$ is a translation it follows from Proposition 2.17 that $R \circ \rho$ is a glide-reflection.

The proof of the second case is left to the reader. q.e.d.

2.5 THE MAIN THEOREMS

We are now ready to prove the two theorems that constitute the complete classification of all the rigid motions of the Euclidean plane.

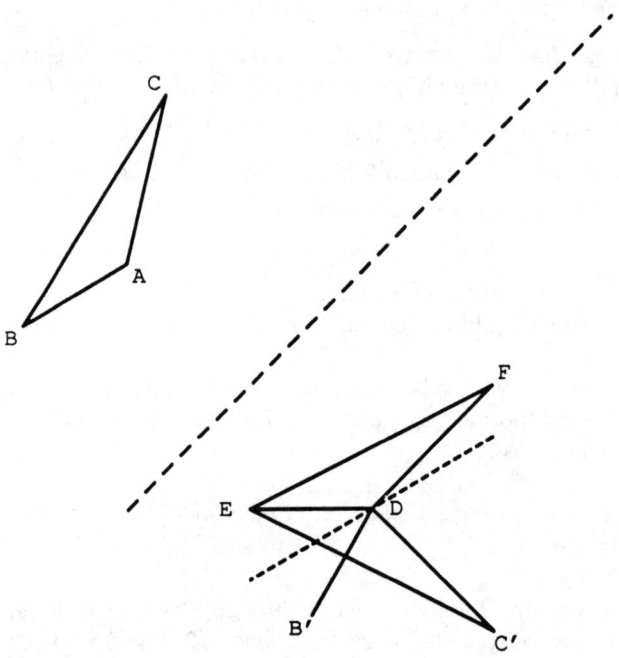

FIGURE 2.9

Proposition 2.19. *Suppose $\triangle ABC \cong \triangle DEF$. Then there exists a sequence of no more than three reflections such that the composition of these reflections maps the points A, B, C onto D, E, F, repsectively.*

PROOF: Clearly there is a map f_1, which is either a reflection or the identity, and which maps A onto D (Fig. 2.9). Set $B' = f_1(B)$ and let f_2 be that map which is either the reflection or the identity, and which maps B' onto E. Since $DB' = AB = DE$ it follows that f_2 fixes D. At this point the reader should note that the composition $f_2 \circ f_1$ maps A onto D and B onto E. Let $C' = f_2 \circ f_1(C)$. If $C' = F$, then f_1, f_2 is the desired sequence of reflections. Otherwise, $\triangle DEC' \cong \triangle ABC \cong \triangle DEF$, and so $\rho_{DE}(C') = F$. It is now easily verified that $\rho_{DE} \circ \rho_m \circ \rho_k$ does indeed maps the points A, B, C, onto D, E, F, respectively. q.e.d.

The proof of this chapter's first main theorem now follows directly from Theorem 2.3 and Proposition 2.19.

Theorem 2.20. *Every rigid motion is the composition of at most three reflections.* □

Since we have already described how all the special rigid motions defined above interact when composed, it is not surprising any more that we can actually classify all the rigid motions.

Theorem 2.21. *Every rigid motion is either a translation, a rotation, or a glide-reflection.*

PROOF: If a rigid motion is a reflection then it is by definition a glide-reflection. If it is the composition of two reflections, then by Propositions 2.10 and 2.11 it must be either a translation or a rotation. If it is the composition of three reflections, then the composition of the first two is either a translation or a rotation, and hence, by Propostions 2.17 and 2.18, the composition of all three is a glide-reflection. q.e.d.

The following corollary may seem to be anticlimactic at this point. Nevertheless, it describes an important property of rigid motions.

Corollary 2.22. *Every rigid motion has an inverse.* □

The following examples all refer to Fig. 2.10 in which triangles ABC, XAB, XBY, YBD are congruent equilateral triangles and the points H, I, J bisect the segments AC, AB, XB, respectively, and τ denotes the translation τ_{AB}.

Example 2.23. Determine $f = R_{B,\pi/3} \circ \tau$. Since $\tau(A) = B$ it follows that $f(A) = R_{B,\pi/3}(B) = B$. Hence, by Proposition 2.14 f is a rotation of angle $\pi/3$ that carries A to B. This yields $f = R_{X,\pi/3}$.

48 THE POINCARÉ HALF-PLANE

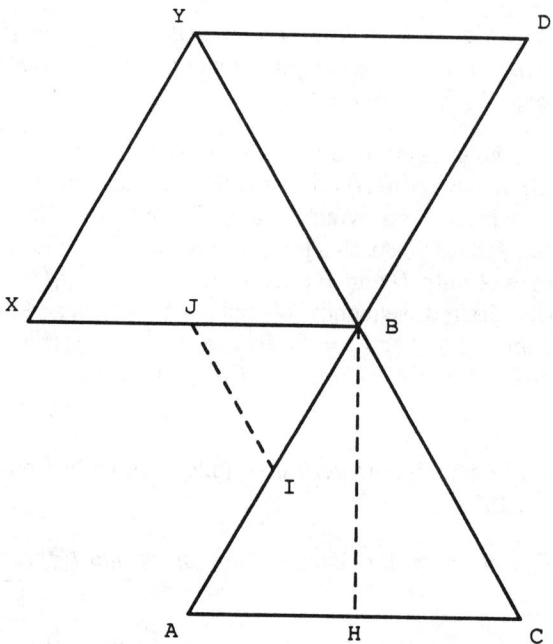

FIGURE 2.10

Example 2.24. Determine $f = R_{C,\pi/3} \circ \tau$. Since $R_{C,\pi/3}(B) = A$, it follows that $f(A) = A$. Hence, by Proposition 2.13 f is a rotation of angle $\pi/3$ that fixes A. Thus, $f = R_{A,\pi/3}$.

Example 2.25. Determine $f = \tau \circ R_{B,\pi/3}$. Here $f(B) = D$. Hence, by Proposition 2.13, f is a rotation of angle $\pi/3$ that maps B to D. Thus, $f = R_{Y,\pi/3}$.

Example 2.26. Determine $f = R_{B,\pi/3} \circ \rho_{AB}$. By Proposition 2.16 f is a glide-reflection and it is easy to see that $f(A) = C$ and $f(B) = B$. Since f has a fixed point it must be a pure reflection. Since the axis of this reflection must bisect the segment AC it follows that $f = \rho_{BH}$.

Example 2.27. Determine $f = \rho_{BC} \circ \tau_{AB}$. By Proposition 2.17 f is a glide-reflection. Since $f(A) = B$ and $f(B) = X$ it follows that the axis of f bisects both the line segments AB and BX. The reader can now easily verify that $f = \gamma_{IJ}$.

2.6 RIGID MOTIONS AND ABSOLUTE GEOMETRY

Just as many of the theorems of Euclidean geometry also hold in the more general context of absolute geometry, much of the information that was derived in this chapter about the Euclidean rigid motions is also valid in this wider context. This fact will only become crucial in Chapter 9 where the non-Euclidean rigid motions are investigated in detail. Nevertheless, the readers should be made aware at this point of the following observations.

The proofs of 2.1-2.4 made no use, either explicit or implicit, of the notion of parallelism, and hence *every rigid motion is determined by its effect on any triple of noncollinear points* in **absolute geometry** too.

Whereas translations and glide-reflections rely on the notion of parallelism for their very definition, such is not the case for rotations and reflections. In particular, the latter only require the notion of a perpendicular bisector, and the existence of such a line is guaranteed by Euclid's Propositions 10 and 11, both of which are propositions of absolute geometry. Thus 2.19 and 2.20 hold in the wider context too, and so *every rigid motion of any* **absolute geometry** *is the composition of at most three reflections*.

2.6 EXERCISES

1. Let A, B, C be the vertices of the triangle in Fig. 2.10. Use the method of Example 2.13 to find the centers of the following rotations:
   a) $R_{A,\pi/2} \circ R_{B,\pi/2}$ b) $R_{B,\pi/2} \circ R_{A,\pi/2}$ c) $R_{C,\pi} \circ R_{A,\pi/2}$
   d) $R_{A,\pi/2} \circ \tau_{BC}$ e) $R_{A,\pi/2} \circ \tau_{CA}$ f) $\tau_{CA} \circ R_{A,\pi/2}$.
2. Prove that if P is a point on the line m and q is any angle, then both $R_{P,\theta} \circ \rho_m$ and $\rho_m \circ R_{P,\theta}$ are reflections. What are their axes?
3. Given a figure consisting of two points P and Q, sketch a construction of the fixed point of $\tau_{PQ} \circ R_{P,\pi/4}$.
4. Given a figure consisting of three points A, B, C, sketch a construction of the fixed point of $\tau_{BC} \circ R_{A,2\pi/3}$.
5. Show that $\rho_m \circ R_{C,\theta} \circ \rho_m = R_{C,-\theta}$ if the point C is on the line m.
6. Let ABC be a clockwise triangle with oriented interior angles α, β, γ at the vertices A, B, C, respectively. Prove that $R_{C,2\gamma} \circ R_{B,2\beta} \circ R_{A,2\alpha} = Id$.
7. If k, m, n are the perpendicular bisectors of the sides AB, BC, CA respectively, of $\triangle ABC$, show that their composition $\rho_k \circ \rho_m \circ \rho_n$ is a reflection. What is the fixed line of this reflection?
8. Show that $\rho_k \circ \rho_m \circ \rho_n = \rho_n \circ \rho_m \circ \rho_k$ whenever the line k, m, n are either concurrent or have a common perpendicular.
9. Let P and Q be any two distinct points. Prove that the composition $R_{Q,\pi} \circ \rho_{PQ} \circ R_{P,\pi}$ is a glide-reflection and find its axis.
10. Show that the composition of the reflections in the three angle bisectors of a triangle is a reflection in a line that is perpendicular to one side of the triangle.
11. Let $ABCD$ be a square whose diagonals BD and AC intersect at the point P. Identify the following:
    a) $\rho_{BA} \circ \rho_{BD}$ b) $\rho_{AC} \circ \rho_{AB}$ c) $\rho_{AB} \circ \rho_{AC}$
    d) $R_{B,\pi/2} \circ R_{D,\pi/2}$ e) $R_{B,\pi} \circ R_{A,\pi}$ f) $R_{P,\pi/2} \circ \gamma_{DC}$
    g) $R_{D,\pi/2} \circ \gamma_{DC}$ h) $\gamma_{DA} \circ \gamma_{DC}$ i) $\gamma_{CB} \circ \gamma_{DC}$
    j) $\tau_{AC} \circ \gamma_{CB}$.

12. Prove Proposition 2.2.
13. Prove Proposition 2.6.
14. Prove Proposition 2.8.
15. Prove Proposition 2.11.
16. Complete the proof of Proposition 2.14.
17. Prove Lemma 2.16.
18. Complete the proof of Proposition 2.17.
19. Complete the proof of Proposition 2.18.
20. Let $A(a_1, a_2)$ and $B(b_1, b_2)$ be fixed points. Explain why the plane transformation $Q = f(P)$ that maps the point $P(x, y)$ to the point $Q(x', y')$ where

$$x' = x + b_1 - a_1$$
$$y' = y + b_2 - a_2$$

is in fact the translation τ_{AB}.

21. Let θ be an angle. Show that the plane transformation $Q = f(P)$ that maps the point $P(x, y)$ to the point $Q(x', y')$ where

$$x' = x \cos \theta - y \sin \theta$$
$$y' = x \sin \theta + y \cos \theta$$

is in fact the rotation $R_{O,\theta}$.

22. Let θ be an angle, and let m be the straight line through the origin with inclination θ to the positive x axis. Show that the plane transformation $Q = f(P)$ that maps the point $P(x, y)$ to the point $Q(x', y')$ where

$$x' = x \cos 2\theta + y \sin 2\theta$$
$$y' = x \sin 2\theta - y \cos 2\theta$$

is in fact the reflection ρ_m.

23. Express the rotation $R_{A,\pi/2}$, where $A = (0, 1)$, as a coordinate transformation of the type described in Exercises 20–22.
    (Hint: note that $R_{A,\pi/2} = \tau \circ R_{O,\pi/2} \circ \tau^{-1}$ where $\tau = \tau_{AB}$.)
24. Express the reflection ρ_m as a coordinate transformation where m is the line $y = 2x + 1$.
25. If f is any rigid motion and τ is any translation, identify the rigid motion $\tau \circ f \circ \tau^{-1}$.
26. If f is any rigid motion and R is any rotation, identify the rigid motion $R \circ f \circ R^{-1}$.
27. If f is any rigid motion and ρ is any reflection, identify the rigid motion $\rho \circ f \circ \rho^{-1}$.

A *symmetry* of a plane figure is a rigid motion that fixes that figure.

28. Describe the symmetries of the equilateral triangle.
29. Describe the symmetries of an isosceles triangle which is not equilateral.
30. Describe the symmetries of the square.
31. Describe the symmetries of a rectangle which is not a square.
32. Describe the symmetries of the regular n-gon.

The following exercises call for the use of a graphics package such as those available on Mathematica or Maple.

33. Write a script that translates lines, triangles, and circles.
34. Write a script that rotates lines, triangles, and circles.
35. Write a script that reflects lines, triangles, and circles in a given line.
36. Write a script that performs glide-reflections of given lines, triangles, and circles.

3

Inversions

3.1 AN INTERESTING NON-RIGID TRANSFORMATION

We now introduce the inversion – a transformation of the plane that is not rigid but will prove very useful in the sequel. These inversions resemble the rigid motions of the plane in that they transform straight lines and circles into straight lines and/or circles. We begin with a strengthened version of Proposition 32 of Book III of Euclid's *Elements*.

Proposition 3.1. *Let AB be a chord of a circle and let AT be any ray from A. Then the line AT is tangent to the circle if and only if $\angle BAT$ is equal to the angle at the circumference subtended by the intercepted arc.* □

The proof of Proposition 32 can be found in Appendix A. The proof of the converse is relegated to Exercise 1.

Proposition 3.2. *Let P be a point outside a given circle q, let PT be a tangent with contact point T, and let PAB be a secant with chord AB. Then*

$$PA \cdot PB = PT^2.$$

PROOF: By the previous proposition $\angle ATP = \angle PBT$ (Figure 3.1). Since $\angle TPA$ is common to both the triangles TPA and BPT, it follows that these two triangles are similar, and hence

$$\frac{PA}{PT} = \frac{PT}{PB},$$

and the proposition follows immediately. q.e.d.

52 THE POINCARÉ HALF-PLANE

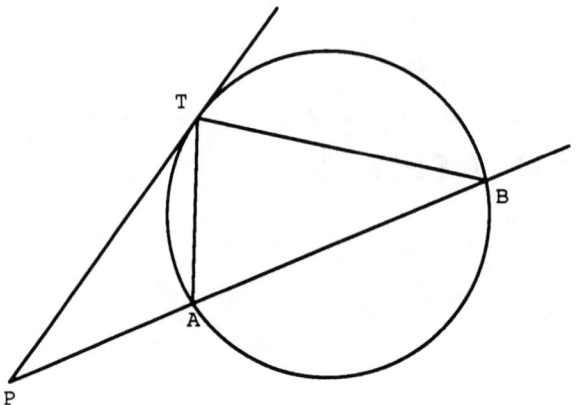

FIGURE 3.1.

Given a circle q with center C and radius k, the two points P, P' are said to be *symmetrical with respect to q* if

a) C, P, P' are collinear with C outside the segment PP'

and

b) $CP \cdot CP' = k^2$.

It is clear that for a fixed circle q, the point P is symmetrical with the point P' if and only if the point P' is symmetrical with the point P. Moreover a point P is symmetrical with itself if and only if it lies on the circumference of q. Note that no point is symmetrical to C, and C is the only such point. Given such a point P inside the circle, its corresponding symmetrical point P' can be easily found as follows. Let QR be the chord of the circle q that contains P and is perpendicular to the radius CP (Figure 3.2). The point P' is the intersection of the line tangent to q at Q with the (extended) radius CP.

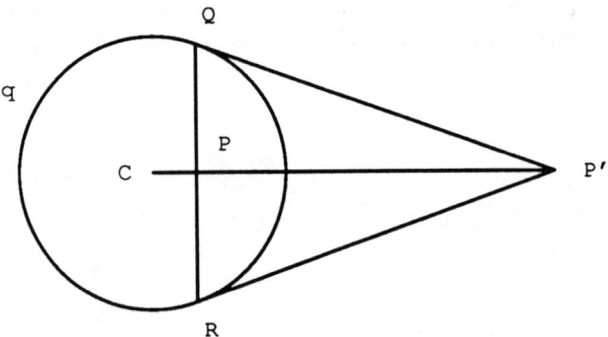

FIGURE 3.2.

Let C be a fixed point and k a positive real number. The *inversion* $I_{C,k}$ is a function such that
$$I_{C,k}(P) = P'$$
where P and P' are symmetrical with respect to the circle with center C and radius k. The inversion $I_{C,k}$ is undefined at the point C, and so, strictly speaking it is not really a transformation of the plane. The reader should view this as merely a nuisance rather that a genuine difficulty. There are several ways of getting around this problem and we chose the easiest one of simply ignoring the inaccuracy and modifying the notion of a transformation to include inversions as well. Note that inversions, like reflections, are involutions.

It is clear that the fixed points of $I_{C,k}$ are exactly those that constitute the circle centered at C with radius k. Note that with the exception of the point C itself, the points of any straight line m through C are mapped by $I_{C,k}$ back onto m. Hence we shall say that $I_{C,k}$ maps such a line m onto itself, even though this is not completely correct, since $I_{C,k}$ is not defined at C. It is also easy to see that every circle centered at C is mapped by $I_{C,k}$ onto another circle centered at C.

Theorem 3.3. *The inversion $I_{C,k}$ maps*

a) *straight lines containing C onto themselves,*
b) *straight lines not containing C onto circles through C,*
c) *circles through C onto straight lines not containing C,*
d) *circles not through C onto circles not through C,*

When the inversion transforms a circle into a straight line, or vice versa, the straight line that joins C to the center of the circle is perpendicular to the given straight line. When the inversion transforms a circle into a circle, their centers are collinear with C.

PROOF:
  a) This is clear.
  b) Let m be a straight line not containing C, and let H be a point of m such that CH is perpendicular to m (Figure 3.3). Let P be an arbitrary point of m and let
$$H' = I_{C,k}(H) \quad \text{and} \quad P' = I_{C,k}(P).$$
Since $CH \cdot CH' = k^2 = CP \cdot CP'$ it follows that
$$\frac{CH'}{CP} = \frac{CP'}{CH}.$$
Moreover, since $\angle HCP$ is common to both $\triangle HCP$ and $\triangle P'CH'$, it follows that these two triangles are similar, and hence
$$\angle H'P'C = \angle PHC = \frac{\pi}{2}.$$

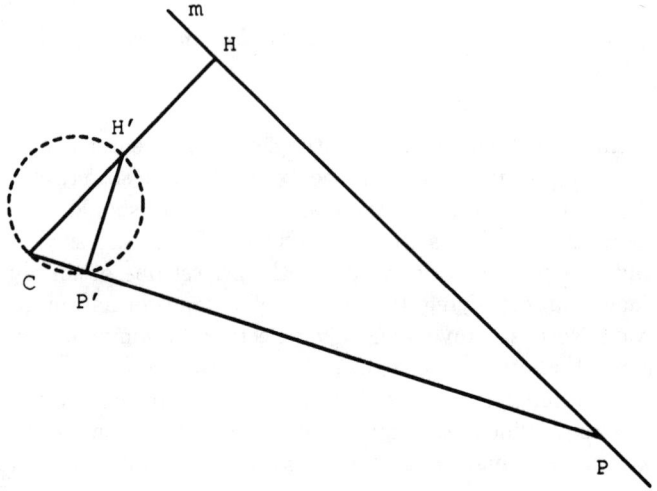

FIGURE 3.3.

Since the position of H' is independent of that of P and P', it follows that the locus of P' is the circle that has CH' as its diameter.

c) It is clear that in the above proof the line m can be chosen so that CH' coincides with any line segment that terminates at C. Thus any circle that contains C is the image of some line m not through C. Since the inversion $I_{C,k}$ is an involution we are done.

d) Let p be a circle not containing C, and let P be an arbitrary point on p. Let DE be a diameter of p whose extension contains C, and let the given inversion $I_{C,k}$ map the points D, E, P onto the points D', E', P', respectively (Figure 3.4). Since

$$CP \cdot CP' = CD \cdot CD' = CE \cdot CE' = k^2,$$

it follows that

$$\frac{CD}{CP'} = \frac{CP}{CD'}$$

and

$$\frac{CE}{CP'} = \frac{CP}{CE'}.$$

Since the angle $\angle DCP$ is common to the four triangles DCP, $D'CP'$, ECP, $P'CE'$, it follows that the first two are similar to each other, as are the last two. Consequently,

$$\angle CDP = \angle CP'D' \quad \text{and} \quad \angle CEP = \angle CP'E'.$$

Using the facts that the exterior angle of a triangle is equal to the sum of the two interior angles that are not adjacent to it, and that the angle subtended by

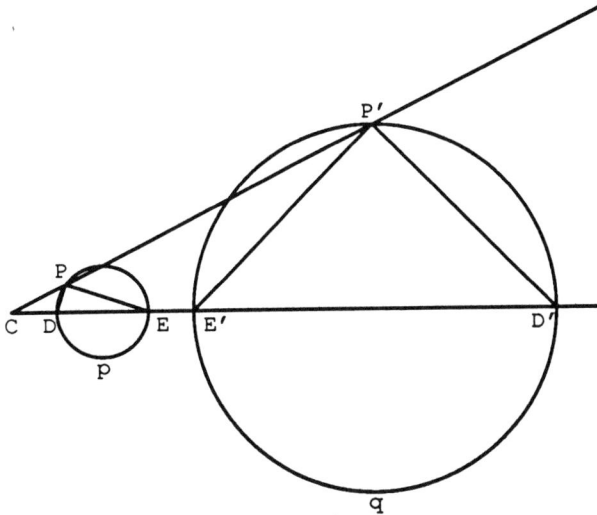

FIGURE 3.4.

a diameter at the circumference is $\frac{\pi}{2}$, we get

$$\angle E'P'D' = \angle CP'D' - \angle CP'E' = \angle CDP - \angle CEP = \angle DPE = \frac{\pi}{2}.$$

Since the position of D' and E' is independent of that of P and P', it follows that the locus of P' is a circle that has $D'E'$ as its diameter. q.e.d.

Example 3.4. Let $I = I_{O,4}$, where O denotes the origin, let m denote the line $x = -4$, and let q denote the circle with center $(0,2)$ and radius 1 (Fig. 3.5). Then, by part ii of the above theorem, $I(m)$ is a circle that contains the origin. Since I fixes $(-4,0)$, this circle must contain $(-4,0)$. By the same theorem's last assertion, the center of $I(m)$ lies on the x axis. Thus $I(m)$ is the circle of radius 2 that is centered at $(-2,0)$.

Turning to the circle q, it follows from the same theorem that $I(q)$ is a circle that is bisected by the Y axis. The intercepts of q with the Y axis are $(0,1)$ and $(0,3)$ and these are transformed by I to the points $(0,16)$ and $(0,\frac{16}{3})$. Hence these two points are the endpoints of a diameter of $I(q)$.

Example 3.5. What is the inversion $I_{C,k}$ that transforms the circle of radius 2 that is centered at the origin into the line $y = 6$? It follows from Theorem 3.3 that C must be one of the points $(0, \pm 2)$. Since C cannot separate a point from its image, it must be $(0,-2)$. Finally, since C transforms the point $(0,2)$ to the point $(0,6)$, it follows that

$$k = \sqrt{[2-(-2)] \cdot [6-(-2)]} = 4\sqrt{2}.$$

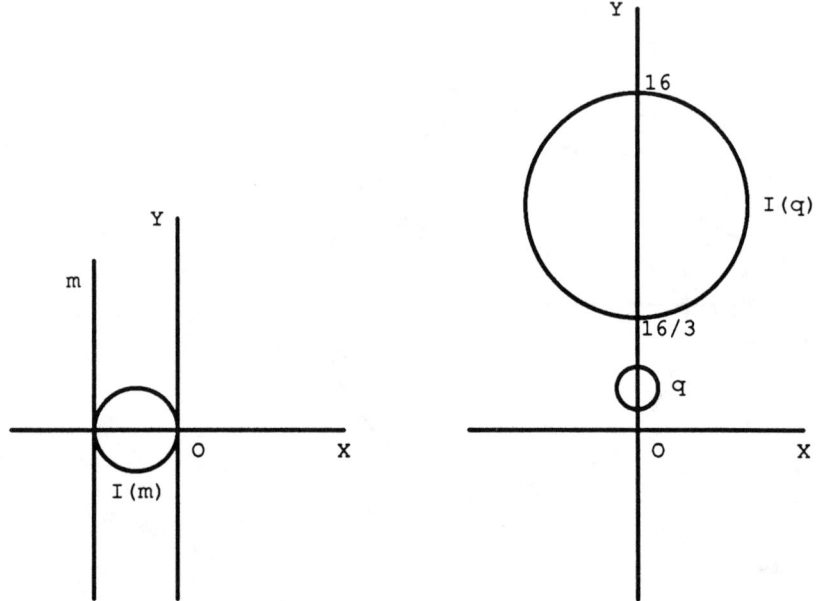

FIGURE 3.5.

Inversions are not rigid motions. A pair of points that are both very close to the center of the inversion, and so also close to each other, will be transported by the inversion into a pair a points that are very far from each other. However, inversions do share a very important property with rigid motions. Both types of transformations also preserve measures of angles. It is so intuitively clear that the rigid motions preserve the measures of angles that we will not belabor the point any further except to say that this follows from the three sides congruence theorem which is Proposition 8 of Euclid. That inversions preserve the measures of angles is indeed a remarkable coincidence. Note that inversions do *not* preserve the angles themselves, since straight lines are in general inverted into arcs. However, inversions do transform any given angle into another angle that has the same measure. A word of caution is required here. It is clear that inversions reverse the orientation of angles. Thus when we say that they preserve angles, we mean that the *unsigned* measure is preserved, although the sign is reversed. Transformations that preserve measures of angles are said to be *conformal,* and we shall study many more in great detail in later chapters. In more advanced contexts, the term "conformal" is reserved for transformations that preserve both the measure and the orientation (sign) of the angle. For the sake of simplifying the terminology we shall use this word in the broader sense in this book.

Since inversions do not transform rectilineal angles into rectilineal angles,

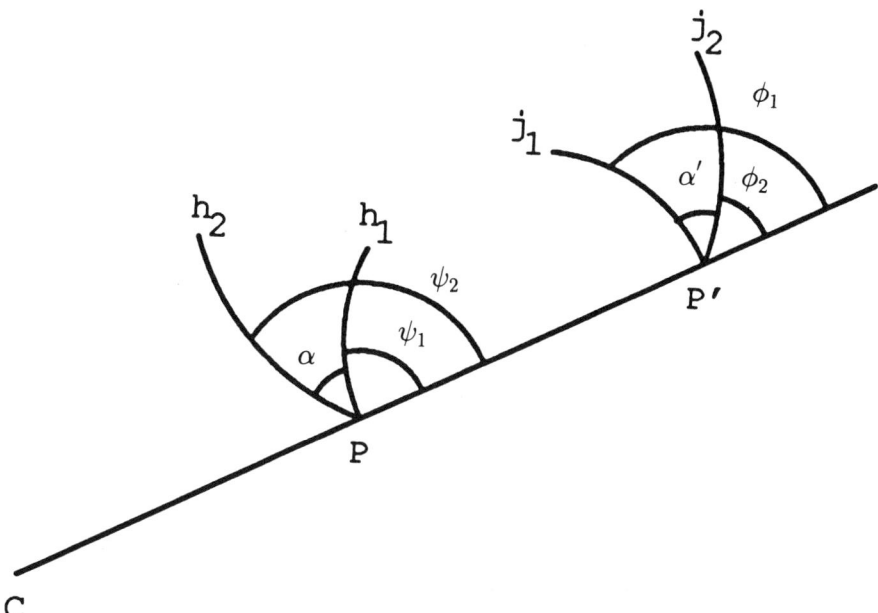

FIGURE 3.6.

we must understand the term angle here to apply to the configuration formed by any two intersecting curves, much the same as Euclid had in mind in his Definition 8. The measure of such an angle is the measure of the rectilineal angle formed by the tangents to these two curves at their point of intersection.

Theorem 3.6. *Inversions are conformal transformations of the plane.*

PROOF: Let $I_{C,k}$ be an inversion. Place a polar coordinate system with its origin at C and its initial ray along the positive x-axis. Let h denote the curve

$$r = f(\theta) \quad \alpha \leq \theta \leq \beta.$$

The inversion $I_{C,k}$ then maps h to the curve h' given by

$$r = F(\theta) = \frac{k^2}{f(\theta)} \quad \alpha \leq \theta \leq \beta.$$

Suppose now that the angle α at P has sides h_i given by the equations $r = f_i(\theta)$ for $i = 1, 2$ respectively (Fig. 3.6). Suppose further that the given inversion $I_{C,k}$ maps P, h_1, and h_2 onto P', j_1, and j_2 respectively. Then the image curves j_i have the equations

$$r = F_i(\theta) = \frac{k^2}{f_i(\theta)}, \quad i = 1, 2.$$

58 THE POINCARÉ HALF-PLANE

For any point P on any curve $r = f(\theta)$ the angle from the radius vector CP to the tangent line at P is given by

$$\tan \psi = \frac{r}{r'}.$$

Consequently

$$\tan \phi_1 = \frac{F_1}{F_1'} = \frac{k^2/f_1}{-k^2 f_1'/f_1^2} = -\frac{f_1}{f_1'} = -\tan \psi_1.$$

Hence, $\phi_1 = \pi - \psi_1$ and similarly $\phi_2 = \pi - \psi_2$. This, in turn yields

$$\alpha' = \phi_1 - \phi_2 = (\pi - \psi_1) - (\pi - \psi_2) = \psi_2 - \psi_1$$
$$= -\alpha$$

q.e.d.

Two circles are said to be *orthogonal* if they intersect and their respective tangents at a point of intersection are perpendicular to each other. Since the tangent to a circle is known to be perpendicular to the radius through the point of intersection, it follows that two circles are orthogonal if and only if they intersect and the tangent to each circle at their point of intersection also passes through the center of the other circle.

The following proposition will prove very useful later when we discuss non-Euclidean circles.

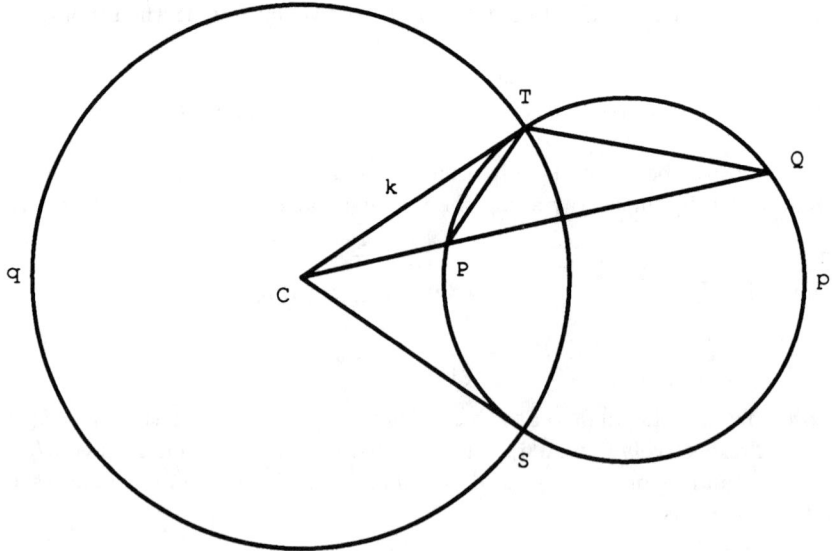

FIGURE 3.7.

Proposition 3.7. *Let q be a circle with center C and radius k, and let p be any other circle. Then the inversion $I_{C,k}$ fixes the circle p if and only if the circles p and q are orthogonal.*

PROOF: Let p be a circle orthogonal to q and suppose they intersect at a point T. Let P be an arbitrary point of the circle p, and let Q be the other point where the secant line CP intersects the circle p. By Proposition 3.2,
$$CP \cdot CQ = CT^2 = k^2,$$
and hence $I_{C,k}(P) = Q$. In other words, the inversion $I_{C,k}$ maps the circle p onto itself.

Conversely, suppose the circle p is fixed by $I_{C,k}$. Since $I_{C,k}$ interchanges points inside and outside q, it follows that p must intersect q in two points, say S and T. Since the inversion $I_{C,k}$ fixes both p and the straight lines CS and CT, it follows that the points S and T are both fixed by $I_{C,k}$. Let P be any other point of p, and let Q be the other intersection of the secant CP with the circle p. Since
$$CP \cdot CQ = CT^2 = k^2,$$
it follows that
$$\frac{CP}{CT} = \frac{CT}{CQ},$$
and hence triangles CPT and CTQ are similar. Consequently,
$$\angle CTP = \angle CQT,$$
and so, by Proposition 3.1, the line CT is tangent to the circle p. Thus, the tangents to p and q through T are perpendicular to each other, and so the circles p and q are orthogonal to each other. q.e.d.

3.2 AN APPLICATION OF INVERSIONS (OPTIONAL)

Inversions have many interesting properties and have been studied in great detail. Of particular significance is the fact that they transform straight lines into circles and vice versa. This gives them a primary role in the investigation of ruler and compass constructions. The readers are referred to Eves' and Pedoe's books for more details on this topic. In this text we limit ourselves to showing how the inversion of a circle into a straight line can be used to reduce a difficult problem about the former into a triviality about the latter. First, of course, a lemma is needed.

Lemma 3.8. *Let the inversion $I_{C,k}$ transform the points P and Q into the points P' and Q' respectively. Then*
$$P'Q' = \frac{k^2 PQ}{CP \cdot CQ}.$$

60 THE POINCARÉ HALF-PLANE

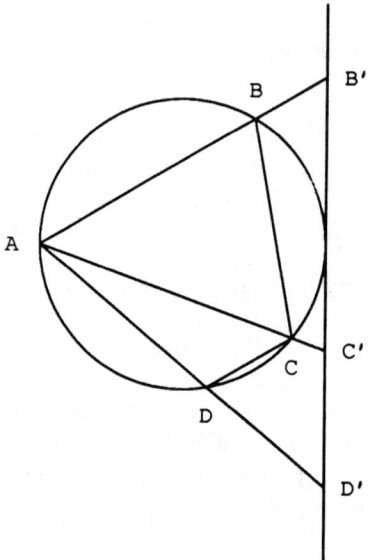

FIGURE 3.8.

PROOF: Assume first that the points C, P, Q are not collinear. From the definition of inversions we conclude that

$$CP \cdot CP' = k^2 = CQ \cdot CQ'.$$

Hence

$$\frac{CP}{CQ} = \frac{CQ'}{CP'}$$

and so $\triangle CPQ$ and $\triangle CQ'P'$ are similar. Consequently,

$$\frac{P'Q'}{PQ} = \frac{CQ'}{CP} = \frac{CQ' \cdot CQ}{CP \cdot CQ} = \frac{k^2}{CP \cdot CQ},$$

from which the desired result follows immediately.

As this is the only case which will be used in the proof of the following theorem, the remaining case, where the points C, P, Q are collinear is relegated to Exercise 12. q.e.d.

Proposition 3.9. *In a cyclic quadrilateral the product of the diagonals is equal to the sum of the products of the two pairs of opposite sides.*

PROOF: Let $ABCD$ be a cyclic quadrilateral inscribed in a circle of diameter k (Fig. 3.8). By Theorem 3.3 $I_{A,k}$ inverts this circle into a line tangent to the circle, and the points B, C, D, into points B', C', D' on that line. It is clear that

$B'C' + C'D' = B'D'$ whence, by the above lemma,

$$\frac{k^2 BC}{AB \cdot AC} + \frac{k^2 CD}{AC \cdot AD} = \frac{k^2 BD}{AB \cdot AD}$$

or

$$BC \cdot AD + CD \cdot AB = BD \cdot AC.$$

q.e.d.

3.3 EXERCISES

1. Prove Proposition 3.1.
2. Prove that the points P and P' of Figure 3.2 are indeed symmetrical with respect to the circle q.
3. If O denotes the origin, to what point or curve does the inversion $I_{O,5}$ transform the sets below?
   a) The point (-2,3).
   b) The point (6,8).
   c) The line $y = 2x$.
   d) The line $x + y = 5$.
   e) The line $y = 5$.
   f) The line $y = x + 10$.
   g) The circle centered at O with radius 3.
   h) The circle centered at O with radius 10.
   i) The circle centered at (0,3) with radius 1.
   j) The circle centered at (0,3) with radius 5.
   k) The circle centered at (0,3) with radius 10.
   l) The circle centered at (15,0) with radius 2.
   m) The circle centered at (15, 0) with radius 13.
4. For each of the following pairs of curves, decide whether there exists an inversion that transforms one onto the other. Identify the inversion if it exists.
   a) The y axis and the straight line $x = 2$.
   b) The circle $x^2 + y^2 = 16$ and the straight line $x = 2$.
   c) The circle $x^2 + y^2 = 16$ and the straight line $x = 4$.
   d) The circle $x^2 + y^2 = 16$ and the straight line $x = 8$.
   e) The circles $x^2 + y^2 = 16$ and $x^2 + y^2 = 100$.
   f) The circles $x^2 + y^2 = 16$ and $(x - 4)^2 + y^2 = 16$.
   g) The circles $x^2 + y^2 = 16$ and $(x - 34)^2 + y^2 = 900$.
5. Let p be a fixed circle and let P be a fixed point not on p. Show that there exists another point P' such that every circle containing P and orthogonal to the circle p also contains the point P'.
6. Let p be a fixed circle and let P be a fixed point. Show that the locus of the centers of all the circles that contain P and are orthogonal to p is a straight line.
7. Let I be an inversion and let p be a circle such that $I(p)$ is also a circle. When do p and $I(p)$ have unequal radii?
8. Let p and q be circles with unequal radii. Prove that there is an inversion that transforms p onto q.
9. Two circles p, q touch at a point T, and a variable circle through T cuts p and q orthogonally in P and Q respectively. Prove that PQ passes through a fixed point.
10. Suppose A, B, C, D are four points on a circle. If T is the point of contact of a circle containing A and B with another circle containing C and D, show that the locus of T is a circle.

11. Let two circles p and q intersect in A and B, and let the diameters of p and q through B cut q and p in C and D respectively. Show that the line AB passes through the center of the circle containing B, C, and D.
12. Complete the proof of Lemma 3.8.
13. Let A, B be any two points, let $\tau = \tau_{AB}$, and let k be any positive number. Prove that
$$\tau \circ I_{A,k} \circ \tau^{-1} = I_{B,k}.$$
14. Let R be a rotation and let $I_{A,k}$ be any inversion. Identify $R \circ I_{A,k} \circ R^{-1}$.
15. Let I be any inversion and let f be any rigid motion. Is $I \circ f \circ I^{-1}$ necessarily a rigid motion?

The following exercises call for the use of a graphics package such as those available on Mathematica or Maple.

16. Write a script that inverts an arbitrary line in an arbitrary circle.
17. Write a script that inverts an arbitrary circle in an arbitrary circle.

4

The Hyperbolic Plane

4.1 THE HYPERBOLIC DISTANCE

Imagine a two dimensional universe, with a superimposed Cartesian coordinate system, in which the x-axis is infinitely cold. Imagine further, that as the objects of this universe approach the x-axis, the drop in temperature causes them to contract. Thus the inhabitants of this strange land will find that it takes them less time to walk along a horizontal line from $A(0,1)$ to $B(1,1)$ (Figure 4.1) than it takes to walk along a horizontal line from $C(0,.5)$ to $D(1,.5)$. Since their rulers contract just as much as they do,

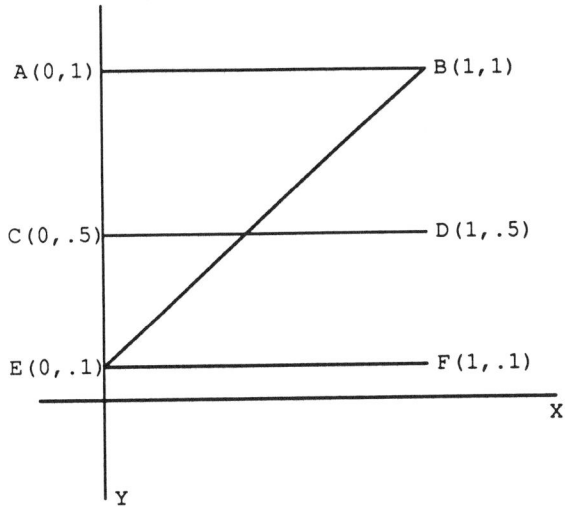

FIGURE 4.1.

this observation will not seem at all paradoxical to them. If we assume further that the contraction is such that the outside observer sees the length of a given object as being inversely proportional to its distance from the x-axis, then the inhabitants will find that walking from A to B takes half the time of walking from C to D, and one tenth the time of walking from $E(0,.1)$ to $F(1,.1)$.

We shall later see that the nature of the x-axis is such as to make impossible any communication between the lower and the upper half-planes. Consequently we restrict our attention to the upper half-plane and refer to it as the *hyperbolic plane*. It is also known as the *Poincaré upper half-plane*. The exact differences between these two names will be discussed in Section 4.3 and again in Chapter 15. It is very important to note that what distinguishes the hyperbolic plane from the Euclidean one is not the underlying space, but the way distance is measured in this common underlying space. Loosely speaking, the hyperbolic distance is defined in the above paragraph as

$$\text{hyperbolic length} = \frac{\text{Euclidean length}}{y},$$

with the constant of proportionality being chosen as 1. The reader is warned against taking this as a formal definition of hyperbolic length. It does indeed tell us that the hyperbolic length of the line segment joining A to B is $1/1 = 1$ and that the hyperbolic length of the line segment joining E to F is $1/.1 = 10$. However, it does not provide an immediate way of calculating the hyperbolic lengths of, say, the line segments joining E to A or to B. To deal with these lines we make use of the infinitesimal calculus. Let $P(x, y)$, $Q(x + dx, y)$, and $R(x + dx, y + dy)$ be an infinitesimal triangle. The Euclidean length of the infinitesimal line segment PR is of course

$$\text{Euclidean length of PR} = \sqrt{dx^2 + dy^2}.$$

Accordingly, we define, again loosely, the hyperbolic length of the infinitesimal line segment PR as

$$\text{hyperbolic length of PR} = \frac{\sqrt{dx^2 + dy^2}}{y} \qquad (1)$$

This definition turns the computation of hyperbolic lengths into calculus exercises. For example, along the line segment joining E to A, dx is identically zero, and hence,

$$\text{hyperbolic length of EA} = \int_{.1}^{1} \frac{dy}{y} = \ln 1 - \ln .1$$

$$= \ln 10 \approx 2.303.$$

Turning to the slanted line EB, we note that its equation is

$$y - .1 = \frac{1 - .1}{1 - 0}(x - 0)$$

or

$$y = .9x + .1.$$

Differentiation now yields the equation

$$dy = .9\,dx.$$

If these values of y and dy are substituted into (1) and integration is again used to add the hyperbolic lengths of the infinitesimal line segments along EB we get

$$\text{hyperbolic length of EB} = \int_0^1 \frac{\sqrt{dx^2 + (.9\,dx)^2}}{.9x + .1}$$

$$= \sqrt{1.81} \int_0^1 \frac{dx}{.9x + .1} = \left. \frac{\sqrt{1.81}}{.9} \ln(.9x + .1) \right|_0^1$$

$$\approx 3.442.$$

It should now be clear to the reader that (1) provides us with a method for expressing the hyperbolic length of any curve as a definite integral. We define the *hyperbolic length* of an arbitrary curve γ as the expression

$$\int_\gamma \frac{\sqrt{dx^2 + dy^2}}{y} \qquad (2)$$

For example the hyperbolic length of the segment of the parabola $y = x^2$ that joins the points (2,4) and (3,9) is

$$\int_2^3 \frac{\sqrt{dx^2 + (2x\,dx)^2}}{x^2} = \int_2^3 \sqrt{\frac{1 + 4x^2}{x^2}}\,dx.$$

While this integral does not yield itself easily to the standard methods of integration taught in first year calculus, it can be evaluated, with arbitrary accuracy, by Simpson's rule. Thus, we indeed do have a method for evaluating the hyperbolic length of an arbitrary curve in the hyperbolic plane. The following two propositions are offered as further examples that will turn out to be very useful in the sequel.

Proposition 4.1. *Let q be a circle with center $C(c,0)$ and radius r. If P and Q are points of q such that the radii CP and CQ make angles α and β ($\alpha < \beta$) respectively, with the positive x-axis, then the*

$$\text{hyperbolic length of arc } PQ = \ln \frac{\csc \beta - \cot \beta}{\csc \alpha - \cot \alpha}.$$

PROOF: If t is the angle from the positive x-axis to the radius through an arbitrary point (x,y) on q, then

$$x = c + r\cos t$$
$$y = r\sin t.$$

Consequently

$$dx = -r\sin t\, dt \quad \text{and} \quad dy = r\cos t\, dt,$$

and the hyperbolic length of the arc PQ is

$$\int_\alpha^\beta \frac{\sqrt{(-r\sin t\, dt)^2 + (r\cos t\, dt)2}}{r\sin t} = \int_\alpha^\beta \frac{r\, dt}{r\sin t}$$
$$= \int_\alpha^\beta \csc t\, dt = \ln\frac{\csc\beta - \cot\beta}{\csc\alpha - \cot\alpha}.$$

q.e.d.

The above propostion implies that the hyperbolic length of a circular arc that is centred on the x-axis depends only on the inclination of the radii through the arc's endpoints to the positive x-axis. This verifies the next corollary.

Corollary 4.2. *If PQ and $P'Q'$ are arcs of Euclidean circles which share the same center on the x axis and such that each of the triples C, P, P' and C, Q, Q' is collinear, then the two arcs PQ and $P'Q'$ have the same hyperbolic length.* □

The hyperbolic length of vertical line segments is also easily computed by the same means that were used to compute the hyperbolic length of the line segment EA of Figure 4.1 above.

Proposition 4.3. *The hyperbolic length of the Euclidean line segment joining the points $P(a,y_1)$ and $Q(a,y_2)$, $0 < y_1 \leq y_2$, is*

$$\ln\frac{y_2}{y_1}.$$

PROOF: See Exercise 16. □

4.2 HYPERBOLIC STRAIGHT LINES

Let us return to the examples that were developed at the beginning of this chapter. It was calculated there that the line segments AB, EA, and EB have hyperbolic lengths 1, 2.303, and 3.442, respectively. Since

$$3.442 > 1 + 2.303$$

it follows that in the hyperbolic plane the Euclidean straight line joining two points does not provide the shortest hyperbolic distance between those two points. This, of course, poses the very natural problem of identifying the curve whose hyperbolic length is the shortest amongst all the curves that join a given pair of points in the hyperbolic plane. Such curves are called *geodesic segments*, and their identification is surprisingly, and thankfully, quite easy.

Theorem 4.4. *The geodesic segments of the hyperbolic plane are either*

a) *arcs of Euclidean semicircles that are centered on the x-axis,*

or

b) *segments of Euclidean straight lines that are perpendicular to the x-axis.*

PROOF: Let $P(x_1, y_1)$ and $Q(x_2, y_2)$ be any two points of the hyperbolic plane, and let g be a curve joining them. Two cases must be considered.

CASE I: $x_1 \neq x_2$. In this case the Euclidean line segment PQ is not perpendicular to the x-axis. Let $C(c,0)$ be the x-intercept of the perpendicular bisector to PQ. Place a polar coordinate system so that its origin coincides with C and its initial ray points in the same direction as the positive x-axis. Relative to this polar coordinate system let the geodesic segment g be a part of the curve whose equation is $r = f(\theta)$, and let the coordinates of P and Q be (r_P, α) and (r_Q, β), respectively (Fig. 4.2). By (1) the hyperbolic length of g is

$$\int_g \frac{\sqrt{dx^2 + dy^2}}{y}.$$

As these polar coordinates are related to the defining Cartesian coordinates by the equations

$$x = c + r \cos \theta$$

$$y = r \sin \theta,$$

it follows that

$$\frac{dx}{d\theta} = \frac{dr}{d\theta} \cos \theta + r \frac{d \cos \theta}{d\theta} = r' \cos \theta - r \sin \theta$$

$$\frac{dy}{d\theta} = \frac{dr}{d\theta} \sin \theta + r \frac{d \sin \theta}{d\theta} = r' \sin \theta + r \cos \theta.$$

68 THE POINCARÉ HALF-PLANE

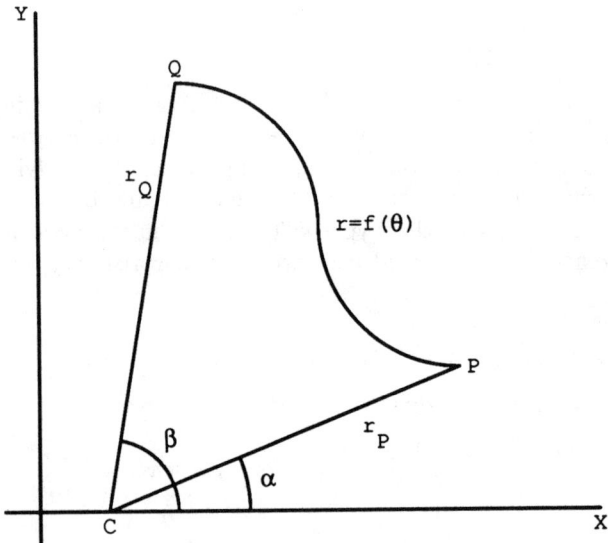

FIGURE 4.2.

Consequently,

$$dx^2 + dy^2 = (r'\cos\theta - r\sin\theta)^2 d\theta^2 + (r'\sin\theta + r\cos\theta)^2 d\theta^2$$
$$= [r'^2(\cos^2\theta + \sin^2\theta) + 2rr'(\cos\theta\sin\theta - \sin\theta\cos\theta) + r^2(\sin^2\theta + \cos^2\theta)] d\theta^2$$
$$= (r'^2 + r^2) d\theta^2.$$

Substituting this back into expression (1) we conclude that the hyperbolic length of g is

$$\int_\alpha^\beta \frac{\sqrt{r'^2 + r^2}}{r\sin\theta} d\theta \geq \int_\alpha^\beta \frac{\sqrt{r^2}}{r\sin\theta d\theta} = \int_\alpha^\beta \csc\theta\, d\theta$$
$$= \ln\frac{\csc\beta - \cot\beta}{\csc\alpha - \cot\alpha}.$$

Since we already know by Proposition 4.1 that the last expression above happens to be the hyperbolic length of the arc of the Euclidean circle with radius $CP = CQ$, it follows that this arc is the geodesic segment joining P and Q.

CASE II: $x_1 = x_2$. Let g have equation $x = f(y)$. If dx/dy is denoted by f', then the hyperbolic length of g is

$$\int_{y_1}^{y_2} \frac{\sqrt{f'^2 dy^2 + dy^2}}{y} = \int_{y_1}^{y_2} \frac{\sqrt{f'^2 + 1}}{y} dy \geq \int_{y_1}^{y_2} \frac{dy}{y} = \ln\frac{y_2}{y_1}.$$

This last expression, however, was already shown in Proposition 4.3 to be the hyperbolic length of the Euclidean line segment PQ. Hence the Euclidean line segment joining P and Q is also the geodesic segment joining them.

q.e.d.

A Euclidean semicircle that is centered on the x-axis and lies in the upper half-plane is called a *geodesic* or a *hyperbolic straight line*. The same name also applies to Euclidean half-lines that begin at the x-axis, are perpendicular to it, and also lie in the upper half-plane. Whenever it becomes necessary to distinguish between them, we shall speak of the former as *bowed* geodesics and of the latter as *straight*. This terminology will be used notwithstanding the fact that both types of geodesics are straight relative to the hyperbolic way of measuring distance in the sense that every segment of one of these lines is the shortest curve joining its endpoints. It is important to note that in such a hyperbolic universe light, which by a well known physical principle must follow geodesics rather than Euclidean straight lines, would necessarily travel along these hyperbolic geodesics, be they bowed or straight. Thus, the inhabitant of a hyperbolic universe would agree with Euclid's Definition 4 which says that every segment lies evenly with its endpoints.

Any two points of the upper half-plane can be joined by a geodesic. Points that have the same abscissa are joined by a straight geodesic. Points that have different abscissas are joined by a bowed geodesic whose center (as a Euclidean semicircle) is the intersection of the x-axis with the perpendicular bisector of the Euclidean line segment joining the given points.

The *hyperbolic distance* between the points P and Q, denoted by $h(P, Q)$, is the hyperbolic length of the geodesic joining them.

Example 4.5. Find the hyperbolic length of the geodesic segment joining the points $A(8,4)$ and $B(0,8)$ of Fig. 4.3.

The center C of the geodesic joining A and B is obtained by finding the intersection of the perpendicular bisector to the Euclidean line segment AB with the x-axis. Since the Euclidean straight line joining A and B has slope

$$\frac{4-8}{8-0} = -\frac{1}{2}$$

and the midpoint M of the Euclidean line segment AB has coordinates

$$\left(\frac{0+8}{2}, \frac{8+4}{2}\right) = (4, 6)$$

it follows that this perpendicular bisector has equation

$$y - 6 = 2(x - 4)$$

70 THE POINCARÉ HALF-PLANE

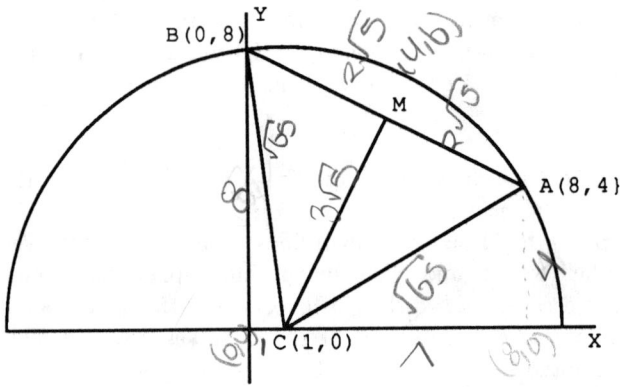

FIGURE 4.3.

and its x-intercept is $(1,0)$. Since $CA = CB = \sqrt{(8-1)^2 + (4-0)^2} = \sqrt{65}$ it follows that the hyperbolic length of the geodesic segment joining A and B is

$$\ln \frac{\frac{\sqrt{65}}{8} - \left(\frac{-1}{8}\right)}{\frac{\sqrt{65}}{4} - \frac{7}{4}} = 1.450....$$

Example 4.6. Find the hyperbolic midpoint M of the geodesic joining the points $A(a,p)$ and $B(a,q)$.

Since the geodesic joining the points A and B is straight, it follows from Proposition 4.3 that M has coordinates (a,b) where b satisfies the equation

$$\frac{b}{p} = \frac{q}{b}.$$

Thus $M = (a, \sqrt{pq})$.

The reader is referred to Exercise 6 for the description of a method for finding the hyperbolic midpoint of an arbitrary hyperbolic geodesic segment.

4.3 HYPERBOLIC ANGLES

Thus far we have defined a new distance in the plane and found out what its "straight lines" are. We next turn to measurement of angles in this new geometry. In agreement with Euclid's use of the word, an angle is allowed to have arbitrary curves as its sides. The Euclidean measure of such an angle is defined to be the usual measure of the rectilineal angle formed by the two tangents to the sides of the angle at its vertex. Note that in this more general context it may easily happen that non-congruent angles have the same measure. In fact, all the angles at the origin one of whose sides is the x-axis and the other the curve $y = x^n$, $n = 2, 3, 4, ...$, have the same measure of 0. In contrast with the hyperbolic distance and geodesics, the task of defining *the hyperbolic measure of an angle* turns out to be quite easy. The hyperbolic measure of an angle is identical with its Euclidean measure.

If the readers feel that this definition is too facile, they are probably quite right. The fact of the matter is that geometrical measures cannot be arbitrarily defined. They must possess the properties specified in Euclid's Common Notions. Moreover, for lengths of geodesic segments and measures of rectilineal angles (those formed by geodesics) one would expect equality in measure to entail congruence. Namely, geodesic segments of equal lengths should be congruent, and the same must hold for rectilineal angles of equal measure. It is now clear that at this point we must provide for congruences in the hyperbolic plane.

4.4 HYPERBOLIC RIGID MOTIONS

As was demonstrated in Chapter 2, the Euclidean rigid motions enjoy two properties:
  i) they are generated by reflections,
  ii) they are classified into reflections, translations, rotations, and glide-reflections.

In this chapter we shall content ourselves with defining the hyperbolic analogs of reflections. In the next chapter it will be argued that property i above also holds for all hyperbolic rigid motions. The classification of all the hyperbolic motions is known but is too complicated for the purposes of this book. Instead, an elegant algebraic description will be provided in Chapter 9.

Theorem 4.7. *The following transformations of the hyperbolic plane preserve both hyperbolic lengths and measures of angles:*
  a) inversions $I_{C,k}$ where C is on the x-axis.
  b) reflections ρ_m where m is perpendicular to the x-axis,
  c) translations τ_{AB} where AB is parallel to the x-axis,

PROOF: We already know by Theorem 3.4 that all inversions preserve the Euclidean measures of angles. It also follows from the SAS congruence theorem that all Euclidean rigid motions also preserve the Euclidean measures of angles. Since the hyperbolic measure of an angle coincides with its Euclidean measure it follows that all the transformations in question do indeed preserve the hyperbolic measures of angles. We now go on to prove their rigidity.

a) Let $I_{C,k}$ be the given inversion. Place a polar coordinate system with its origin at C and its initial ray along the positive x-axis. Let h denote the curve

$$r = f(\theta) \qquad \alpha \le \theta \le \beta.$$

The inversion $I_{C,k}$ then maps h to the curve h' given by

$$r = F(\theta) = \frac{k^2}{f(\theta)} \qquad \alpha \le \theta \le \beta.$$

As we saw in the proof of Theorem 4.4, the hyperbolic length of h' is

$$= \int_{h'} \frac{\sqrt{r'^2 + r^2}}{r \sin \theta} d\theta = \int_a^\beta \frac{\sqrt{F'^2 + F^2}}{F \sin \theta} d\theta$$

$$= \int_\alpha^\beta \frac{\sqrt{(-k^2 f'/f^2)^2 + (k^2/f)^2}}{k^2 \sin \theta / f} d\theta = \int_\alpha^\beta \frac{\sqrt{f'^2 + f^2}}{f \sin \theta} d\theta$$

$$= \int_h \frac{\sqrt{r'^2 + r^2}}{r \sin \theta} d\theta = \text{hyperbolic length of } h.$$

Thus the given inversion does indeed preserve hyperbolic lengths.

b) Let the line m have equation $x = c$. It is clear that the two points (x_1, y_1) and (x_2, y_2) are symmetrical with respect to the this line if and only if

$$c = \frac{x_1 + x_2}{2} \quad \text{and} \quad y_1 = y_2.$$

If γ is any curve parametrized as $[u(t), v(t)]$, $a \le t \le b$, then $\rho_m(\gamma)$ has parametrization $[2c - u(t), v(t)]$, $a \le t \le b$. Now

along γ : $\qquad dx = u'(t) dt \qquad dy = v'(t) dt$

along $\rho_m(\gamma)$: $\qquad dx = -u'(t) dt \qquad dy = v'(t) dt$.

Consequently both γ and $\rho_m(\gamma)$ have the common value

$$\int_a^b \frac{\sqrt{u'^2(t) + v'^2(t)}}{v(t)} dt$$

as their hyperbolic length.

c) Let τ be the translation such that $\tau(x, y) = (x + h, y)$ for some fixed number h. If γ is any curved parametrized as $[u(t), v(t)]$, $a \le t \le b$, then $\tau(\gamma)$ has parametrization $[u(t) + h, v(t)]$, $a \le t \le b$. Now along both γ and $\tau(\gamma)$ we have

$$dx = u'(t) dt \qquad dy = v'(t) dt.$$

Consequently both γ and $\tau(\gamma)$ have the common value
$$\int_a^b \frac{\sqrt{u'^2(t)+v'^2(t)}}{v(t)}\,dt$$
as their hyperbolic length. q.e.d.

Corollary 4.8. *All the transformations listed in the statement of Theorem 4.7 carry geodesics onto geodesics.* □

At this point it may not be clear that hyperbolic rectilineal angles that have the same measure are necessarily hyperbolically congruent, or that the same holds for geodesic segments. The establishment of these facts has been assigned to the reader as Exercises 14 and 15. In the next chapter we shall demonstrate that all hyperbolically rectilineal *right* angles are hyperbolically congruent, and so the reader may wish to postpone attempting these problems until then.

4.5 RIEMANNIAN GEOMETRY (OPTIONAL)

The reader is now acquainted with two geometries. Euclidean geometry and the hyperbolic geometry of the upper half-plane. Even on the half-plane that they share these geometries are different in the sense that they endow the same curves with different lengths. This makes it clear that the notion of distance is actually relative in the sense that the same underlying space can be endowed with different notions of length. Each such notion gives rise a new geometry with its own properties and theorems.

The recognition that the measurement of length on a surface is to a great extent independent of the manner in which the surface is defined is fairly explicit in the works of Gauss, and we shall have something to say about this in Chapter 12. However, the credit for completely emancipating the notion of length goes to G. F. B. Riemann (1826–1866) who proposed in the year of 1854 a very general modification of planar distances. He suggested that mathematicians study geometries in which the length of an arbitrary curve γ lying in the plane would be given by the expression

$$\int_\gamma \sqrt{E\,dx^2 + 2F\,dx\,dy + G\,dy^2} \qquad (3)$$

wherein E, F, and G are functions of x and y that satisfy the constraints

$$E > 0, \quad G > 0, \quad EG - F^2 > 0. \qquad (4)$$

The integrand of (3) needs some clarification. Let γ be a plane curve with parametric equations

$$x = x(t), \quad y = y(t) \quad a \le t \le b.$$

Then the integrand of (3) should be thought of as an abbreviation for the unobjectionable, though more complicated looking, integral

$$\int_a^b \sqrt{E\left(\frac{dx}{dt}\right)^2 + 2F\frac{dx}{dt}\frac{dy}{dt} + G\left(\frac{dy}{dt}\right)^2}\, dt.$$

For example, when

$$E = G = \frac{1}{y^2} \quad \text{and} \quad F = 0$$

expression (3) reduces to the hyperbolic length defined in (2). When

$$E = G = 1 \quad \text{and} \quad F = 0,$$

and γ is any curve parametrized as $y = f(x)$, $x = x$, then

$$\int_\gamma \sqrt{dx^2 + dy^2} = \int_a^b \sqrt{1 + \left(\frac{dy}{dx}\right)^2}\, dx$$

which the reader should recognize as the Euclidean length of the curve γ.

The constraints (4) are dictated by the requirement that any reasonable definition of length must provide for a *positive* length along any nondegenerate curve. Exercise 22 demonstrates what can go wrong when these constraints are violated. This is not to say that integrals of type (3) which do not satisfy these constraints are of no interest. Far from it, as is made clear by the role that the integral

$$\int_\gamma \sqrt{dx^2 - dy^2}$$

plays in the special theory of relativity. However, this integral cannot then be interpreted as a length in any conventional sense.

Be that as it may, in the sequel we shall restrict ourselves to length integrals that do satisfy constraint (4). When discussing these integrals it has become customary to focus on the expression

$$E\, dx^2 + 2F\, dxdy + G\, dy^2$$

which is called a *Riemann metric*. The special cases

$$dx^2 + dy^2 \quad \text{and} \quad \frac{dx^2 + dy^2}{y^2}$$

are called the *Euclidean metric* and the *Poincaré metric*, respectively. Before going on to some specific examples, it is important to note that many metrics are only defined in part of the plane. Thus the Poincaré metric is clearly undefined on the x axis, and, as we have seen, it is customary to restrict its domain to the upper half-plane.

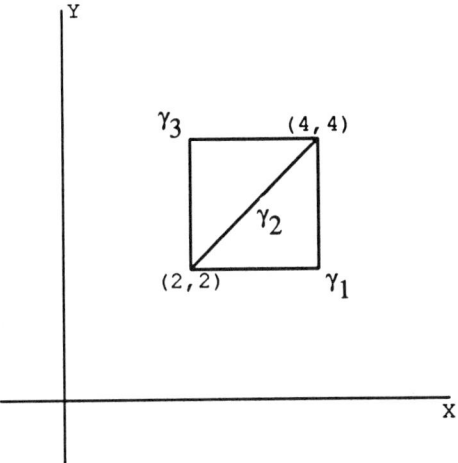

FIGURE 4.4.

Example 4.9. Consider the Riemann metric $y\,dx^2 + 2x\,dxdy + x^2\,dy^2$. Since $EG - F^2 = x^2y - x^2 = x^2(y-1)$, we can make sure that the necessary constraint is satisfied by restricting attention to the quarter plane defined by the inequalities $x > 0$ and $y > 1$. We shall compute the length of three curves joining the points (2,2) and (4,4) relative to this metric. (Figure 4.4).

Let γ_1 denote the polygonal line obtained by first joining (2,2) to (4,2) and then joining (4,2) to (4,4). Along the first of these line segments $dy = 0$ and $y = 2$; along the second, $dx = 0$ and $x = 4$. Consequently the length of γ_1 relative to this metric is

$$\int_2^4 \sqrt{2\,dx^2} + \int_2^4 \sqrt{4^2\,dy^2} = \sqrt{2}\int_2^4 dx + 4\int_2^4 dy$$

$$= 2\sqrt{2} + 8 = 10.83....$$

Let γ_2 denote the Euclidean straight line segment joining (2,2) to (4,4). Along this curve $y = x$ and $dy = dx$ and hence its length, relative to the given metric is

$$\int_2^4 \sqrt{3x + x^2}\,dx = 8.47....$$

Finally, let γ_3 denote the polygonal line that first joins (2,2) to (2,4) and then goes on to (4,4). Along the first of these line segments $dx = 0$ and $x = 2$; along the second $dy = 0$ and $y = 4$. Hence the length of this this curve relative to the given metric is

$$\int_2^4 \sqrt{4\,dy^2} + \int_2^4 \sqrt{4\,dx^2} = 8.$$

In principle, one should be able to mimic our discussion of the Poincaré metric for any Riemannian metric. Every such metric has geodesics. The differential equations that characterize the geodesics of a metric are known, but they are in general too intractable to admit an explicit solution. Thus even such a seemingly simple question as

If P and Q are two points in the domain of a certain Riemann metric, is there only one geodesic segment joining them?

can turn out to be very difficult. Nevertheless, this topic, and its generalization to spaces with an arbitrary number of dimensions has blossomed into a major mathematical discipline known as Riemannian geometry. The reasons for its success are both the intrinsic interest of the field and its applicability to other scientific disciplines both inside and outside mathematics. We shall return to this topic in Chapter 12 to describe one of its most important theorems.

The relativization of length and distance resulted in other consequences for mathematics above and beyond the proliferation of geometries. Chief among those has been the creation of the new discipline of topology which has assumed as its object the study of those properties of space that are completely independent of the notion of distance. After all, if distance is relative, then it is in some sense only "superficial" and it does make sense to look for concepts that strike "deeper" into the nature of space. We shall not do so in this book and the reader is referred to the Chinn and Steenrod's book for an elementary introduction to this topic, and to Massey's book for a more advanced treatment.

4.6 EXERCISES

1. Compute the hyperbolic length of the following curves joining the points (0,1) and (1,2):
   a) $y = x + 1$ b) $y = x^2 + 1$ c) $y = x^3 + 1$.
2. Find the points that divide the geodesic segment joining (0,10) and (0,1) into portions the ratio of whose hyperbolic lengths are:
   a) $1:1$ b) $2:1$ c) $3:2$.
3. Find the hyperbolic distance between each of pair of the three points (-1,2), (3,2), and (0,1).
4. Find the inversion that transforms the bowed geodesic from (0,0) to (4,0) to the straight geodesic $x = 10$.
5. Find the inversion which transforms the bowed geodesic from (-4,0) to (2,0) to the bowed geodesic from (10,0) to (20,0).

6. Let A and B be two points joined by a bowed geodesic segment g. Let $C(c,0)$ be a point on the x-axis and $M(c,y)$ be the point of g above C. Then M is the hyperbolic midpoint of g if $\angle BCM = \angle ACM$. (Hint: examine the combined effect of the reflection ρ_{CM} and the inversion $I_{C,CM}$ on g.)
7. Show that if the points $P(x_1, y_1)$ and $Q(x_2, y_2)$ are contained in a bowed geodesic with Euclidean center $(c,0)$ and radius r, then

$$h(P, Q) = \left| \ln \frac{(x_1 - c - r)y_2}{y_1(x_2 - c - r)} \right|.$$

8. Describe a Euclidean method for constructing some hyperbolic isosceles triangles.
9. Describe a Euclidean method for constructing some hyperbolic right triangles.
10. Given a hyperbolic geodesic g and a point P on g, describe a Euclidean method for constructing a hyperbolic geodesic through P that is orthogonal to g.
11. Given a hyperbolic geodesic g and a point P not on g, describe a Euclidean method for constructing a hyperbolic geodesic through P that is orthogonal to g.
12. Given a hyperbolic geodesic segment, describe a Euclidean method for constructing its perpendicular bisector.
13. Given any two distinct points of the hyperbolic plane, show that there exists a hyperbolic reflection that carries one onto the other.
14. Let AB and CD be geodesic segments of equal length. Prove that there is a hyperbolic rigid motion that maps A and AB onto C and CD, respectively.
15. Let g and h be a pair of intersecting geodesics, and let g' and h' be another pair of intersecting geodesics such that $\angle(g, h) = \angle(g', h')$. Prove that there is a hyperbolic rigid motion that maps g and h onto g' and h', repsectively.
16. Prove Proposition 4.3.
17. Compute the lengths of the three curves of Example 4.9 relative to the metric $x\,dx^2 + 2\,dx\,dy + y\,dy^2$.
18. The Riemann metric

$$\frac{dx^2 + dy^2}{1 - x^2 - y^2}$$

has the interior of the unit disk as its domain. For any point (a,b) inside the unit disk compute, relative to this metric, the length of the Euclidean line segment that joins it to the origin. What happens to this distance as the point (a,b) approaches the circumference of the unit disk?
19. If (a,b) is any point inside the unit disk, show that the Euclidean line segment joining it to the origin is a geodesic relative to the metric of problem 18.
20. The Riemann metric

$$\frac{(1-y^2)\,dx^2 + 2xy\,dx\,dy + (1-x^2)\,dy^2}{1 - x^2 - y^2}$$

has the interior of the unit disk as its domain. For any point (a,b) inside the unit disk compute, relative to this metric, the length of the Euclidean line segment that joins it to the origin. What happens to this distance as the point (a,b) approaches the circumference of the unit disk?
21. If (a,b) is any point inside the unit disk, show that the Euclidean line segment joining it to the origin is a geodesic relative to the metric of problem 20.
22. Evaluate $\int_\gamma \sqrt{dx^2 - dy^2}$ where g is the Euclidean straight line segment that joins the points (1,1) and (2,2).

78 THE POINCARÉ HALF-PLANE

23. Show that if a and b are two positive real numbers then the geodesics of the metric $adx^2 + bdy^2$ are Euclidean straight lines. Generalize this exercise.

The following exercises call for the use of a computer package such as Mathematica or Maple.

24. Write a script that draws the hyperbolic geodesic joining an arbitrary pair of points.
25. Write a script that computes the hyperbolic distance between any two points.
26. Write a script that finds the hyperbolic midpoint of the geodesic segment joining any two points. (See Exercise 6 above.)
27. Write a script that, given any point and any geodesic, will draw the geodesic through the given point that is orthogonal to the given geodesic.
27. Write a script that computes the hyperbolic length of any curve.
28. Write a script that computes the length of any curve relative to any Riemann metric.

5

Euclidean Versus Hyperbolic Geometry

5.1 EUCLID'S POSTULATES REVISITED

We now reconsider Euclid's five postulates one at a time and show that the first four remain valid in the context of hyperbolic geometry, whereas the parallel postulate does not.

Postulate 1. *To draw a straight line from any point to any point.*

That this postulate also holds in the hyperbolic plane was already noted following the proof of Theorem 4.4.

Postulate 2. *To produce a finite straight line continuously in a straight line.*

Let g be any geodesic containing the two points P and Q. We will show that if P, while moving along the geodesic g, approaches the x-axis, then $h(P,Q)$ becomes indefinitely large. We assume here that g is a bowed geodesic centered at $C(c,0)$, leaving the case of the straight geodesics as Exercise 18. Let the coordinates of P and Q relative to a standard polar coordinate system be placed at C be (r_P, α) and (r_Q, β) respectively. Then, by Proposition 4.1,

$$\lim_{\alpha \to 0} h(P,Q) = \lim_{\alpha \to 0} \ln \frac{\csc \beta - \cot \beta}{\csc \alpha - \cot \alpha}$$
$$= \lim_{\alpha \to 0} \ln[(\csc \beta - \cot \beta)(\csc \alpha + \cot \alpha)]$$

which is clearly infinite since $\csc 0 = \cot 0 = \infty$.

Postulate 3. *To describe a circle with any center and radius.*

In a sense the existence of circles in the upper half-plane is quite obvious. Given any point C, a positive real number r, and any ray (half-geodesic) g emanating

80 THE POINCARÉ HALF-PLANE

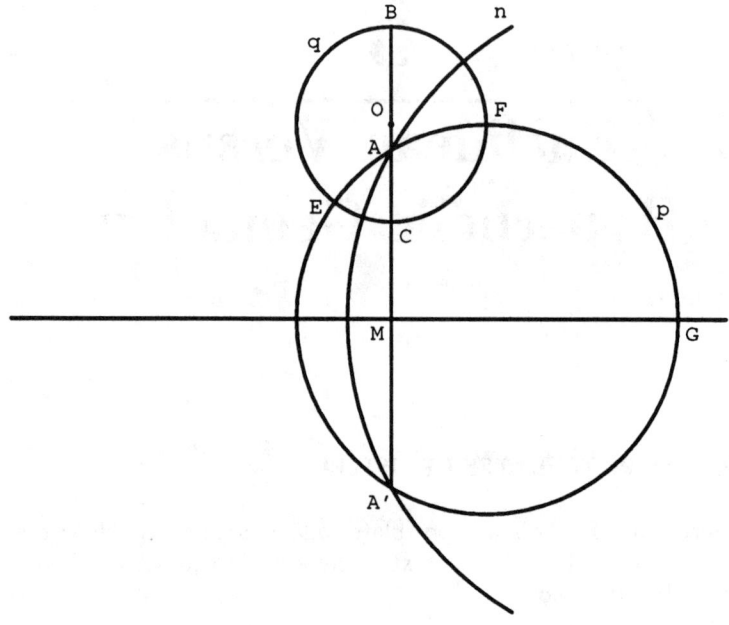

FIGURE 5.1

from C, it follows from Postulate 2 that there is a point P_g on g that is at a hyperbolic distance of r from C. The locus of all such point P_g is the hyperbolic circle with center C and hyperbolic radius r. The following theorem, however, may come as a great surprise to the reader.

Theorem 5.1. *Every Euclidean circle in the upper half-plane is also a hyperbolic circle.*

Proof: Let q be a Euclidean circle with center O and diameter BC perpendicular to the x-axis (Fig. 5.1). If q is to be proven to be also a hyperbolic circle, then common sense dictates that its hyperbolic center A should be the hyperbolic midpoint of the segment BC. It follows from Example 4.5 that

$$AM = \sqrt{BM \cdot CM}. \tag{1}$$

We will now show that every geodesic p through A divides q into two hyperbolically congruent parts. (When this is accomplished, we will of course be very close to our goal.) This hyperbolic congruence will be the hyperbolic reflection I_p, namely the inversion that leaves every point of p fixed. Proposition 3.5 states that this can only happen provided that the circles p and q are orthogonal. However, and this is what makes this proof work, p and q

are orthogonal if and only if the hyperbolic reflection in the original circle q fixes the circle p. Thus, it is now necessary to prove that the inversion $I_{O,k}$, where k is the Euclidean radius of q, fixes the circle p.

The inversion $I_{O,k}$ of course fixes the intersections E and F of p and q. Since $I_{O,k}(p)$ is necessarily a circle containing E and F, it now suffices to show that $A' = I_{O,k}(A)$ is also on p. Since p has its center on the x-axis, this is tantamount to showing that A and A' are symmetrical with respect to the x-axis. Now, if $O = (\,\cdot\,, b)$, then $B = (\,\cdot\,, b+k)$, $C = (\,\cdot\,, b-k)$, and, by (1) above, $A = (\,\cdot\,, \sqrt{b^2 - k^2})$. Hence the ordinate of A' equals

$$OM - OA' = b - \frac{k^2}{OA} = b - \frac{k^2}{b - \sqrt{b^2 - k^2}}$$

$$= b - \frac{k^2}{b - \sqrt{b^2 - k^2}} \cdot \frac{b + \sqrt{b^2 - k^2}}{b + \sqrt{b^2 - k^2}}$$

$$= b - \frac{k^2(b + \sqrt{b^2 - k^2})}{b^2 - (b^2 - k^2)} = b - \frac{k^2(b + \sqrt{b^2 - k^2})}{k^2}$$

$$= b - (b + \sqrt{b^2 - k^2}) = -\sqrt{b^2 - k^2},$$

which is the negative of the ordinate of A. Hence A' is on both p and $I_{O,k}(p)$ which implies that $p = I_{O,k}(p)$. This means that p and q are orthogonal and so $q = I_p(q)$. In other words, every hyperbolic straight line through A divides q into two hyperbolically congruent parts.

We conclude the proof by showing that the hyperbolic length of the arc AE is a constant in the sense that it is independent of the position of p, as long as p passes through A. This is accomplished by producing a hyperbolic reflection that transforms AE onto AC. Let G be either of the intersections of p with the x axis, and let n be the the bowed geodesic centered at G and passing through A. Since $I_n(A) = A$ and $I_n(A') = A'$, it follows that $I_n(p) = BM$. However, by the first part of the proof, $I_n(q) = q$, and hence I_n transforms E, which is the intersection of q and p, onto C which is the intersection of q and BM. Note that were we to chose the other intersection to play the role of G then E would actually be transformed to B. In either case the same conclusion is reached: the hyperbolic length of the arc AE is constant, and so q is also a hyperbolic circle centered at A.

q.e.d.

Proposition 5.2. *If a circle has Euclidean center (h, k) and a Euclidean radius r, then it has the hyperbolic center (H, K), and the hyperbolic radius R, where*

$$H = h, \quad K = \sqrt{k^2 - r^2}, \qquad R = \frac{1}{2} \ln \frac{k + r}{k - r}$$

and
$$h = H, k = K\cosh(R), \qquad r = K\sinh(R).$$

PROOF: Let B and C be, respectively, the points of the circle that lie directly above and below (h, k). It is clear that their coordinates are $(h, k + r)$ and $(h, k - r)$. It then follows from Proposition 4.3 that the hyperbolic diameter of this circle equals
$$\ln \frac{k+r}{k-r}$$
which gives us the desired expression for R. Similar considerations yield the value of K, and the fact that $H = h$ is clear.

Conversely, when the expression for R is inverted it yields
$$\frac{k}{r} = \frac{e^{2R}+1}{e^{2R}-1} = \coth(R).$$
If this expression is solved for k and r simultaneously with $K^2 = k^2 - r^2$ we obtain
$$r^2 = \frac{K^2}{\coth^2(R) - 1} = K^2 \sinh^2(R),$$
and
$$k^2 = K^2 + r^2 = K^2[1 + \sinh^2(R)] = K^2 \cosh^2(R).$$
q.e.d.

Corollary 5.3. *Every hyperbolic circle is also a Euclidean circle.* □

Because of the crucial role that the number π plays in Euclidean geometry, the hyperbolic length of a circle is also of interest. Suppose therefore that q is a circle with Euclidean center $C(h, k)$ and Euclidean radius r. The circle q is then the graph of the equation
$$(x - h)^2 + (y - k)^2 = r^2.$$
Let $P(x, y)$ denote an arbitrary point on the circle, and let t denote the angle from the positive x-axis to the radius CP. Then the circle has the parametric equations:
$$x = h + r \cos t$$
$$y = k + r \sin t.$$
Hence, $dx = -r \sin t \, dt$, $dy = r \cos t \, dt$, and
$$\text{the hyperbolic length of } q = \int_q \frac{\sqrt{dx^2 + dy^2}}{y}$$
$$= 2 \int_{-\pi/2}^{\pi/2} \frac{r \, dt}{k + r \sin t}$$
$$= \left[2r \cdot \frac{2}{\sqrt{k^2 - r^2}} \arctan \frac{k \tan(t/2) + r}{\sqrt{k^2 - r^2}} \right]_{-\pi/2}^{\pi/2}$$

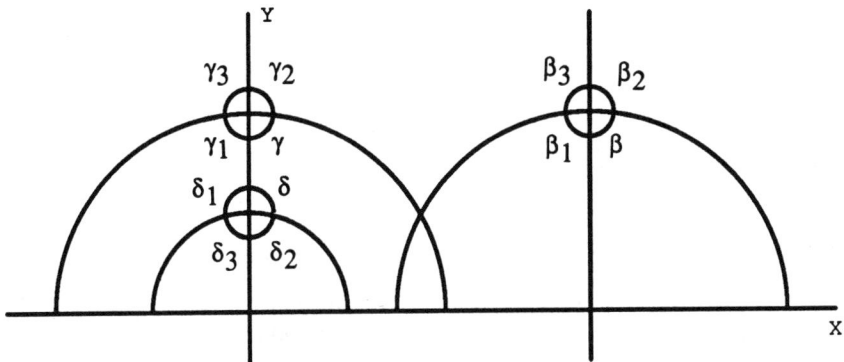

FIGURE 5.2

$$= \frac{4r}{\sqrt{k^2 - r^2}} \left[\arctan \frac{k+r}{\sqrt{k^2 - r^2}} - \arctan \frac{-k+r}{\sqrt{k^2 - r^2}} \right]$$

$$= \frac{4r}{\sqrt{k^2 - r^2}} \left[\arctan \sqrt{\frac{k+r}{k-r}} + \arctan \sqrt{\frac{k-r}{k+r}} \right]$$

$$= \frac{2\pi r}{\sqrt{k^2 - r^2}},$$

because of the trigonometric identity $\arctan x + \arctan \frac{1}{x} = \frac{\pi}{2}$.

Consequently the ratio of the hyperbolic length of a circle to the hyperbolic length of its diameter is

$$\frac{\frac{2\pi r}{\sqrt{k^2 - r^2}}}{\ln \frac{k+r}{k-r}}$$

a quantity that clearly depends on both k and r. In Exercise 8 the readers are requested to evaluate some limiting values of this ratio.

Postulate 4. *That all right angles are equal to one another.*

We will show that every two hyperbolic right angles are in fact congruent. More specifically, given any right angle α we will show that there is a sequence of hyperbolic rigid motions that carry α to the fixed right angle δ of Figure 5.2, whose vertex is assumed to have ordinate 1.

It is clear that the reflection in the y-axis carries the angle δ_1 onto δ and that the inversion $I_{O,1}$ carries the angles δ_2 and δ_3 onto δ and δ_1, respectively. Moreover, if k is the ordinate of the common vertex of angles $\gamma, \gamma_2, \gamma_3$, then the inversion $I_{O,\sqrt{k}}$ carries these angles onto the angles $\delta, \delta_2, \delta_3$ respectively, and it

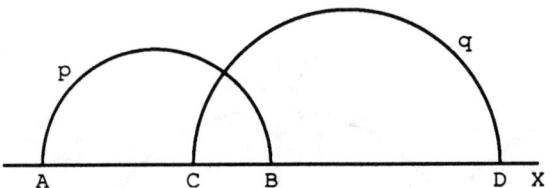

FIGURE 5.3

is now also clear that there is a hyperbolic translation that carries the angles β, β_1, β_2, β_3 onto the respective γ's. These considerations demonstrate that all right angles one of whose sides is a straight geodesic are indeed congruent. Suppose now that α is a right angle both of whose sides p and q are bowed geodesics (Figure 5.3). If d is the Euclidean length of the line segment AB, then by Theorem 3.3 the inversion $I_{A,d}$ maps the geodesic p into a straight geodesic, thus showing that α is congruent to a right angle one of whose sides is a straight geodesic. Since those angles were already shown to be hyperbolically congruent to δ we are done. q.e.d.

Playfair's Postulate (Euclid's Postulate 5). *Given a straight line m and a point P not on m, there is a unique straight line that is parallel to m and contains P.*

If we define two non intersecting geodesics as parallel then it is easy to see that Playfair's Postulate no longer holds in the upper half-plane. In fact, the geodesics p, q, r of Figure 5.4 all pass through the point P and are all parallel to the geodesic g. The reader can no doubt construct many other such parallels. There is of course another way to look at parallel lines. In Euclidean geometry, given a straight line m, the locus of all those points P whose distance from m is a constant d is a pair of straight lines p and p' which are both parallel to m and which lie at a distance d on either of its sides. While the hyperbolic analog of this is not quite so simple, it is interesting enough to merit some discussion. The resulting configuration has an elegance all its own.

First we need to discuss the hyperbolic distance from a point to a geodesic.

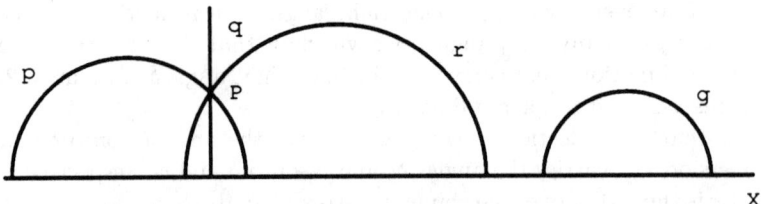

FIGURE 5.4

EUCLIDEAN VERSUS HYPERBOLIC GEOMETRY 85

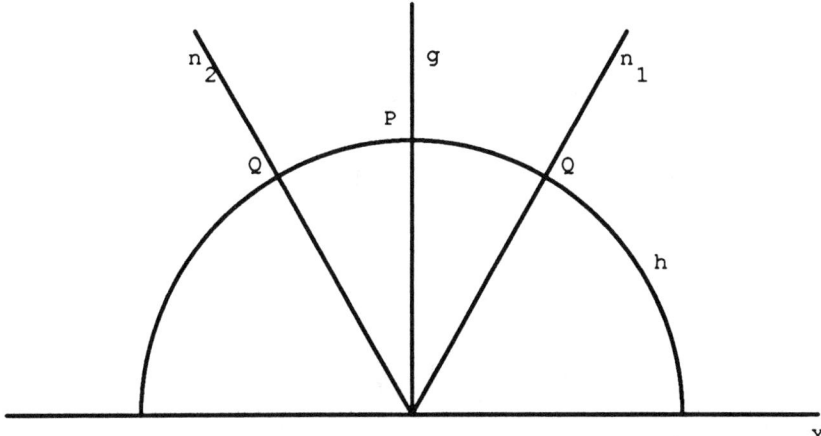

FIGURE 5.5

Given a geodesic g and a point P not on g, it follows from Proposition 12 of Euclid that there is a geodesic h containing P that is perpendicular to g. (The reader who is uncomfortable with this line of reasoning can prove this directly as a statement about orthogonal circles.) It also follows from Euclid's Propositions 17 and 27 that the geodesic h is in fact unique. In complete analogy with Euclidean geometry, the hyperbolic distance of the point P from the geodesic g is the hyperbolic length of the unique geodesic segment that joins P to g and is perpendicular to g.

Suppose first that g is a straight geodesic. Referring to Figure 5.5, it is clear that the bowed geodesics h that are positioned symmetrically about g are all orthogonal to it. Moreover, it follows from Corollary 4.2 that regardless of the position of the point P on the geodesic g, the hyperbolic length of the geodesic segment PQ of h has a constant value, say d. Thus the locus of points P that are at a constant hyperbolic distance from g consists of the two Euclidean straight lines n_1 and n_2, which, while they are not hyperbolic straight lines, may still be said to be parallel to it. Note that it is very tempting to say here that n_1 and n_2 intersect g at infinity.

If g is a bowed geodesic, let A and B be its Euclidean endpoints, and let r be its Euclidean radius (Figure 5.6). By Theorem 3.3 the inversion $I_{A,2r}$ transforms g into a Euclidean ray that emanates from B. Since this ray must also be a geodesic, it follows that g is transformed into the straight geodesic g' that lies directly above B. Let m_1 and m_2 be the above described locus of all the points that lie at a constant hyperbolic distance d from g'. Since the inversion $I_{A,2r}$ is a hyperbolic rigid motion it follows that the curves

$$n_1 = I_{A,2r}(m_1) \text{ and } n_2 = I_{A,2r}(m_2)$$

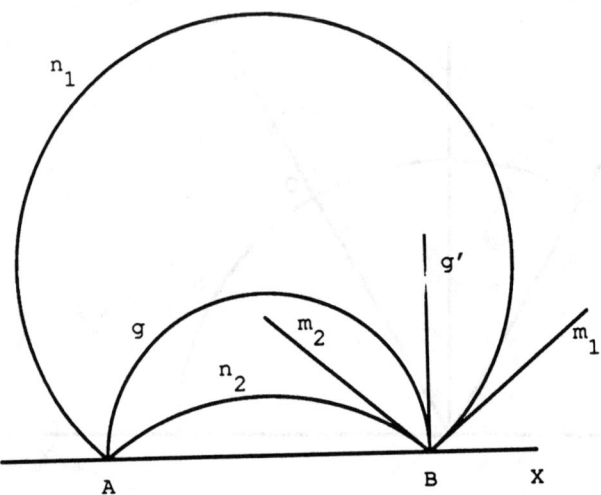

FIGURE 5.6

constitute the locus of all the points that are at a constant hyperbolic distance d from the geodesic g. However, in view of Theorem 3.3 we may now conclude that this locus consists of the two circular arcs joining A and B that are depicted in Figure 5.6. Since the Euclidean straight lines m_1 and m_2 make the same angle with the geodesic g', the same holds for the angles that n_1 and n_2 make with g at the point B. Consequently, if n_1' is the reflection of n_1 in the x-axis, then n_2 and n_1' together form a single Euclidean circle.

Example 5.4. Let us find the curves that lie at the constant hyperbolic distance 1 from the y-axis. It follows from the above considerations that these are two Euclidean rays from the origin. If one ray has inclination $\alpha < \pi/2$ from the positive x-axis, then, by Proposition 4.1,

$$\ln \frac{\csc 90° - \cot 90°}{\csc \alpha - \cot \alpha} = 1.$$

This yields the equations

$$\csc \alpha - \cot \alpha = e^{-1} \quad \text{and} \quad \csc \alpha + \cot \alpha = e$$

from which we conclude that

$$\alpha = \sin^{-1}\left(\frac{2}{e + e^{-1}}\right) \approx 40°.$$

Thus the locus in question consists of the two rays from the origin whose inclinations from the positive and the negative portions of the x-axis are 40°.

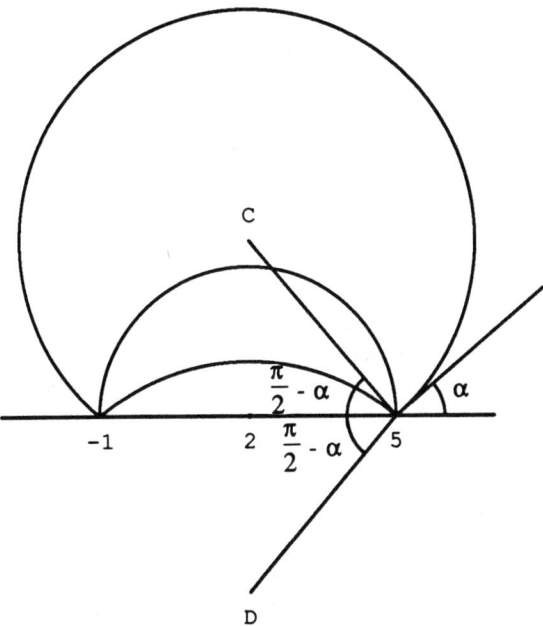

FIGURE 5.7

Example 5.5. Describe the locus of those points that lie at the constant distance of 1 from the bowed geodesic with Euclidean center (2,0) and Euclidean radius 3 (Fig. 5.7).

By the discussion preceding Example 5.4, this locus consists of two arcs of Euclidean circles that pass through the points $(-1,0)$ and $(5,0)$ whose tangents, at those points, make an angle of

$$\alpha = \sin^{-1}\left(\frac{2}{e+e^{-1}}\right) \approx 40°$$

with the x-axis. Since the radius through the point of contact is perpendicular to the tangent, it now follows that the centers C and D of these arcs have coordinates $(2, \pm 3 \tan\left(\frac{\pi}{2} - \alpha\right)) \approx (2, \pm 3.52)$.

5.2 ABSOLUTE GEOMETRY

Since Euclid's Propositions 1–28 depend on the first four postulates only, and since these hold for the hyperbolic plane, it follows that these propostions are also valid in this non-Euclidean context. We shall henceforth feel free to make use of these propositions in the further development of the hyperbolic plane. This can result in a considerable amount of simplification. Consider, for example, the statement that the angles at the base of a hyperbolic triangle are equal. If we tried to prove this directly in terms of the hyperbolic length formula we would be sure to get bogged down in some formidable details. This unpleasantness is obviated by the observation that the proof of the Euclidean analog (which appears in the appendix) is in fact absolute and hence holds in hyperbolic geometry as well.

5.3 HYPERBOLIC GEOMETRY

The upper half-plane together with the Poincaré metric is generally known as the *Poincaré upper half-plane.* It is a concrete model of hyperbolic geometry which, strictly speaking, is the axiomatic geometry obtained by replacing the uniqueness stipulation of Playfair's postulate with the seemingly more relaxed alternative that through a given point outside a given line there exist more than one line parallel to the given line. There are many other such concrete models of hyperbolic geometry and two more are examined in detail in Chapters 13 and 14.

This distinction between the axiom system and its models notwithstanding, we shall continue to refer to the geodesics of the Poincaré upper half-plane as hyperbolic straight lines for the sake of succinctness. The same convention will be used for other concepts as well, as is done in the next section.

5.4 HYPERBOLIC RIGID MOTIONS

The aggregate of the rigid motions of the hyperbolic plane is considered by many mathematicians to be one of the most interesting of all mathematical structures. Chapter 9 is devoted to a detailed description of these rigid motions, but it so happens that we already have accrued a fair amount of information about them and it might be wise to pause here for the purpose of a summary, even though this information will not be required until we get to Chapter 9.

As was pointed out in the last section of Chapter 2, Propositions 2.1–4 are in fact absolute propositions and hence they are valid in hyperbolic geometry as well. Consequently, *if two hyperbolic rigid motions agree at three hyperbolically noncollinear points, then they agree everywhere in the half-plane.*

Since the definition of translations implicitly assumes the validity of Playfair's Postulate, these Euclidean rigid motions have no clear cut hyperbolic analog. Later, in Chapter 9, we shall nevertheless define hyperbolic translations, but, as the readers will see, these will bear only a very slight resemblance to their Euclidean counterparts. Rotations, on the other hand, are absolute entities and so they are well defined within hyperbolic geometry as well. The analytic description and the visualization of hyperbolic rotations is

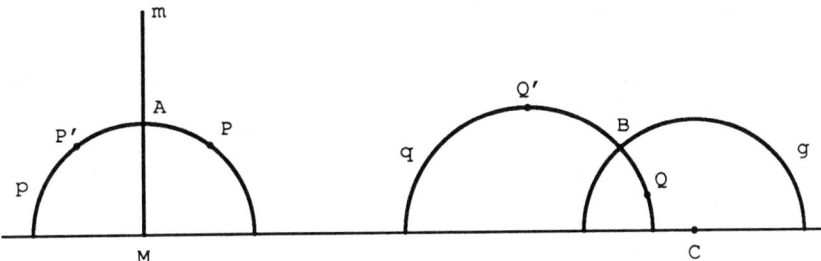

FIGURE 5.8

deferred to Chapter 9. Some further comments will be made below following the discussion of reflections.

A *hyperbolic reflection* is either a Euclidean reflection in a straight geodesic or an inversion centered at some point on the x-axis. We shall now proceed to justify this nomenclature by showing that these hyperbolic rigid motions are indeed the hyperbolic analogs of Euclidean reflections. Euclidean reflections are defined by the property that their axes are the perpendicular bisectors of the line segment joining any point to its image, and we shall show that the hyperbolic reflections possess the same property. Consider first the reflection ρ_m whose axis is the straight geodesic m above the point M on the x-axis (Fig. 5.8). If P is any point of the upper half-plane, let p be the bowed geodesic through P that is centered at M. Since p and m are orthogonal at their intersection A it follows that $\rho_m(p) = p$ and so $P' = \rho_m(P)$ is also on p. Moreover, ρ_m is a hyperbolic rigid motion and so the geodesic segments PA and $P'A$ have equal hyperbolic lengths. Thus, m is the hyperbolic perpendicular bisector of the geodesic segment joining P to its image $P' = \rho_m(P)$.

Next let us examine the hyperbolic reflection that consists of the inversion $I_{C,k}$ where C is some point on the x-axis (Fig. 5.8). Since this inversion fixes every point on the bowed geodesic g that is centered at C and has radius k, we may refer to this geodesic as the axis of $I_{C,k}$. For any point Q let $Q' = I_{C,k}(Q)$ and let q be the geodesic that contains Q and is orthogonal to g at their intersection B. That such a geodesic necessarily exists follows from Proposition 12 of Book I of *The Elements*. By Proposition 3.5, $I_{C,k}(q) = q$ and so Q' too is on q. Moreover, $I_{C,k}$ is a hyperbolic rigid motion and so the geodesic segments QB and $Q'B$ have equal hyperbolic lengths. Thus, g is the hyperbolic perpendicular bisector of the geodesic segment joining Q to its image $Q' = I_{C,k}(Q)$.

Having verified that hyperbolic reflections satisfy the definition of reflections given in Chapter 2, we note that the proof of Proposition 2.10 is absolute and consequently *the composition of two hyperbolic reflections with intersecting axes is a hyperbolic rotation*. The center of this rotation is the intersection of the axes and the angle of the rotation is twice the angle between the axes.

Since glide-reflections assume the existence of translations we cannot say anything about their hyperbolic counterparts at this point.

Proposition 2.19 states that given any two congruent triangles, each can be transformed into the other by means of a series of at most three reflections. The proof we offered is in fact absolute and so this proposition holds in the half-plane as well. More significantly, Theorem 2.20 is now seen to hold in the hyperbolic context. This is so useful that we state it as a new theorem.

Theorem 5.6. *Every hyperbolic rigid motion is the composition of at most three hyperbolic reflections.* □

5.5 EXERCISES

1. Find the hyperbolic center and radius of the Euclidean circle with center (5,4) and radius 3. Find both its hyperbolic and Euclidean circumferences.
2. Find the Euclidean center and radius of the circle with hyperbolic center (5,4) and radius 3.
3. Describe the locus of all the points that are at a constant hyperbolic distance 2 from the positive y-axis.
4. Describe the locus of all the points that are at a constant hyperbolic distance 2 from the geodesic consisting of the semicircle centered at the origin with radius 3.
5. Let L be the hyperbolic circumference of a circle with Euclidean center (a, b) and radius $b/3$. Evaluate
$$\lim_{b \to 0} L.$$
6. Use Exercise 7 of Chapter 4 to obtain an algebraic proof of the fact that every hyperbolic circle is also a Euclidean circle.
7. Prove that the hyperbolic circumference of a circle with hyperbolic radius R is $2\pi \sinh(R)$.
8. If $\pi(q)$ denotes the ratio of the hyperbolic length of the circle q with Euclidean center at (h, k) and Euclidean radius r to the hyperbolic length of its diameter, compute the limit of $\pi(q)$ when
    a) r approaches 0;
    b) $k = 2r$ and both approach 0.
9. If a hyperbolic triangle has three equal angles, are its sides necessarily equal to each other?
10. Does every hyperbolic triangle have a circumscribing circle?
11. Does every hyperbolic triangle have a circumscribed circle?
12. Are the interior angle bisectors of every hyperbolic triangle concurrent?
13. Prove that every hyperbolic rigid motion has an inverse.
14. Discuss the hyperbolic analog of the Euclidean theorem which states that the tangent line to a circle is perpendicular to the radius through the point of contact.
15. Discuss the hyperbolic analog to the Euclidean theorem which relates the central angle of a circle to the angle at the circumference that is subtended by the same arc.

16. Suppose g and h are two geodesics, which, as figures of the Euclidean plane, are tangent to each other.
    a) Prove that the point of tangency is on the x-axis;
    b) Prove that the two geodesics are hyperbolically asymptotic near that point of tangency.
17. Prove or disprove: if a, b, c, are three positive numbers such that the sum of every two exceeds the third, then there exists a hyperbolic triangle the lengths of whose sides are a, b, c, respectively.
18. Show that the hyperbolic distance from the point $(a, 1)$ to the point (a, y) diverges to infinity as
    a) y converges to zero;   b) y diverges to infinity.
19. Suppose that two perpendicular bisectors of a hyperbolic triangle intersect in a point P in the upper half plane. Prove that the third perpendicular bisector also passes through the point P.
20. Suppose that two perpendicular bisectors of a hyperbolic triangle intersect in a point P on the x axis. Prove that the third perpendicular bisector also passes through the point P.
21. Construct a hyperbolic triangle no two of whose perpendicular bisectors intersect.

6
The Angles of the Hyperbolic Triangle

6.1 INTRODUCTION

It will be shown in this chapter that the sum of the angles of every hyperbolic triangle is less than π and that given any three angles α, β, γ whose sum is less than π, there does indeed exist a hyperbolic triangle with α, β, γ as its angles. It is proved that any two such triangles must be hyperbolically congruent. The chapter concludes with a discussion of the new kinds of tilings that are feasible in the hyperbolic plane.

6.2 THE STANDARD POSITION OF A TRIANGLE

A *hyperbolic triangle ABC* consists of three points A, B, C *(vertices)* that do not lie on a single geodesic and the three geodesic segments *(sides)* that join each pair of vertices. Such a triangle is said to be in *standard position* if the vertices A, B, C have coordinates $(0, k)$, (s, t), $(0, 1)$ respectively, where $k > 1$ and $s > 0$.

The following observation will prove very useful in discussing arbitrary hyperbolic triangles.

Proposition 6.1. *Every hyperbolic triangle can be brought into standard position by a hyperbolic rigid motion.*

PROOF: If the hyperbolic $\triangle ABC$ already has its vertices C at $(0,1)$ and B at (s,t) with $s > 0$, but A is at $(0,k)$ with $k < 1$ (Fig. 6.1a), then it is clear that the reflection $I_{0,1}$ will transform it into a $\triangle A'B'C$ that is in standard position.

If the hyperbolic $\triangle ABC$ has both of its vertices $A(0, a)$ and $C(0, c)$ on the y-axis, then, by reflecting this triangle in the y-axis if necessary, we may

94 THE POINCARÉ HALF-PLANE

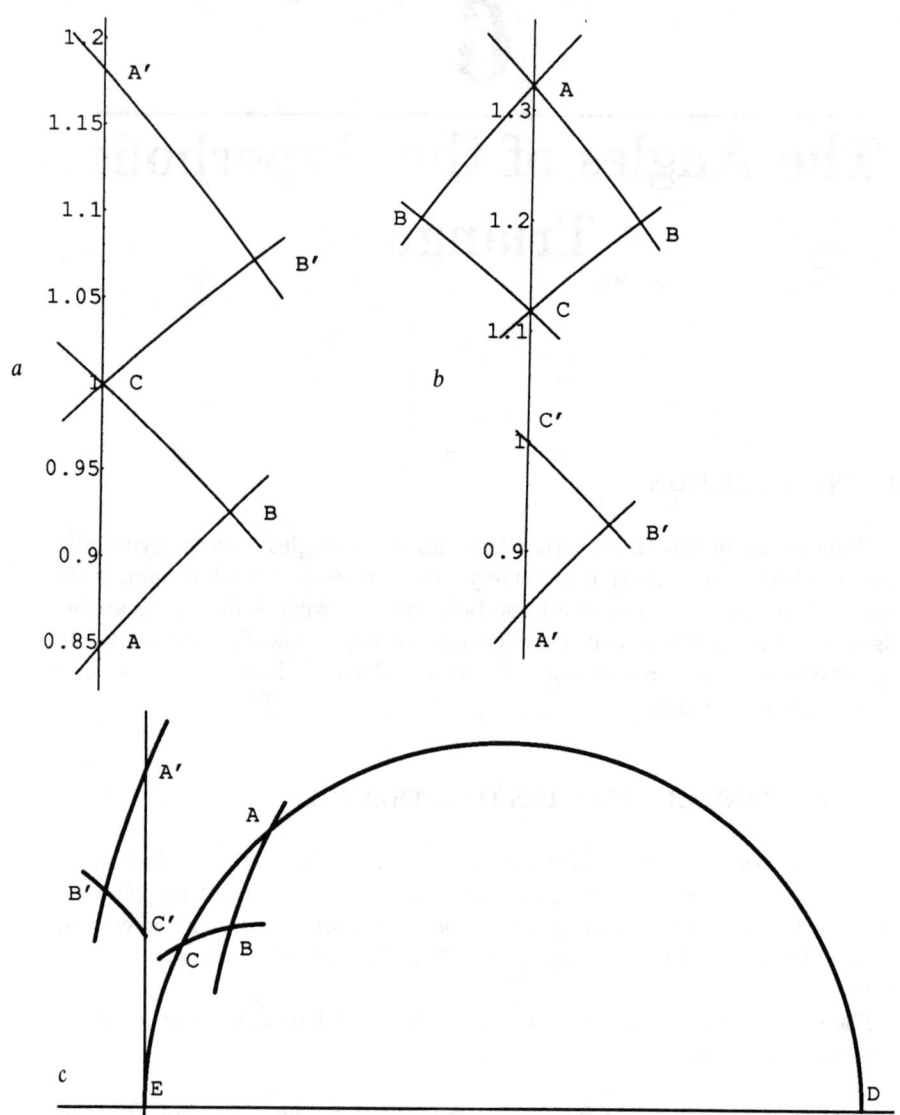

FIGURE 6.1.

assume that $B(s, t)$ has $s > 0$ (Fig. 6.1b). But then the reflection $I_{O,\sqrt{c}}$ transforms $\triangle ABC$ into a triangle of the type discussed above.

Finally, if the hyperbolic triangle $\triangle ABC$ is in an arbitrary position, we may assume that its side AC is a segment of a bowed geodesic g that joins some two points D and E on the x-axis. The fact that horizontal translations are hyperbolic rigid motions allows us to assume that E coincides with the origin O

(Fig. 6.1c). If $D = (d, 0)$ then the inversion $I_{D,d}$ transforms g onto the y-axis, and so the given triangle is transformed into $\triangle A'B'C'$ which has one of its sides on the y-axis. Since we already know that this latter triangle can be brought into standard position, we are done.

q.e.d.

Should the readers attempt to apply this proposition to any specific triangle, such as that with vertices $(0,1)$, $(1,1)$, and $(1,3)$, they will soon discover that the computational details are both considerable and uninteresting. This proposition is a tool that will prove very useful in the proof of some of the subsequent theorems. In and of itself, however, it does not illuminate the nature of the hyperbolic triangle, and so we shall give no specific application of it.

6.3 THE SUM OF THE ANGLES OF THE HYPERBOLIC TRIANGLE

A *geodesic ray* at a point is the analog of a Euclidean ray, namely, it is the name attached to either of the portions of a geodesic that any point on it determines. Given two geodesic rays g and h from some point P, we denote the angle at P from g to h by $\angle(g, h)$.

Proposition 6.2. *At any given point P, let g be a straight geodesic ray and let g_1, g_2, g_3 be bowed geodesics rays centered at C_1, C_2, C_3 respectively (Fig. 6.2). Then*

$$\angle(g, g_1) = \angle DC_1P, \quad \angle(g_1, g_2) = \angle C_1PC_2,$$

and

$$\angle(g_3, g_1) = \pi - \angle C_1PC_3.$$

PROOF: Let D' be a point of g such that P is between D and D', and let PT_1 and PT_2 be the Euclidean tangent lines to g_1 and g_2, respectively, at P. Since the tangent to a circle is perpendicular to the radius through the point of contact, it follows that

$$\angle(g_1, g_2) = \angle T_1PT_2 = \angle T_1PC_1 - \angle T_2PC_1 = \frac{\pi}{2} - \angle T_2PC_1$$
$$= \angle T_2PC_2 - \angle T_2PC_1 = \angle C_1PC_2.$$

Similarly,

$$\angle(g, g_1) = \angle D'PT_1 = \pi - \angle T_1PC_1 - \angle C_1PD = \frac{\pi}{2} - \angle C_1PD = \angle DC_1P$$

and

$$\angle(g_3, g_1) = \angle(g_3, g) + \angle(g, g_1) = \angle PC_3D + \angle DC_1P = \pi - \angle C_1PC_3.$$

q.e.d.

THE POINCARÉ HALF-PLANE

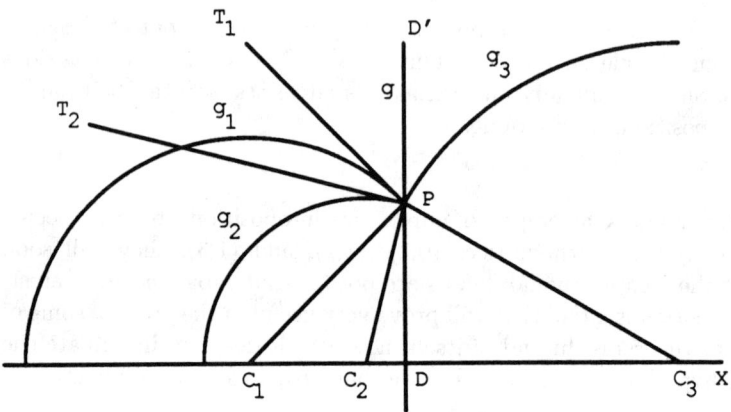

FIGURE 6.2.

This proposition provides us with an explicit method for constructing arbitrary hyperbolically rectilineal angles. We now present two applications of this method.

Example 6.3. Consider the point $P(3,4)$ which lies on the bowed geodesic of radius 5 centered at the origin (Fig. 6.3). Suppose we wish to construct another bowed geodesic which makes an angle of 70° from the given geodesic. It follows from the above proposition that it suffices to find a point $A(a,0)$ on the x-axis such that the Euclidean rectilineal angle $\angle OPA = 70°$. Since $\tan(\angle AOP) = 4/3$ it follows that $\angle AOP \approx 53.1°$. Hence $\angle PAO \approx 56.9°$ and

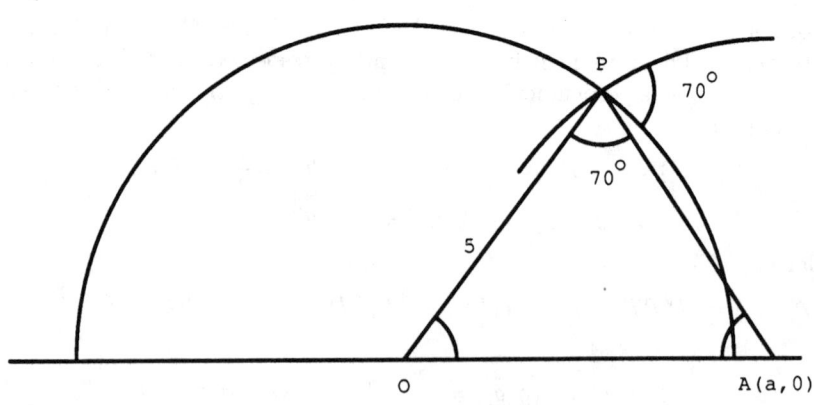

FIGURE 6.3.

THE ANGLES OF THE HYPERBOLIC TRIANGLE 97

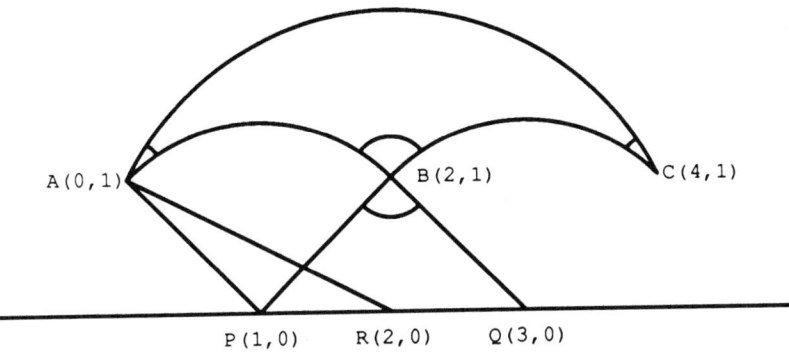

FIGURE 6.4.

so, by the Law of Sines,

$$a = \frac{5 \sin 70°}{\sin(\angle PAO)} \approx 5.6$$

and

$$AP = \frac{5 \sin (\angle AOP)}{\sin (\angle PAO)} \approx 4.8.$$

Example 6.4. What are the angles of the hyperbolic triangle with vertices $A(0,1)$, $B(2,1)$, $C(4,1)$ of Figure 6.4? It is clear that the Euclidean centers of the bowed geodesics AB, BC, and CA are the points $P(1,0)$, $Q(3,0)$, and $R(2,0)$ respectively. Hence,

$$\angle(AC, BC) = \angle(AB, AC) = \angle PAR$$
$$= \cos^{-1}\left(\frac{AP^2 + AR^2 - PR^2}{2AP \cdot AR}\right) = \cos^{-1}\left(\frac{2+5-1}{2\sqrt{2}\sqrt{5}}\right)$$
$$= \cos^{-1}\left(\frac{3\sqrt{10}}{10}\right) \approx 18.4°,$$

and

$$\angle(CB, AB) = \pi - \angle PBQ = \pi - \cos^{-1}\left(\frac{BP^2 + BQ^2 - PQ^2}{2BP \cdot BQ}\right)$$
$$= \pi - \cos^{-1}\left(\frac{2+2-4}{2\sqrt{2}\sqrt{2}}\right) = \pi - \cos^{-1}(0) = 90°.$$

It is worthy of note that the sum of the angles of the hyperbolic triangle of Example 6.4 is considerably less than 180°. We now go on to show that this situation is not an anomaly. It is convenient to begin with right triangles.

98 THE POINCARÉ HALF-PLANE

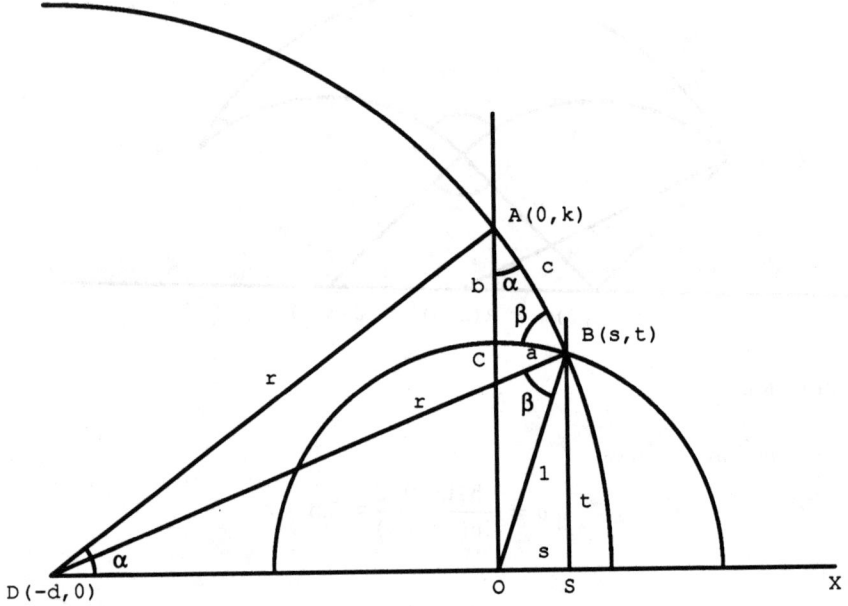

FIGURE 6.5.

Proposition 6.5. *The sum of the angles of any hyperbolic right triangle is less than π.*

PROOF: Let C denote the vertex of the triangle where the right angle resides. By Proposition 6.1 we may assume that the given triangle is in standard position with $C(0,1)$, $A(0,k)$, and $B(s,t)$, as depicted in Fig. 6.5. Since $\angle BCA$ is a right angle it follows that the side BC is a segment of the bowed geodesic that is centered at the origin O and that side AC is a segment of the positive y-axis. Let $D(-d,0)$ be the center of the bowed geodesic containing the hypotenuse AB, and let SB be the straight geodesic containing B. By Proposition 6.2

$$\alpha = \angle CAB = \angle ODA, \quad \beta = \angle ABC = \angle DBO.$$

We will show that $\alpha < \frac{\pi}{2} - \beta$. Since by Proposition 17 of Euclid both α and β are acute angles, it suffices to show that $\sin\alpha < \sin(\frac{\pi}{2} - \beta) = \cos\beta$. However, the reader can easily verify the equivalence of the following assertions.

$$\sin\alpha < \cos\beta,$$
$$\frac{k}{r} < \frac{r^2 + 1^2 - d^2}{2r} = \frac{k^2 + 1}{2r},$$

THE ANGLES OF THE HYPERBOLIC TRIANGLE 99

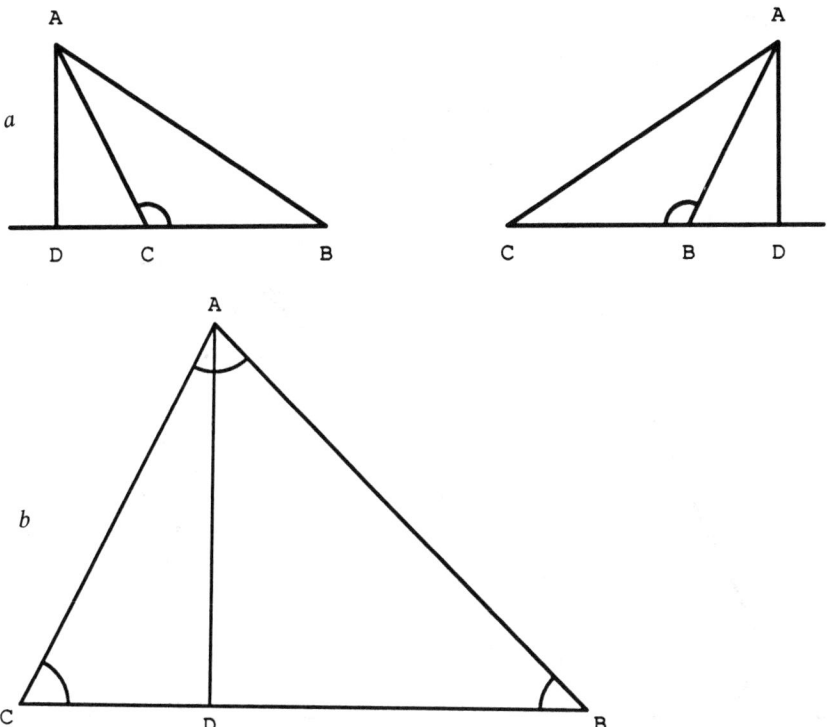

FIGURE 6.6.

$$2k < k^2 + 1,$$
$$0 < (k-1)^2.$$

Since the last assertion is valid as long as A and C are distinct, we have proven that $\alpha + \beta < \frac{\pi}{2}$, and so the sum of the angles of a hyperbolic right triangle is indeed less than π.

q.e.d.

Theorem 6.6. *The sum of the angles of any hyperbolic triangle is less than π.*

PROOF: We first show that every triangle must have an internal altitude. Suppose that AD is an external altitude of $\triangle ABC$ (Fig. 6.6a). Then, by Proposition 16 of Euclid either

$$\angle BCA > \angle CDA = \frac{\pi}{2} \quad \text{or} \quad \angle ABC > \angle ADB = \frac{\pi}{2}.$$

Thus, if the altitude to side BC is external, one of the angles at B and C must be obtuse. However, by Proposition 17 of Euclid the angles at two of the

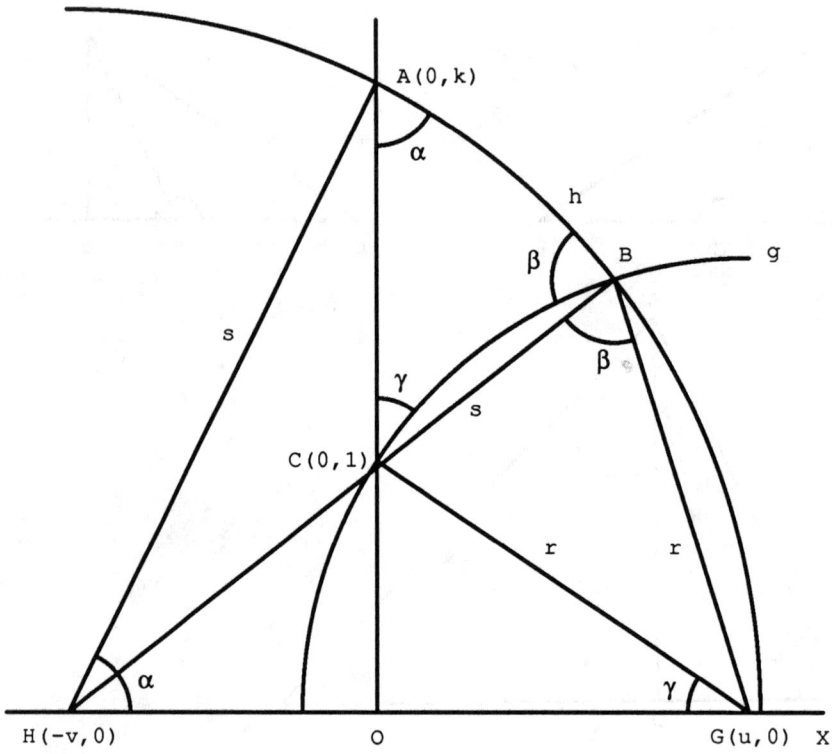

FIGURE 6.7.

vertices of any triangle are acute, and hence the altitude from the third vertex must be internal. Thus, given a hyperbolic triangle ABC, we may assume that the altitude AD is internal (Fig. 6.6b). It then follows from the previous proposition that

$$\angle ABC + \angle BAD < \frac{\pi}{2} \quad \text{and} \quad \angle BCA + \angle CAD < \frac{\pi}{2}.$$

Consequently,

$$\angle ABC + \angle BCA + \angle CAB = \angle ABC + \angle BCA + \angle BAD + \angle CAD$$
$$< \frac{\pi}{2} + \frac{\pi}{2} = \pi.$$

q.e.d.

In Euclidean geometry the sum of the angles of every triangle is π. Moreover, this is the only constraint on the angles of the triangle. In other words, given any three angles whose sum is π, they are indeed the angles of some Euclidean triangle. Similarly, Theorem 6.6 above constitutes the only constraint on the angles of the hyperbolic triangle.

Theorem 6.7. *Given any three angles whose sum is less than π, they are indeed the angles of some hyperbolic triangle.*

PROOF: Let α, β, γ, be three arbitrary positive angles such that $\alpha + \beta + \gamma < \pi$. Let us see what conditions must be satisfied by a hyperbolic $\triangle ABC$ in standard position so that $\angle CAB = \alpha$, $\angle ABC = \beta$, and $\angle BCA = \gamma$. As depicted in Fig. 6.7, let $G(u, 0)$ and $H(-v, 0)$ be the Euclidean centers of the bowed geodesics BC and AB, respectively, and let r and s be their respective Euclidean radii. By Proposition 6.2

$$\angle CGO = \gamma, \quad \angle OHA = \alpha, \quad \angle HBG = \beta.$$

The trigonometry of the Euclidean triangles GCO, AHO, and BGH yields the constraints

$$u = r \cos \gamma \qquad (1)$$

$$v = s \cos \alpha \qquad (2)$$

and

$$(u + v)^2 = r^2 + s^2 - 2rs \cos \beta \qquad (3)$$

It follows from the Theorem of Pythagoras that $r^2 = u^2 + 1$ which, in conjunction with (1) yields

$$r = \csc \gamma \quad \text{and} \quad u = \cot \gamma.$$

When these values of u, r, and v are substituted in (3) we get

$$(\cot \gamma + s \cos \alpha)^2 = \csc^2 \gamma + s^2 - 2s \cos \beta \csc \gamma$$

which simplifies to the quadratic

$$s^2 \sin^2 \alpha - 2s(\cos \alpha \cot \gamma + \cos \beta \csc \gamma) + 1 = 0 \qquad (4)$$

It is also clear that if ABC is a hyperbolic triangle in standard position that satisfies these constraints, then its angles are indeed α, β, and γ. The quadratic (4) has the discriminant

$$4(\cos \alpha \cot \gamma + \cos \beta \csc \gamma)^2 - 4 \sin^2 \alpha.$$

We now show that this discriminant is necessarily positive, thus guaranteeing that the quadratic equation (4) does indeed have solution in s. Since $\alpha + \beta + \gamma < \pi$, we have

$$\alpha + \gamma < \pi - \beta.$$

As the cosine function is monotone decreasing in the first two quadrants this yields the following sequence of equivalent statements:

$$\cos(\alpha + \gamma) > \cos(\pi - \beta) = -\cos \beta,$$

$$\cos \alpha \cos \gamma - \sin \alpha \sin \gamma > -\cos \beta,$$

$$\cos \alpha \cos \gamma + \cos \beta > \sin \alpha \sin \gamma > 0,$$

$$\cos\alpha\cot\gamma + \cos\beta\csc\gamma > \sin\alpha > 0$$
$$(\cos\alpha\cot\gamma + \cos\beta\csc\gamma)^2 > \sin^2\alpha.$$

The last of these inequalities clearly establishes the positivity of the discriminant of (4). Hence this quadratic has two real solutions for any given positive angles α, β, γ such that $\alpha + \beta + \gamma < \pi$. If we set $v = s\cos\alpha$, then s, v, $u = \cot\gamma$, and $r = \csc\gamma$ satisfy Eqs. (1–3) and so the desired triangle does indeed exist.

q.e.d.

The above proof contains a blueprint for the construction of hyperbolic triangles with specified angles. To illustrate this we shall construct a hyperbolic equilateral triangle in standard position each of whose angles measures 45°. If the angles α, β, γ of Fig. 6.7 all measure 45° then

$$r = \csc 45° = \sqrt{2} \quad \text{and} \quad u = \cot 45° = 1.$$

Moreover, the quadratic (4) now reduces to

$$s^2 - 2s(2 + \sqrt{2}) + 2 = 0$$

one of whose roots is

$$2 + \sqrt{2} + \sqrt{(2+\sqrt{2})^2 - 2} \approx 6.52$$

From this we also obtain

$$v = s\cos 45° = \frac{2 + \sqrt{2} + \sqrt{(2+\sqrt{2})^2 - 2}}{\sqrt{2}} \approx 4.61.$$

This gives us sufficient information to redraw Fig. 6.7 so that the three angles α, β, γ measure 45° each. The resulting hyperbolic equilateral triangle appears in Fig. 6.8.

6.4 A NEW CONGRUENCE THEOREM

Let ABC be any hyperbolic triangle with angles α, β, γ

$$\angle CAB = \alpha, \quad \angle ABC = \beta, \quad \text{and } \angle BCA = \gamma.$$

Place this triangle in standard position, and let r and s be as in Fig. 6.7. It then follows that C and A have coordinates $(0, 1)$ and $(0, s\sin\alpha)$ respectively. By Proposition 4.3 the hyperbolic length of the side AC is

$$|\ln(s\sin\alpha)|. \tag{5}$$

Since s was obtained by solving the quadratic Eq. (4) whose coefficients depend on α, β, γ alone, it would seem that there may be two triangles in standard position that are determined by these angles. In fact, there is only one. To see this, let s_1 and s_2 denote the two possible values of s. As these are the roots of

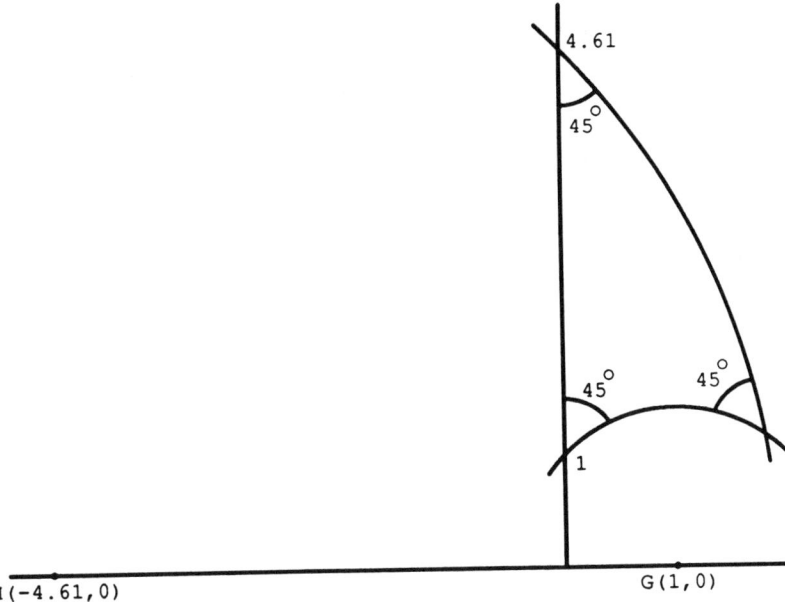

FIGURE 6.8.

the quadratic (4) it follows that

$$s_1 s_2 = \frac{1}{\sin^2 \alpha}$$

and hence

$$s_2 \sin \alpha = \frac{1}{s_1 \sin \alpha}.$$

However, $|\ln(1/x)| = |\ln x|$, and so it follows that the length of side AC, as given by (5), is completely determined by the angles α, β, γ. Since this argument could have been applied to any side of the given hyperbolic triangle ABC, it follows that every hyperbolic triangle is completely determined by its angles. This is of course in marked contrast with the state of affairs in Euclid's geometry.

THEOREM 6.8. *If two hyperbolic triangles have their respective angles equal, then they are hyperbolically congruent.*

□

It follows from expression (5) that the hyperbolic length of each side of the hyperbolic equilateral triangle of Fig. 6.8 is

$$\ln \frac{2 + \sqrt{2} + \sqrt{(2+\sqrt{2})^2 - 2}}{\sqrt{2}} = 1.528\ldots.$$

104   THE POINCARÉ HALF-PLANE

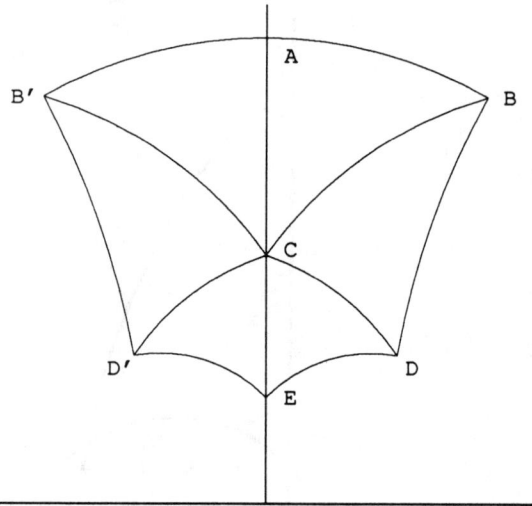

FIGURE 6.9.

It is clear that this observation can be extended to a formula that expresses the length of the side of the arbitrary hyperbolic triangle as a function of its angle (Exercise 7). We shall not pursue this train of thought here as the trigonometry of the hyperbolic triangle is discussed in great detail in Chapter 8.

6.5   REGULAR TESSELATIONS (OPTIONAL)

It is common knowledge that only square, hexagonal, and triangular tiles can

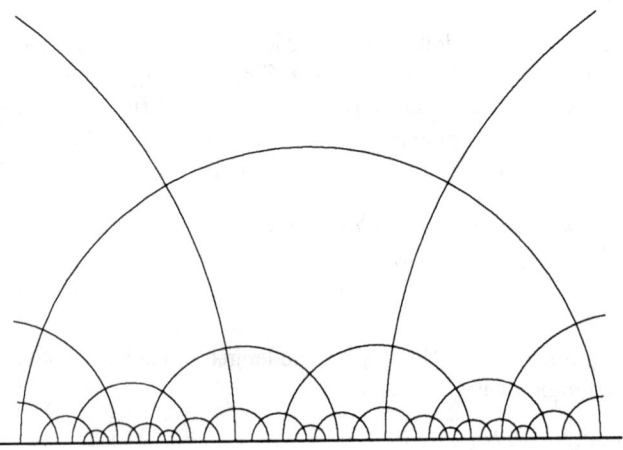

FIGURE 6.10.

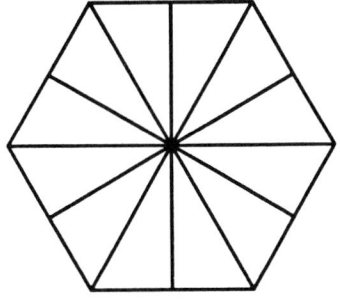
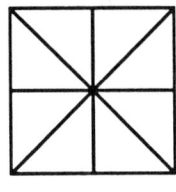

FIGURE 6.11.

be used in the construction of the floors of bathrooms and kitchens. This observation assumes that one is restricted to a tiling by congruent pieces whose common shape is that of a regular polygon. The reason for the limitation to squares and hexagons is that the corners of the other regular polygons cannot be fitted together in the required manner. The corners of the pentagon have an angle of 108°, so that four of them, meeting at a common vertex, must overlap, whereas three are not sufficient to cover the full angle of 360°. On the other hand, the angles of regular n-gons with $n > 6$, all exceed 120°, the angle of the regular hexagon, and so they too cannot be fitted together to cover the floor without overlapping.

Non-Euclidean geometry, on the other hand, allows for a much larger selection of tiles. In fact, given any integer $n > 6$, there are infinitely many different ways of tiling the hyperbolic plane with regular n-gons. The reason for this is that for each such n there exist infinitely many non-congruent regular n-gons each of which can be used for such a tiling.

THEOREM 6.9. *For each pair of integers $k > n > 2$ there is a hyperbolic regular n-gon with angle $(n-2)\pi/k$.*

PROOF: It will suffice to show that for each such pair k, n there exists an isosceles hyperbolic triangle with angles
$$\frac{2\pi}{n}, \quad \frac{(n-2)\pi}{2k}, \quad \frac{(n-2)\pi}{2k}.$$
For, once these triangles are available, the required regular n-gon is obtained by placing n copies of this triangle around a common vertex in much the same way that a Euclidean regular n-gon can be constructed out of n isosceles triangles each of which has angles $2\pi/n, (n-2)\pi/2n, (n-2)\pi/2n$.

The existence of the required isosceles triangle follows from the following arithmetical observations. Since $k > n > 2$ it follows that
$$(n^2 - 2n) < k(n-2)$$

or
$$2k + n^2 - 2n < kn$$
or
$$\frac{2}{n} + \frac{n-2}{k} < 1$$
or
$$\frac{2\pi}{n} + \frac{(n-2)\pi}{2k} + \frac{(n-2)\pi}{2k} < \pi.$$

Hence, Theorem 6.7 guarantees the existence of the required triangle. That this triangle is isosceles follows from Proposition 6 of Euclid.

q.e.d.

For example, if $k = 2n - 4$, the resulting hyperbolic n-gon will have a right angle as each of its angles. In particular, this means that the hyperbolic contractor has the option of tiling the kitchen floor with pentagonal tiles so that exactly four tiles meet at each corner! Unfortunately, this option may not be available to this contractor in "practice". It follows from Theorem 6.8 that once the angle of the regular n-gon has been fixed, its size has been fixed too, and that size might not be convenient. In fact, that size might conceivably exceed that of the kitchen to be tiled.

Example 6.10. Let us construct a right angled hyperbolic regular pentagon. We begin by constructing a triangle with angles $\alpha = 90°$, $\beta = 45°$, and $\gamma = 36°$ in standard position. For this triangle (Figure 6.7) $u = \cot 36°$, $r = \csc 36°$, $v = 0$, and s is a solution of the quadratic $s^2 - s\sqrt{2}\csc 36° + 1 = 0$ and hence $s = 1.8717\ldots$. This is sufficient information to draw the hyperbolic triangle ABC of Fig. 6.9. Let $\triangle AB'C$ be the reflection of $\triangle ABC$ in the y-axis. Then the geodesic BD is the inversion of BB' in BC, and $B'D'$ is the inversion of BB' in $B'C$. The geodesic DC is the inversion of $B'C$ in BC, and the geodesic $D'C$ is the inversion of BC in $B'C$. Finally, DE is the inversion of BD in DC, and $D'E$ is the inversion of $B'D'$ in $D'C$.

In the above discussion we have only demonstrated that the argument used to limit the number of regular tiles of Euclidean geometry does not apply to the hyperbolic plane. On that basis we assumed that it is indeed possible to tile the complete hyperbolic plane with, say, right angled regular pentagons. After all, it would seem that all one has to do is to lay down a tile, fit as many tiles as possible around it, and thus go on indefinitely. The fact that the corners are all right angled would seem to preclude the possibility of any overlaps occuring. Thus Fig. 6.10 contains a tiling of the upper half-plane by a right angled regular pentagon. The careful reader, however, will note that this process assumes that the growing tiled portion of the plane is always hyperbolically convex and that it will never sprout branches that might meet and improperly overlap at some far away point. The following theorem, stated without proof, guarantees that this will not happen.

Theorem 6.10: *Let D be a hyperbolic triangle with angles π/l, π/m, and π/n. Then the triangles obtained by inverting (or reflecting) $\triangle$ in each of its sides and then inverting each of the images in each of its sides, and so on indefinitely, cover the hyperbolic plane without gaps or improper overlaps.*

$\square$

It may not be immediately evident to the reader that this theorem does indeed guarantee the existence of the aforementioned regular hyperbolic tilings. It is a fact, however, that every regular polygon, whether in the Euclidean or in the hyperbolic plane, can be decomposed into triangles of this type. Fig. 6.11 illustrates how this can be done for the square and the regular hexagon of the Euclidean plane. The square consists of eight triangles with angles $\pi/2$, $\pi/4$, $\pi/4$, and the regular hexagon consists of twelve triangles with angles of $\pi/2$, $\pi/3$, $\pi/6$. Successive reflections of each of these triangles in its sides results in a triangular tiling of the plane. Finally, when these triangles are properly grouped in groups of containing 8 or 12 triangles we obtain a square or hexagonal tiling of the plane.

This is admittedly a roundabout method for finding a polygonal tiling of the Euclidean plane, especially in view of the fact that it can produce only square and hexagonal tilings. The advantage of this approach is that it produces many such tilings of the hyperbolic plane. Specifically, the hyperbolic regular n-gon with angle $2\pi/m$ at each vertex can be decomposed into $2n$ congruent triangles with angles $\pi/2$, π/m, π/n, and, by the above theorem this triangle can be used to construct a tiling of the hyperbolic plane by the regular n-gon with angle $2\pi/m$. The actual construction of such tilings is quite easy with any graphics package such as those in Mathematica and Maple. The tiling that appears in Fig. 6.10 took the right angled hyperbolic regular pentagon of Fig. 6.9 and inverted it in its sides. This process was repeated several times to produce this pentagonal tiling.

The interested readers are referred to the articles by Poincaré and Maskit, as well as the books by Caratheodory and Beardon for proofs of Theorem 6.10 and/or related theorems.

6.6 EXERCISES

1. Construct a hyperbolically rectilineal angle of 45° whose vertex is (0,1) and one of whose sides is the y-axis.
2. Construct a hyperbolically rectilineal angle of 60° whose vertex is the point (1,1) and one of whose sides is the bowed geodesic that contains (1,1) and is centered at the origin.

3. Construct a hyperbolic equilateral triangle each of whose angles is $\pi/6$. What is the length of its side?
4. Determine the angles of the hyperbolic triangle whose vertices are $A(1,2)$, $B(3,2)$, and $C(7,2)$.
5. Determine the angles of the hyperbolic triangle whose vertices are $A(1,2)$, $B(3,2)$, and $C(7,4)$.
6. Describe a Euclidean method for bisecting any hyperbolically rectilineal angle.
7. Find a formula for the hyperbolic length a_α of the side of the hyperbolic equilateral triangle with angle α. Evaluate the limits of a_α as α approaches 0 and as α approaches $\frac{\pi}{3}$.
8. Prove that if $0 < \alpha < 2\pi$ then there is a hyperbolic quadrilateral the sum of whose angles is α.
9. Prove or disprove: given any four angles whose sum is less than 2π, they are the angles of some hyperbolic quadrilateral (this is difficult).
10. Find a necessary and sufficient condition for any n angles to be the angles of a hyperbolic n-gon (this is difficult).
11. Explain why there is no such figure as a hyperbolic square. Comment on parallelograms, rhombuses and regular polygons.
12. Let b_α be the hyperbolic length of the side of the equilateral hyperbolic quadrilateral each of whose angles is α. Evaluate the limits of b_α as α approaches 0 and $\frac{\pi}{2}$.
13. Construct a hyperbolic triangle no two of whose altitudes meet. (Hint construct a hyperbolic triangle whose altitudes are segments of vertical geodesics.)
14. Show that every parallelogram can be used to tile the Euclidean plane.
15. Show that every triangle can be used to tile the Euclidean plane.
16. Show that every quadrilateral can be used to tile the Euclidean plane.
17. Explain why the hyperbolic equilateral triangle each of whose angles measures 50° cannot be used to tile the hyperbolic plane.
18. Generalize Exercise 17.
19. Construct a tiling of the upper half-plane by right angled hexagons.

The following exercises call for the use of a computer package such as those available on Mathematica or Maple.

20. Write a script that places a given hyperbolic triangle in standard position.
21. Write a script that computes the angles of a given hyperbolic triangle.
22. Write a script that computes the lengths of the sides of the hyperbolic triangle whose interior angles are α, β, and γ.

7
Hyperbolic Area

7.1 THE GENERAL DEFINITION OF AREA

The definition of hyperbolic area is considerably harder to motivate than that of hyperbolic length. This task will be facilitated by a review of Euclidean area.

The naive approach to areas takes the square all of whose sides have length 1 as the unit for areas. Since a rectangle with integral sides of lengths a and b obviously contains ab such units, it is reasonable to define the rectangle's area as ab unit squares. This definition is then extended to cover first rectangles with rational dimensions, and then rectangles with arbitrary dimensions.

The next step is to observe that by cutting off one appropriate corner of a parallelogram and pasting it onto the opposite side, the parallelogram can be converted into a rectangle of equal area. This leads us to assign bh as the area of the parallelogram with base b and altitude h. Since each parallelogram is split by its diagonal into two congruent triangles, we can now conclude that the area of the triangle with base b and altitude h should be $bh/2$ unit squares. Every polygonal region can be split into triangles by means of appropriately drawn diagonals, and so we now have a method for computing the area of an arbitrary polygon. Calculus can then be used to extend this approach to regions with curved boundaries.

Since squares do not exist in hyperbolic geometry, this strategy fails here. Of course, one could replace the unit square with some sort of a unit equilateral triangle, but a little reflection will convince the readers that the execution of this approach still leads to considerable complications. Consider, for example, what would happen if the hyperbolic equilateral triangle all of whose angles are 45° were chosen to provide the unit area measurement. How would one go about computing the hyperbolic area of the equilateral triangle all of whose angles are 44°?

Many of today's high school geometry texts employ a more sophisticated axiomatic strategy in their treatment of areas. This usualy involves some version of the following four axioms:

Existence: Every polygonal region has an area that is a positive real number;

Invariance: Congruent polygons enclose regions with equal areas.

Additivity: If the polygonal region R is the union of two polygonal regions S and T which overlap only on their boundaries, then the area of R equals the sum of the areas of S and T.

Rectangle: The area of the rectangle is the product of the lengths of its base and altitude.

This strategy also stipulates the existence of rectangles which are not available in the hyperbolic plane. In view of the failure of these two approaches, it behoves us to reexamine Euclid's definition of area. This definition, it was claimed in Chapter 1, is presented axiomatically in the Common Notions. In other words, Euclid simply assumes that a reasonable notion of area exists, where the reasonableness of the notion is made precise by the requirements of the Common Notions. The grounds for his assumption were presumably provided by his everyday experience. Some Greeks had fields and these fields had sizes. The *plethron,* a square of approximately 10,000 square feet was their unit for measuring these areas. Unfortunately we are very lacking in such an experience to guide us in the context of the half-plane.

However, there is a formula for the Euclidean area that does have a very natural hyperbolic analog. If R is any region in the Euclidean plane, then its Euclidean area is given by the double integral

$$\iint_R dx\, dy.$$

The rationale behind this formula is that the region R can be decomposed into infinitesimal rectangles of dimensions dx and dy. In view of the definition of hyperbolic distances given in Chapter 4 it is plausible to speculate that the hyperbolic area of such an infinitesimal rectangle is

$$\frac{dx}{y} \frac{dy}{y}.$$

and hence that the *hyperbolic area* of the region R, denoted by $ha(R)$, is

$$\iint_R \frac{dx\, dy}{y^2} \tag{1}$$

We now show that this poorly motivated definition does indeed yield a reasonable notion of area. Specifically, we show that this notion of area satisfies all of Euclid's Common Notions.

Common Notion 1 simply requires that two regions that have the same hyperbolic area as a third region have the same hyperbolic area as each other. This is indeed clear since if R, S, T are regions such that $ha(R) = ha(T)$ and $ha(S) = ha(T)$ then it follows on logical grounds alone that $ha(R) = ha(S)$.

The verification of Common Notion 2, which stipulates that when equals are added to equals the results are equal, requires some mathematics. It is clear from his subsequent proofs that what Euclid had in mind when he spoke of addition is a situation wherein two polygons are being juxtaposed so that their overlap, if any, is restricted to their boundaries. If R and S are two such regions and $R \cup S$ denotes their union (sum), then it follows from the general properties of integrals that

$$ha(R \cup S) = \iint_{R \cup S} \frac{dx\,dy}{y^2} = \iint_R \frac{dx\,dy}{y^2} + \iint_S \frac{dx\,dy}{y^2} = ha(R) + ha(S).$$

Common Notion 3, which says that when equals are subtracted from equals the results are equal can be disposed of in a similar manner and will not be belabored.

Common Notion 4 requires that congruent regions have equal areas. In other words, it must be shown that the above defined hyperbolic area is invariant under hyperbolic rigid motions. It is already known that all such rigid motions are the composition of several reflections in straight geodesics and inversions centered on the x-axis, and hence it suffices to show that these transformations preserve hyperbolic areas. The case of reflections in straight geodesics is left to the reader as Exercise 11. Since every horizontal translation is the composition of two such reflections, it may be concluded that they too preserve hyperbolic area. Thus it only remains to demonstrate the validity of Common Notion 4 for an inversion $I_{C,k}$ that is centered at the point $C(c, 0)$. Let R be the given region and let R' be its image under this inversion (Fig. 7.1). Assume further that a polar coordinate system has been placed at the origin with its initial ray along the positive x-axis. The transformation from polar to cartesian coordinates is given by

$$x = c + r\cos\theta, \quad y = r\sin\theta$$

with Jacobian

$$\frac{\partial(x, y)}{\partial(r, \theta)} = \frac{\partial x}{\partial r}\frac{\partial y}{\partial \theta} - \frac{\partial x}{\partial \theta}\frac{\partial y}{\partial r} = (\cos\theta)(r\cos\theta) - (-r\sin\theta)(\sin\theta)$$

$$= r.$$

The inversion $I_{C,k}$ carries the general point (r, θ) of R into a point (ρ, θ) of R'. Hence, in terms of polar coordinates, the inversion is given by the transformation

$$\rho = \frac{k^2}{r}, \quad \theta = \theta$$

THE POINCARÉ HALF-PLANE

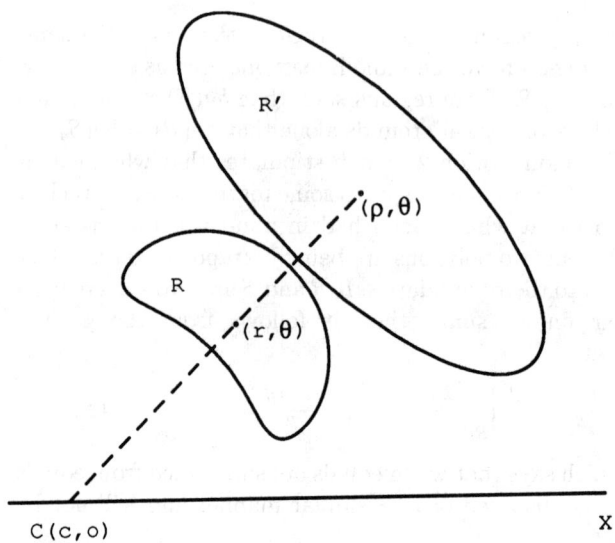

FIGURE 7.1

with Jacobian

$$\frac{\partial(\rho,\theta)}{\partial(r,\theta)} = \frac{\partial \rho}{\partial r}\frac{\partial \theta}{\partial \theta} - \frac{\partial \rho}{\partial \theta}\frac{\partial \theta}{\partial r} = \left(-\frac{k^2}{r^2}\right)(1) - (0)(0) = -\frac{k^2}{r^2}.$$

The relevant integrals then transform as follows:

$$ha(R') = \iint_{R'} \frac{dxdy}{y^2} = \iint_{R'} \frac{\rho d\rho d\theta}{\rho^2 \sin^2\theta}$$

$$= \iint_R \frac{\dfrac{k^2}{r}\dfrac{k^2}{r^2} drd\theta}{\dfrac{k^4}{r^2}\sin^2\theta} = \iint_R \frac{rdrd\theta}{r^2 \sin^2\theta}$$

$$= \iint_R \frac{dxdy}{y^2} = ha(R).$$

Thus definition (1) of hyperbolic area is indeed invariant under hyperbolic rigid motions.

Finally, Common Notion 5 requires that the area of the whole be greater than that of a part. Since definition (1) endows every bona fide two dimensional region with a positive hyperbolic area, the validity of this Common Notion follows from that of Common Notion 3.

It should be of interest to compute the hyperbolic areas of some regions and

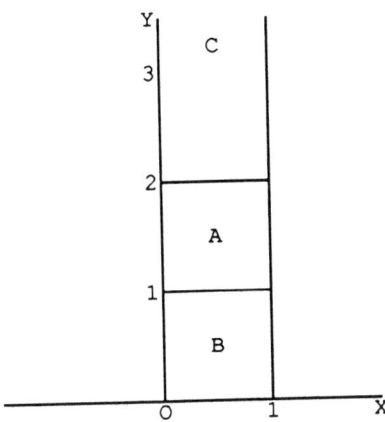

FIGURE 7.2

to compare these to their Euclidean counterparts. Thus the hyperbolic area of the square A of Fig. 7.2 is

$$\int_1^2 \int_0^1 \frac{dxdy}{y^2} = \int_1^2 \frac{dy}{y^2} = \frac{1}{2}$$

compared with a Euclidean area of 1. On the other hand, the hyperbolic area of the square B is

$$\int_0^1 \int_0^1 \frac{dxdy}{y^2} = \int_0^1 \frac{dy}{y^2} = \infty,$$

compared with the finite Euclidean area of 1. Finally, the hyperbolic area of the infinite strip C is

$$\int_2^\infty \int_0^1 \frac{dxdy}{y^2} = \int_2^\infty \frac{dy}{y^2} = \frac{1}{2},$$

compared with its infinite Euclidean area. That the infinite region C has a finite hyperbolic area is not as paradoxical as may seem at first glance. This phenomenon is analogous to the fact that the Euclidean area of the infinite region between the graph of $y = x^{-2}$ and the x-axis to the right of $x = 1$ has the finite value

$$\int_1^\infty \frac{dx}{x^2} = 1.$$

Hyperbolically speaking, the strip C narrows down at a fast enough rate to cause it to have finite hyperbolic area.

114 THE POINCARÉ HALF-PLANE

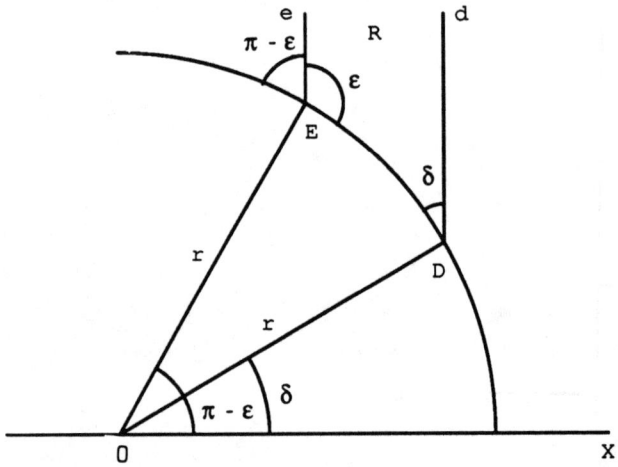

FIGURE 7.3

7.2 THE AREA OF THE HYPERBOLIC TRIANGLE

We shall now show that the above definition of hyperbolic area by means of a double integral results in a very simple formula for the hyperbolic area of an arbitrary hyperbolic triangle. Strangely enough, this formula is even simpler than the well known one for the Euclidean area of a Euclidean triangle. First, a definition is needed.

The *defect* of a hyperbolic triangle with angles α, β, γ is the quantity $\pi - (\alpha + \beta + \gamma)$. It follows from Theorem 6.6 that every hyperbolic triangle has positive defect.

Theorem 7.1. *The area of the hyperbolic triangle is equal to its defect.*

The proof of this theorem is preceded by a lemma that is interesting in and of itself.

Lemma 7.2. *Let DE be a segment of a bowed geodesic. If d and e are the straight geodesics above D and E respectively, and if R is the portion of the infinite strip between d and e that lies above DE, then*

$$ha(R) = \pi - \delta - \epsilon$$

where δ and ϵ are the angles interior to R at D and E respectively.

PROOF: We assume without loss of generality that the geodesic containing DE has Euclidean center O and radius r (Fig. 7.3). By Proposition 6.2

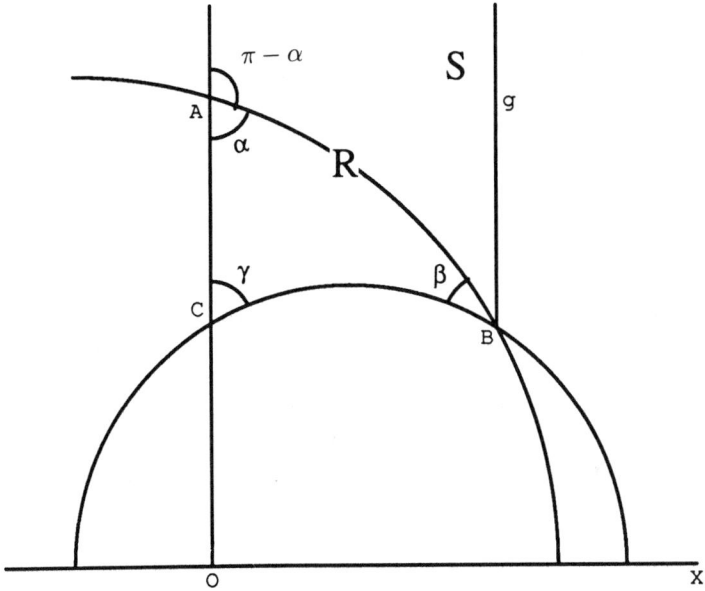

FIGURE 7.4

$\angle XOD = \delta$ and $\angle XOE = \pi - \epsilon$. The geodesic DE has the Euclidean equation

$$x^2 + y^2 = r^2,$$

and so the required area is given by

$$\int_{r\cos(\pi-\epsilon)}^{r\cos\delta} \int_{\sqrt{r^2-x^2}}^{\infty} \frac{dy\,dx}{y^2}.$$

When this double integral is evaluated as an iterated integral we get

$$\int_{-r\cos\epsilon}^{r\cos\delta} \left[-y^{-1} \right]_{\sqrt{(r^2-x^2)}}^{\infty} dx = \int_{-r\cos\epsilon}^{r\cos\delta} \frac{dx}{\sqrt{r^2-x^2}}$$

$$\arcsin\frac{x}{r} \Big]_{-r\cos\epsilon}^{r\cos\delta} = \arcsin(\cos\delta) - \arcsin(-\cos\epsilon)$$

$$= \frac{\pi}{2} - \delta + \frac{\pi}{2} - \epsilon = \pi - \delta - \epsilon.$$

q.e.d.

PROOF of Theorem 7.1: Let ABC be a hyperbolic triangle in standard position, and let g be the straight geodesic through B (Figure 7.4). Let R denote the portion of the infinite strip between g and the y-axis that lies above the geodesic segment BC, and let S denote the portion of the same strip that lies above the geodesic AB. Finally, let α, β, γ denote the angles of the triangle at the vertices A, B, C, repectively. Then, by the above lemma,

$$ha(\triangle ABC) = ha(R) - ha(S)$$
$$= \pi - \gamma - \angle(g, BC) - [\pi - (\pi - \alpha) - \angle(g, AB)]$$
$$= \pi - \gamma - \alpha - [\angle(g, BC) - \angle(g, AB)]$$
$$= \pi - \alpha - \beta - \gamma.$$

q.e.d.

When the areas of specific triangles are computed it is necessary to measure the relevant angles in terms of radians rather than degrees. This is necessitated by the use of calculus in the proof of Lemma 7.2. Thus, the hyperbolic area of the hyperbolic equilateral triangle each of whose angles measures $45°$ is $\pi - 3\pi/4 = \pi/4 = .78....$ Similarly, the hyperbolic area of the hyperbolic triangle with vertices $(0,1)$, $(2,1)$, $(4,1)$ of Example 6.4 is $\pi - \pi/2 - 2\cos^{-1}(3\sqrt{10}/10) = .92....$

Corollary 7.3. *Hyperbolically congruent triangles have equal hyperbolic areas.*

$\square$

The point we are trying to make with this corollary is that its proof does not depend on the sophisticated notion of the Jacobian employed in the previous section. Thus, this obvious corollary yields an alternate and more elementary method for proving that hyperbolic rigid motions do not change the hyperbolic areas of hyperbolic polygons. This method is based on the observation that any hyperbolic (non self-intersecting) polygon can be expressed as the union and difference of a finite number of hyperbolic triangles. Unfortunately, this fact, while plausible and correct, is not easily proven.

7.3 EXERCISES

1. Find a formula for the hyperbolic area of a hyperbolic quadrilateral.
2. Find a formula for the hyperbolic area of an arbitrary hyperbolic polygon.
3. Find the hyperbolic area of the hyperbolic triangle with vertices $(1, 2)$, $(3, 2)$, $(7, 2)$.
4. Find the hyperbolic area of the hyperbolic triangle with vertices $(0, 1)$, $(0, 2)$, and $(2, 1)$.
5. Find the hyperbolic area of the Euclidean rectangle with vertices $(0, 1)$, $(0, 3)$, $(5, 3)$, $(5, 1)$.
6. Find the hyperbolic area of the Euclidean square S with vertices (a, b), $(a + h, b)$, $(a, b + h)$, $(a + h, b + h)$. Find the limit of the ratio of hyperbolic area of S to its Euclidean areas as h approaches 0.
7. Find the limit of the hyperbolic area of the square of Exercise 6 as
    i) b approaches 0,   ii) b diverges to positive infinity,

iii) h diverges to positive infinity.
8. Prove that the hyperbolic area of a circle with hyperbolic radius R is $2\pi[\cosh(R) - 1]$. (Hint: place a polar coordinate system at the Euclidean center of the circle and feel free to refer to an integral table.)
9. Find an expression for the hyperbolic area of the Euclidean triangle with vertices $(0, 1)$, $(a, 1)$, (b, c).
10. Find the Euclidean area of a hyperbolic triangle in standard position and compare it to its hyperbolic area.
11. Prove formally, in terms of Jacobians, that reflections in straight geodesics preserve hyperbolic areas.
12. Prove formally, in terms of Jacobians, that horizontal translations preserve hyperbolic areas.

8

The Trigonometry of the Hyperbolic Triangle

8.1 THE TRIGONOMETRY OF HYPERBOLIC LINE SEGMENTS

It is the purpose of this chapter to convince the readers that the geometry of the hyperbolic plane is just as concrete as that of the Euclidean plane. We aim to accomplish this by developing the trigonometry of this non-Euclidean geometry. The derivation of the hyperbolic Laws of Sines and Cosines by Bolyai and Lobachevsky marks a very important turning point in the evolution of the hyperbolic plane. Up to that point non-Euclidean geometry had the status of a mere logical exercise which was part of an attempted *reductio ad absurdum* proof of the Parallel Postulate. The quotation that appears at the end of Section 11.3 makes it clear that it was the amazing resemblance between the hyperbolic and spherical trigonometries that convinced Lobachevsky of the validity of his geometry and of the independence of the Parallel Postulate. The author also believes that both the intricacy and the apparent internal consistency of the rich hyperbolic trigonometry had much to do with the acceptance of hyperbolic geometry as a valid alternate geometry. The readers are referred to Chapter 15 for a more detailed discussion of these issues.

We begin this section by reexamining the configuration that proved so useful in the study of hyperbolic areas.

Proposition 8.1. *Let AB be a bowed geodesic segment, and let AE and BD be the straight geodesics above A and B, respectively. If $\alpha = \angle EAB$, $\beta = \angle ABD$ and $c = h(A,B)$, then*

$$\text{i)} \quad \sinh c = \frac{\cos \alpha + \cos \beta}{\sin \alpha \sin \beta}$$

120   THE POINCARÉ HALF-PLANE

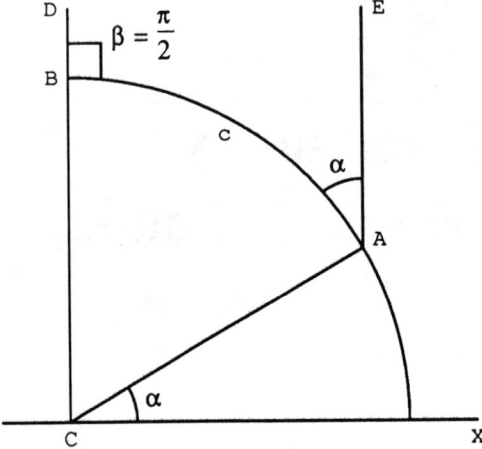

FIGURE 8.1.

$$\text{ii)} \quad \cosh c = \frac{1 + \cos \alpha \cos \beta}{\sin \alpha \sin \beta}$$

$$\text{iii)} \quad \tanh c = \frac{\cos \alpha + \cos \beta}{1 + \cos \alpha \cos \beta}$$

PROOF: Assume first that $\beta = \frac{\pi}{2}$ (Fig. 8.1). In that case, if C is the point on the x-axis directly below B and D, then by Proposition 6.2, $\angle XCA = \alpha$, and by Proposition 4.1,

$$c = \ln \frac{\csc \beta - \cot \beta}{\csc \alpha - \cot \alpha} = \ln \frac{\sin \alpha}{1 - \cos \alpha}$$

Consequently,

$$2 \sinh c = e^c - e^{-c} = \frac{\sin \alpha}{1 - \cos \alpha} - \frac{1 - \cos \alpha}{\sin \alpha}$$

$$= \frac{\sin^2 \alpha - 1 + 2 \cos \alpha - \cos^2 \alpha}{\sin \alpha (1 - \cos \alpha)}$$

$$= \frac{2 \cos \alpha (1 - \cos \alpha)}{\sin \alpha (1 - \cos \alpha)} = 2 \cot \alpha.$$

Hence,

$$\sinh c = \cot \alpha \quad \text{when } \beta = \frac{\pi}{2}.$$

Since

$$\cosh^2 c - \sinh^2 c = 1 \quad \text{and} \quad \csc^2 \alpha - \cot^2 \alpha = 1,$$

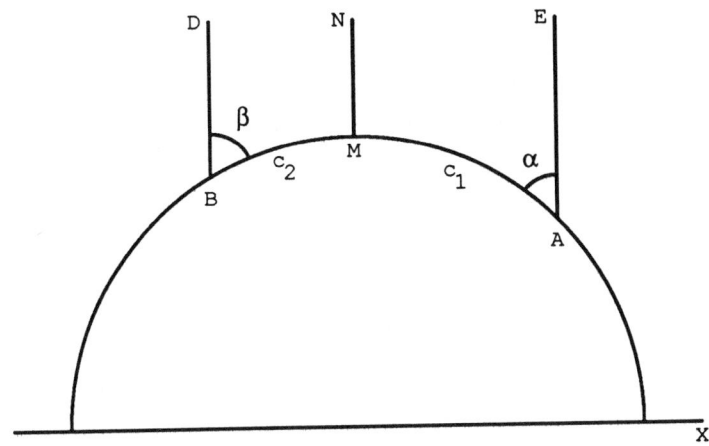

FIGURE 8.2.

it follows that

$$\cosh c = \csc \alpha \quad \text{and} \quad \tanh c = \cos \alpha \quad \text{when} \quad \beta = \frac{\pi}{2}.$$

Turning now to the general case (Fig. 8.2), let M be the point on the geodesic containing AB that has the largest ordinate. We restrict our discussion to the case where M is in between A and B, leaving the alternate case, where M is outside the hyperbolic segment AB as Exercise 2.

If MN is the straight geodesic directly above M, then $\angle AMN = \angle NMB = \frac{\pi}{2}$. Consequently, if $c_1 = h(A, M)$ and $c_2 = h(B, M)$, then, when M lies between A and B,

$$\sinh c = \sinh(c_1 + c_2) = \sinh c_1 \cosh c_2 + \cosh c_1 \sinh c_2$$

$$= \cot \alpha \csc \beta + \csc \alpha \cot \beta = \frac{\cos \alpha + \cos \beta}{\sin \alpha \sin \beta}.$$

This concludes the proof of part i of the proposition. Parts ii and iii are left to the reader (Exercise 1).

q.e.d.

Just like Proposition 6.1, this technical proposition is merely a tool that will facilitate the proofs of the more interesting theorems to follow.

The introductory case where $\beta = \frac{\pi}{2}$ in the above proof is of sufficient interest to merit an explicit statement.

Corollary 8.2. *If in Proposition 8.1 $\beta = \frac{\pi}{2}$, then*

$$\sinh c = \cot \alpha, \quad \cosh c = \csc \alpha, \quad \tanh c = \cos \alpha.$$

□

8.2 HYPERBOLIC RIGHT TRIANGLES

The Theorem of Pythagoras is considered by many mathematicians to be the most important single theorem of mathematics. Many advanced theorems and concepts of geometry, algebra, and number theory are either based on, or else motivated by this simple proposition. The following theorem is the hyperbolic analog of the Theorem of Pythagoras.

Theorem 8.3. *Let ABC be a hyperbolic triangle with a right angle at C. If a, b, c, are the hyperbolic lengths of the sides opposite A, B, C, respectively, then*

$$\cosh c = \cosh a \cosh b.$$

PROOF: Let ABC be in standard position (Fig. 8.3). Let r denote the radius and $D(-d, 0)$ the Euclidean center of the geodesic containing AB. Let a and b be the interior angles of $\triangle ABC$ at A and B, and let β_1 be the angle between the

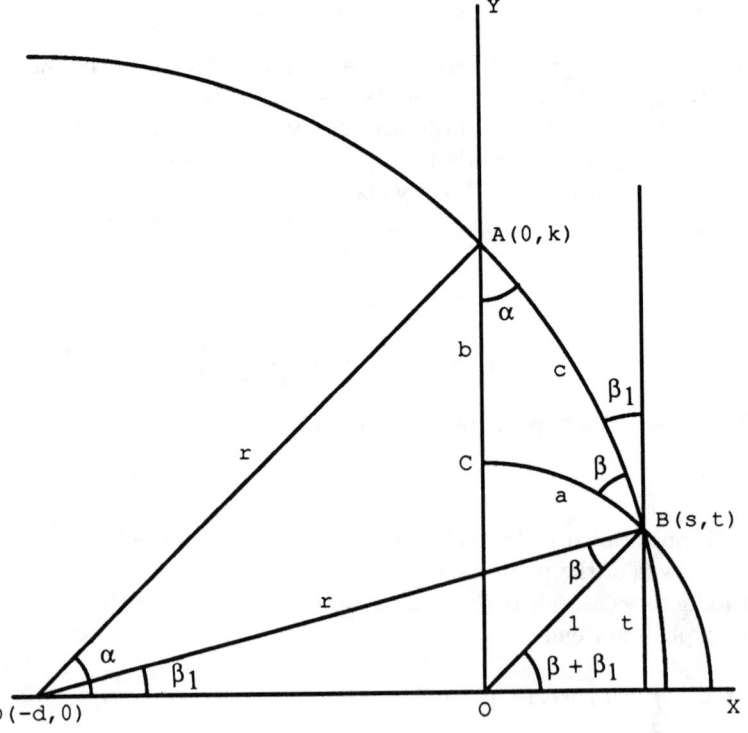

FIGURE 8.3.

geodesic segment AB and the straight geodesic above B. By Proposition 6.2
$$\angle XDA = \alpha, \quad \angle XDB = \beta_1, \quad \angle XOB = \beta + \beta_1,$$
$$\angle DBO = \beta.$$

By Corollary 8.2,
$$\cosh a = \csc(\beta + \beta_1) = \frac{1}{t}.$$

By Proposition 4.3, $b = \ln k$, and so
$$\cosh b = \frac{e^b + e^{-b}}{2} = \frac{k + \frac{1}{k}}{2} = \frac{k^2 + 1}{2k}.$$

By Propostion 8.1,
$$\cosh c = \frac{1 + \cos \beta_1 \cos(\pi - \alpha)}{\sin \beta_1 \sin(\pi - \alpha)}$$
$$= \frac{1 - \frac{d + s\,d}{r}\frac{}{r}}{\frac{tk}{rr}} = \frac{r^2 - d^2 - ds}{kt}$$
$$= \frac{k^2 - ds}{kt} \tag{1}$$

However, by the Euclidean Theorem of Pythagoras,
$$s^2 + t^2 = 1, \quad (s + d)^2 + t^2 = r^2 = d^2 + k^2,$$
and hence
$$2sd + d^2 = r^2 - 1,$$
or
$$sd = \frac{r^2 - d^2 - 1}{2} = \frac{k^2 - 1}{2}.$$

Substituting this value of sd into (1) now yields
$$\cosh c = \frac{k^2 - \frac{k^2 - 1}{2}}{kt} = \frac{k^2 + 1}{2kt}$$
$$= \cosh a \cosh b.$$

q.e.d.

It follows from the hyperbolic Theorem of Pythagoras that if c is the hyperbolic length of the hypotenuse of the hyperbolic right triangle both of whose sides have length 1, then
$$\cosh c = (\cosh 1)^2$$

and so
$$c = \cosh^{-1}((\cosh 1)^2)$$
$$= \ln[(\cosh 1)^2 + \sqrt{((\cosh 1)^2)^2 - 1}] = 1.51....$$
We note that this quantity somewhat exceeds $\sqrt{2} = 1.41...$ which is the Euclidean length of the hypotenuse of the Euclidean counterpart of this right triangle. Nevertheless, the two lengths are fairly close. The following considerations may clarify the relationship between the Euclidean and the hyperbolic Pythagorean theorems. Recall that the exponential function has the infinite series expansion
$$e^x = 1 + x + \frac{x^2}{2!} + \frac{x^3}{3!} + ...$$
Hence, if terms of degree four or higher are ignored, we obtain
$$\cosh x = \frac{1}{2}[e^x + e^{-x}] \approx 1 + \frac{x^2}{2}.$$
But then, if fourth degree terms are still ignored, the statement
$$\cosh c = \cosh a \cosh b$$
is equivalent to the statement
$$1 + \frac{c^2}{2} = \left(1 + \frac{a^2}{2}\right)\left(1 + \frac{b^2}{2}\right)$$
or, upon simplification,
$$c^2 = a^2 + b^2.$$

Thus the Euclidean and the hyperbolic Pythagorean Theorems may be said to agree for infinitely small triangles.

The various parts of the following proposition are easily proved by the same technique that was used to prove the hyperbolic Theorem of Pythagoras, and the details are left to the reader.

Proposition 8.4. *In the hyperbolic right triangle ABC with its right angle at C, let α and β denote the angles at A and B, respectively, and let a, b, c, be the hyperbolic lengths of the sides opposite A, B, C, respectively. Then*

i) $\tanh a = \sinh b \tan \alpha$ and $\tanh b = \sinh a \tan \beta$

ii) $\sinh a = \sinh c \sin \alpha$ and $\sinh b = \sinh c \sin \beta$

iii) $\tanh b = \tanh c \cos \alpha$ and $\tanh a = \tanh c \cos \beta$

iv) $\cosh b \sin \alpha = \cos \beta$ and $\cosh a \sin \beta = \cos \alpha$

v) $\cosh c = \cot \alpha \cot \beta$

□

8.3 THE GENERAL HYPERBOLIC TRIANGLE

Generally speaking, a Euclidean triangle is completely determined by any three of its three sides and three angles. An exception to this rule is the fact that a Euclidean triangle is determined only up to similarity when only its three angles are given. The formulas that allow us to compute all the parts of the triangle from the three given data are known as the Law of Cosines and the Law of Sines. We state them here without proof.

Theorem 8.5'. *In the Euclidean triangle ABC let α, β, γ denote the angles at A, B, C, and let a, b, c, denote the Euclidean lengths of the sides opposite A, B, C, respectively. Then,*

i) $\quad \cos A = \dfrac{b^2 + c^2 - a^2}{2bc}$

ii) $\quad \dfrac{\sin \alpha}{a} = \dfrac{\sin \beta}{b} = \dfrac{\sin \gamma}{c}.$

□

In the hyperbolic plane this exception does not occur and a triangle is completely determined by its angles. Parts i and iii of the following theorem are the hyperbolic analogs of the Euclidean Laws of Cosines and Sines. Part ii has no Euclidean analog, but it is, in an obvious though informal sense, the dual of part i.

Theorem 8.5. *In the hyperbolic triangle ABC let α, β, γ denote the angles at A, B, C, and let a, b, c, denote the hyperbolic lengths of the sides opposite A, B, C, respectively. Then,*

i) $\quad \cos \alpha = \dfrac{\cosh b \cosh c - \cosh a}{\sinh b \sinh c}$

ii) $\quad \cosh a = \dfrac{\cos \beta \cos \gamma + \cos \alpha}{\sin \beta \sin \gamma}$

iii) $\quad \dfrac{\sin \alpha}{\sinh a} = \dfrac{\sin \beta}{\sinh b} = \dfrac{\sin \gamma}{\sinh c}.$

PROOF: We begin with part i. Suppose first that the hyperbolic altitude d from vertex A to side BC falls inside the triangle (Fig. 8.4). If this altitude splits the angle α into α_1 and α_2, and the side a into a_1 and a_2, then the application of the $\cos(x+y)$, $\cosh(x+y)$ formulas, parts ii and iii of Theorem

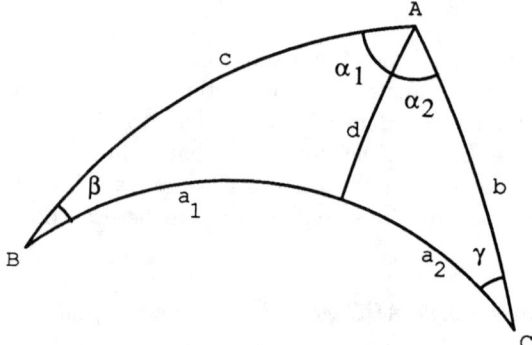

FIGURE 8.4.

8.4, as well as the hyperbolic Theorem of Pythagoras, yield

$$\cos \alpha = \cos(\alpha_1 + \alpha_2) = \cos \alpha_1 \cos \alpha_2 - \sin \alpha_1 \sin \alpha_2$$

$$= \frac{\tanh d}{\tanh c} \frac{\tanh d}{\tanh b} - \frac{\sinh a_1}{\sinh c} \frac{\sinh a_2}{\sinh b}$$

$$= \frac{\cosh b \cosh c \tanh^2 d - \sinh a_1 \sinh a_2}{\sinh b \sinh c}$$

$$= \frac{\cosh b \cosh c \, (1 - \operatorname{sech}^2 d) - \sinh a_1 \sinh a_2}{\sinh b \sinh c}$$

$$= \frac{\cosh b \cosh c - \dfrac{\cosh b}{\cosh d} \dfrac{\cosh c}{\cosh d} - \sinh a_1 \sinh a_2}{\sinh b \sinh c}$$

$$= \frac{\cosh b \cosh c - (\cosh a_1 \cosh a_2 + \sinh a_1 \sinh a_2)}{\sinh b \sinh c}$$

$$= \frac{\cosh b \cosh c - \cosh(a_1 + a_2)}{\sinh b \sinh c}$$

$$= \frac{\cosh b \cosh c - \cosh a}{\sinh b \sinh c}.$$

The proof of the case where the altitude d falls outside the triangle is relegated to Exercise 11.

Part ii is now proven in a similar manner (but with different formulas from Theorem 8.4). Assuming that the altitude d is internal,

$$\cosh a = \cosh(a_1 + a_2) = \cosh a_1 \cosh a_2 + \sinh a_1 \sinh a_2$$

$$= \frac{\cos \alpha_1}{\sin \beta} \frac{\cos \alpha_2}{\sin \gamma} + \frac{\tanh d}{\tan \beta} \frac{\tanh d}{\tan \gamma}$$

$$= \frac{\cos\alpha_1 \cos\alpha_2 + (1 - \operatorname{sech}^2 d)\cos\beta\cos\gamma}{\sin\beta\sin\gamma}$$

$$= \frac{\cos\beta\cos\gamma + \cos\alpha_1\cos\alpha_2 - \dfrac{\cos\beta}{\cosh d}\dfrac{\cos\gamma}{\cosh d}}{\sin\beta\sin\gamma}$$

$$= \frac{\cos\beta\cos\gamma + \cos\alpha_1\cos\alpha_2 - \sin\alpha_1\sin\alpha_2}{\sin\beta\sin\gamma}$$

$$= \frac{\cos\beta\cos\gamma + \cos\alpha}{\sin\beta\sin\gamma}.$$

Exercise 12 deals with the case of an external altitude.
We now turn to part iii. An application of part i to triangle ABC yields

$$\frac{\sin^2\alpha}{\sinh^2 a} = \frac{1 - \cos^2\alpha}{\sinh^2 a}$$

$$= \frac{\sinh^2 b \sinh^2 c - (\cosh b \cosh c - \cosh a)^2}{\sinh^2 a \sinh^2 b \sinh^2 c}$$

$$= \frac{(1 - \cosh^2 b)(1 - \cosh^2 c) - (\cosh b \cosh c - \cosh a)^2}{\sinh^2 a \sinh^2 b \sinh^2 c}$$

$$= \frac{1 - \cosh^2 a - \cosh^2 b - \cosh^2 c + 2\cosh a \cosh b \cosh c}{\sinh^2 a \sinh^2 b \sinh^2 c}.$$

This last expression, however, is symmetrical in a, b, and c, and hence it follows that

$$\frac{\sin^2\alpha}{\sinh^2 a} = \frac{\sin^2\beta}{\sinh^2 b} = \frac{\sin^2\gamma}{\sinh^2 c}$$

Since all the quantities α, β, γ are angles between 0 and π, their sines are all positive. Similarly, since the quantities a, b, c are positive, so are their hyperbolic sines. Consequently,

$$\frac{\sin\alpha}{\sinh a} = \frac{\sin\beta}{\sinh b} = \frac{\sin\gamma}{\sinh c}.$$

q.e.d.

Example 8.6. By Theorem 6.5 there exists an equilateral hyperbolic triangle whose angles are each $\pi/4$. If the common length of the sides of the triangle is a, then by part ii of the above theorem,

$$\cosh a = \frac{\dfrac{1}{2} + \dfrac{\sqrt{2}}{2}}{\dfrac{1}{2}} = 1 + \sqrt{2},$$

and hence

$$a = \cosh^{-1}(1+\sqrt{2}) = \ln[(1+\sqrt{2}) + \sqrt{(1+\sqrt{2})^2 - 1}]$$
$$\approx 1.528.$$

The area of this triangle is of course $\pi - 3\pi/4 = \pi/4$.

Example 8.7. Let α be the angle of the equilateral hyperbolic triangle of side 2. Since $\cosh 2 \approx 3.762$ and $\sinh 2 \approx 3.627$ it follows that

$$\cos \alpha \approx \frac{3.762^2 - 3.762}{3.627^2} \approx .79.$$

Hence $\alpha \approx \cos^{-1} .79 \approx .66$ radians, and the triangle has area approximately $3.14 - 3(.66) = 1.16$.

In Exercises 17,18 the reader is requested to show that the hyperbolic and Euclidean Laws of Sines and Cosines agree when they are applied to infinitesimally small triangles. It follows from this coincidence that infinitesimally small inhabitants of both the hyperbolic and the Euclidean plane cannot distinguish bewtween the properties of the spaces they inhabit.

This could very well apply to us too. We "know" that we inhabit a Euclidean space because the sums of the angles of our triangles is 180°. However, all physical measurements are subject to errors. Moreover, it is clear that we are indeed very small in comparison with the universe. Thus it is quite conceivable that the sums of the angles of our triangles are in fact less than π, but that the triangles within our grasp are so small that our instruments cannot detect the difference. To be somewhat more precise, if these triangles are so small that the *fourth powers* of the lengths of their sides are too small to be detected by measurement then we might very well be hyperbolic creatures.

Ironically, the natives of a Euclidean space can never be sure of what space they inhabit, since the angles of their triangles will always add up to π, but they will always face the possibility that this is due to the fact that their instruments are not accurate enough to detect the defect. On the other hand, the natives of a hyperbolic space, once they have detected a triangle with positive defect, will be certain of the nature of their space.

8.4 EXERCISES

1. Complete the proofs of parts ii and iii of Proposition 8.1 in the case where M is inside the hyperbolic segment AB.
2. Prove Proposition 8.1 in the case where M is outside the hyperbolic segment AB.
3. Find the hyperbolic length of the hypotenuse of the hyperbolic right triangle whose other sides have hyperbolic lengths 3 and 4.

4. Find the hyperbolic length of the hypotenuse of the hyperbolic right triangle whose other sides have hyperbolic lengths .3 and .4.
5. Find the hyperbolic lengths of the legs of the hyperbolic isosceles right triangle whose hypotenuse has hyperbolic length 100.
6. Use the hyperbolic Law of Sines to formulate and prove a hyperbolic version of the Theorem of Menelaus (See Exercise I.20).
7. Use the hyperbolic Theorem of Menelaus to formulate and prove a hyperbolic Theorem of Ceva (See Exercise I.21).
8. Prove that the three medians of the hyperbolic triangle are concurrent.
9. Prove that any two medians of the hyperbolic triangle cut each other into segments whose hyperbolic sines have ratios $2:1$.
10. Prove Proposition 8.4 parts i-v.
11. Prove Theorem 8.5 part i in the case where d is an external altitude.
12. Prove Theorem 8.5 part ii in the case where d is an external altitude.
13. Suppose a hyperbolic equilateral triangle has side a and angle α. Prove that
$$2\cosh\frac{a}{2}\sin\frac{\alpha}{2} = 1.$$
14. Find the area and the lengths of the sides of the hyperbolic right triangle whose acute angles are both $\pi/6$.
15. Find the area and the lengths of the sides of the hyperbolic triangle whose angles are $\pi/6$, $\pi/6$, $\pi/3$.
16. Suppose a triangle has area 1 radian. How large can its sides be? How small can they be?
17. Show that the infinitesimal version of the hyperbolic Laws of Cosines agrees with the Euclidean Law of Cosines.
18. Show that the infinitesimal version of the hyperbolic Law of Sines agrees with the Euclidean Law of Sines.
19. Find the hyperbolic lengths of the sides and diagonals of the hyperbolic regular pentagon each of whose angles is a right angle. Find the hyperbolic areas of the triangle formed by one diagonal and two sides of of this pentagon and of the triangle formed by two diagonals and one side.
20. Prove that the altitude to the base of a hyperbolic right triangle cannot exceed $\ln(1+\sqrt{2})$ in length.
21. Prove that any circle inscribed in a hyperbolic triangle has diameter at most $\ln 3$.
22. If K is the area of a hyperbolic right triangle $\triangle ABC$ in which the right angle is at C, prove that
$$\sin K = \frac{\sinh a \sinh b}{1 + \cosh a \cosh b}.$$
23. Show that if R is the radius of the inscribed circle of a hyperbolic triangle of area K, then
$$\tanh R \geq \frac{1}{2}\sin\left(\frac{K}{2}\right).$$
24. Show that the hyperbolic length of the geodesic joining the midpoints of two of the sides of a hyperbolic triangle is less than half the hyperbolic length of the third side.

9
Complex Numbers and Rigid Motions

9.1 COMPLEX NUMBERS AND THE EUCLIDEAN RIGID MOTIONS

Many surprises have been encountered in this development of the hyperbolic plane. These include the facts that the geodesics of the Poincaré half-plane are Euclidean semicircles, that the hyperbolic circles are in fact also Euclidean circles, and that the Euclidean inversions are hyperbolic rigid motions. Yet possibly the most amazing of the connections brought to light by the Poincaré metric is the role played by complex numbers in this and other geometries. So deep is this connection that mathematicians have come to think of the points of the hyperbolic plane as complex numbers with a positive imaginary part rather than as pairs (x, y) with a positive y. This allows them to apply complex-theoretic information to hyperbolic geometry. Vice versa, non-Euclidean geometrical tools can then be used to resolve problems of the theory of complex variables. The reader will find an explanation of what kind of a light non-Euclidean geometry throws on complex analysis in the historical chapter of this book. At this point suffice it to say that the rigid motions of both the Euclidean and the hyperbolic planes have particularly elegant formulations in terms of complex numbers. The derivation of these formulations is the goal of this chapter.

We remind the reader that with each point (x, y) of the Cartesian plane is associated a complex number

$$z = x + yi$$

where $i^2 = -1$, and that these complex numbers are subject to the four basic arithmetical operations of addition, subtraction, multiplication and division. These operations possess in this new context the same properties of commutativity, associativity and distributivity that they possess in the context of the so called real numbers. Thus,

$$(2+3i) + (1-2i) = (2+1) + (3-2)i = 3+i,$$
$$(2+3i) - (1-2i) = (2-1) + (3+2)i = 1+5i,$$
$$(2+3i)(1-2i) = (2-4i+3i-6i^2) = (2-i+6) = 8-i,$$

and

$$\frac{2+3i}{1-2i} = \frac{2+3i}{1-2i}\frac{1+2i}{1+2i} = \frac{2+4i+3i-6}{1-2i+2i+4} = \frac{-4+7i}{5} = -\frac{4}{5} + \frac{7}{5}i.$$

Let $c = a + bi$ be a fixed complex number and let $z = x + yi$ be an arbitrary complex number. Since $z + c = (x + a) + (y + b)i$ it is clear that the line segment joining z to $z + c$ is both parallel and equal in length to the line joining the origin $O = 0 + 0i$ to the point c. Consequently we have the following lemma.

Lemma 9.1. *If c is any fixed complex number, then the function*

$$f(z) = z + c$$

is a translation of the Euclidean plane. Conversely, every translation of the plane is expressible in this manner. □

We now turn to the Euclidean rotations. Let $z(x, y)$ denote the geometric point that corresponds to the complex number $z = x + yi$. We refer to the length of the line segment Oz as the *modulus* of z, denoted by $|z|$, and to the angle from the positive x axis to the ray Oz as the *argument* of z, denoted by $arg(z)$. It is clear from Figure 9.1 that

$$|z| = \sqrt{x^2 + y^2} \quad \text{and} \quad arg(z) = \arctan\frac{y}{x}.$$

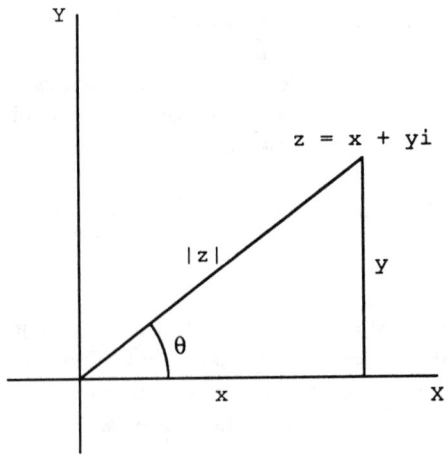

FIGURE 9.1.

If θ is any real number, we define
$$e^{i\theta} = \cos\theta + i\sin\theta.$$
In other words, $e^{i\theta}$ is that complex number of modulus 1 whose argument is θ. The rationale behind this notation is that
$$(\cos\alpha + i\sin\alpha)(\cos\beta + i\sin\beta)$$
$$= \cos\alpha\cos\beta - \sin\alpha\sin\beta + i(\cos\alpha\sin\beta + \sin\alpha\cos\beta)$$
$$= \cos(\alpha+\beta) + i\sin(\alpha+\beta),$$
and so
$$e^{i\alpha}e^{i\beta} = e^{i(\alpha+\beta)}$$
which is, of course, consistent with the multiplication rule for powers with like base.

Let $z = x + yi$ be a complex number and suppose that $r = |z|$ and $\theta = \arg(z)$. Then
$$z = \sqrt{x^2+y^2}\left(\frac{x}{\sqrt{x^2+y^2}} + \frac{yi}{\sqrt{x^2+y^2}}\right)$$
$$= r(\cos\theta + i\sin\theta) = r\,e^{i\theta}.$$

If $w = u + vi$ is another complex number with, say, $|w| = R$ and $\arg(w) = \phi$, then $w = Re^{i\phi}$, and so
$$zw = r\,e^{i\theta}R\,e^{i\phi} = rRe^{i(\theta+\phi)}.$$
In other words
$$|zw| = |z|\,|w| \quad\text{and}\quad \arg(zw) = \arg(z) + \arg(w).$$
Thus, the multiplication of complex numbers has a very geometric interpretation in terms of the modulus and argument. This is summarized in the following proposition.

Proposition 9.2. *The modulus of the product (quotient) of two complex numbers is the product (quotient) of their moduli, and the argument of the product (quotient) is the sum (difference) of their arguments.* □

The portion of this proposition dealing with the argument is often referred to as the argument principle. It follows from this principle that if α is a fixed angle and $z = r\,e^{i\theta}$ is an arbitrary complex number, then the complex number $w = e^{i\alpha}z$ is obtained from z by simply rotating the complex plane by the counterclockwise angle α about the origin. Conversely, every rotation about the origin can be expressed in this manner.

Lemma 9.3. *For any angle α and any complex number c the function*
$$f(z) = e^{i\alpha}(z - c) + c = e^{i\alpha}z + (1 - e^{i\alpha})c$$
is the rotation $R_{c,\alpha}$.

PROOF: The function $f(z)$ is the composition of the translation $z - c$, the rotation $e^{i\alpha}z$ and the translation $z + c$. Hence it is a rotation by the angle α. Since
$$f(c) = e^{i\alpha}(c - c) + c = c,$$
it follows that $f(z) = R_{c,\alpha}$.

q.e.d.

Example 9.4. The 90° counterclockwise rotation of the plane about the point $(0,1)$ has the expression
$$R_{i,\pi/2} = e^{i\pi/2}(z - i) + i = i(z - i) + i = iz + 1 + i$$
in terms of complex numbers. Similarly, the 180° rotation about the point $(2,1)$ has the expression
$$R_{2+i,\pi} = e^{i\pi}(z - 2 - i) + 2 + i = -(z - 2 - i) + 2 + i$$
$$= -z + 4 + 2i.$$
The composition $R_{2+i,\pi} \circ R_{i,\pi/2}$ is now easily computed as
$$f(z) = -(iz + 1 + i) + 4 + 2i = -iz + 3 + i$$
$$= e^{3i\pi/2}z + 3 + i.$$
By Lemma 9.3 we know that this $f(z)$ is a rotation of the form $R_{c,3\pi/2}$. The value of c can be derived from the fact that $f(c) = c$ and hence c satisfies the equation
$$-ic + 3 + i = c.$$
Thus,
$$c = \frac{3+i}{1+i} = \frac{3+i}{1+i} \cdot \frac{1-i}{1-i} = \frac{3 + 1 + i(-3 + 1)}{1 - (-1)} = 2 - i.$$
Hence,
$$R_{2+i,\pi} \circ R_{i,\pi/2} = R_{2-i,3\pi/2}.$$

Finally, we turn to the Euclidean reflections. If $z = x + yi$ is any complex number, then the number $x - yi$ is called its *conjugate* and is denoted by $\bar{z}$. It is easy to see that the conjugate of $\bar{z}$ is z and that every complex number and its conjugate are symmetrical with respect to the x axis. In other words, the

function $f(z) = \bar{z}$ is the reflection ρ_x in the x axis. Note that
$$\overline{e^{i\theta}} = e^{-i\theta}, \quad \overline{z+w} = \bar{z}+\bar{w}, \quad \overline{zw} = \bar{z}\bar{w},$$
$$\arg(\bar{z}) = -\arg z, \quad \text{and} \quad |\bar{z}| = |z|.$$

If m is the line through the origin with inclination θ to the positive x axis, then $R_{O,\theta} \circ \rho_x \circ R_{O,-\theta}$ is the reflection ρ_m in the line m. In terms of complex numbers this composition can be written as
$$e^{i\theta}\overline{e^{-i\theta}z} = e^{2i\theta}\bar{z}.$$

This gives us the complex expression for the reflection in any line through the origin. The following lemma describes all reflections in terms of complex operations.

Lemma 9.5. *If m is any line with inclination α to the positive x axis, and c is a point on m, then the function*
$$f(z) = e^{2i\alpha}\overline{z-c} + c$$
is the reflection in the line m.

PROOF: Let n be the straight line through the origin parallel to m, and let τ be the translation $z+c$ that maps 0 to c. We know that $\tau \circ \rho_n \circ \tau^{-1}$ is a glide reflection. Moreover, this is a glide reflection that fixes every point of the line m, and hence it must in fact be ρ_m. Since
$$\tau \circ \rho_n \circ \tau^{-1}(z) = e^{2i\alpha}\overline{z-c} + c$$
we are done. $\square$

In particular, note that the reflection in the vertical geodesic above the point $(r,0)$ has the expression $-\bar{z} + 2r$.

Example 9.6. Let r be the reflection in the line $y = x - 2$ and let σ be the reflection in the line $x = 5$. These reflections have the complex expressions
$$\rho(z) = e^{2i\pi/4}(\bar{z} - 2) + 2 = i\bar{z} + 2 - 2i$$
$$\sigma(z) = -\bar{z} + 10.$$
Their composition $\rho \circ \sigma$ is the transformation
$$i\overline{(-\bar{z} + 10)} + 2 - 2i = -iz + 2 + 8i$$
$$= e^{3i\pi/2}z + 2 + 8i,$$
which is a 90° clockwise rotation whose center is the solution of the equation
$$-iz + 2 + 8i = z,$$
i.e., the point $5 + 3i$.

The above three lemmas are summarized in the following theorem.

Theorem 9.7. *The rigid motions of the Euclidean plane all have the form*
$$f(z) = e^{i\alpha}z + c \quad \text{or} \quad f(z) = e^{i\alpha}\bar{z} + c,$$
where α is an arbitrary real number and c is an arbitrary complex number. Conversely, every function of either of these forms is a rigid motion of the Euclidean plane.

PROOF: We already know that the translations, rotations and reflections of the Euclidean plane all have these forms. Since the compositions of any two functions of these forms yields another function of one of these forms, the same holds for the glide-reflections.

Conversely, we already know that every function of the form $e^{i\alpha}z + c$ is either a rotation or a translation, and that every function of the form $z + c$ is a translation, and that every function of the form $e^{i\alpha}\bar{z}$ is a reflection. Thus every function of the form $e^{i\alpha}\bar{z} + c$ is the composition of a translation with a reflection and so it is a glide reflection. □

Example 9.8. The transformation $f(z) = i\bar{z} + 1 - i$ is either a reflection or a glide-reflection. We demonstrate that it is in fact a reflection by finding its fixed points, i.e., by showing that there are solutions to the equation
$$i\bar{z} + 1 - i = z.$$
To find these solutions we replace z with $x + iy$ to obtain
$$i(x - iy) + 1 - i = x + iy$$
or
$$ix + y + 1 - i = x + iy.$$
By equating real parts we obtain the equation $x = y + 1$, and by equating imaginary parts the equation $x - 1 = y$ is obtained. Since these two equations are identical it follows that $f(z)$ is the reflection in the line $y = x - 1$.

Example 9.9. The transformation $f(z) = i\bar{z} - 2$ is either a reflection or a glide reflection. Since $f(0) = -2$, the axis of this rigid motion contains the point
$$\frac{0 + (-2)}{2} = -1.$$
Moreover, $f(-1) = -2 - i$, and so $f(z) = \gamma_{-1,-2-i}$.

9.2 HYPERBOLIC RIGID MOTIONS

We now turn to the complex description of the rigid motions of the hyperbolic plane. Of these, the horizontal translations and the reflections in the straight geodesics are of course also Euclidean rigid motions, and as such they can be expressed as

$$f(z) = z + r \quad \text{or} \quad f(z) = -\bar{z} + r,$$

respectively, where r is an arbitrary real number.

The readers must be forwarned here of a minor inaccuracy in which we shall be engaged hereafter. The function $f(z) = z + 1$ is not exactly the same as the hyperbolic rigid motion that consists of moving each point to the right by one unit. These two transformations differ in their domains – one has the entire plane as its domain whereas the other is defined only on half the plane. Nevertheless, for the sake of conciseness, we shall henceforth ignore these distinctions.

Turning to the inversion $I_{O,k}$, note that if $z' = I_{O,k}(z)$ then

$$\arg(z') = \arg(z) \quad \text{and} \quad |z| \, |z'| = k^2.$$

Consequently,

$$z' = I_{O,k}(z) = \frac{k^2}{\bar{z}},$$

since the number $k^2/\bar{z}$ also satisfies the conditions

$$\arg\left(\frac{k^2}{\bar{z}}\right) = \arg(z) \quad \text{and} \quad \left|\frac{k^2}{\bar{z}}\right| |z| = k^2.$$

If $A(a,0)$ is an arbitrary point on the x axis, then, by Exerice 3.13,

$$I_{A,k}(z) = \tau_{OA} \circ I_{O,k} \circ \tau_{AO}(z) = \frac{k^2}{\bar{z} - a} + a.$$

This gives us an analytic description description of all the inversions that are also hyperbolic rigid motions.

Example 9.10. The inversion $I_{O,2}$ has the expression

$$\frac{2^2}{\bar{z}}$$

and so it maps the point $1 + i$ to the point

$$\frac{2^2}{\overline{1+i}} = \frac{4}{1-i} = \frac{4}{1-i} \cdot \frac{1+i}{1+i} = \frac{4+4i}{1+1} = 2 + 2i.$$

On, the other hand, if $A = (3,0)$, then the inversion $I_{A,4}$ has the analytic expression

$$\frac{4^2}{\bar{z} - 3} + 3 = \frac{3\bar{z} + 7}{\bar{z} - 3}.$$

138 THE POINCARÉ HALF-PLANE

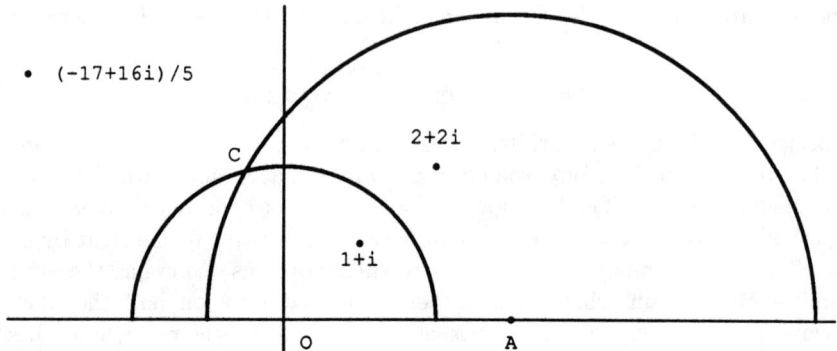

FIGURE 9.2a.

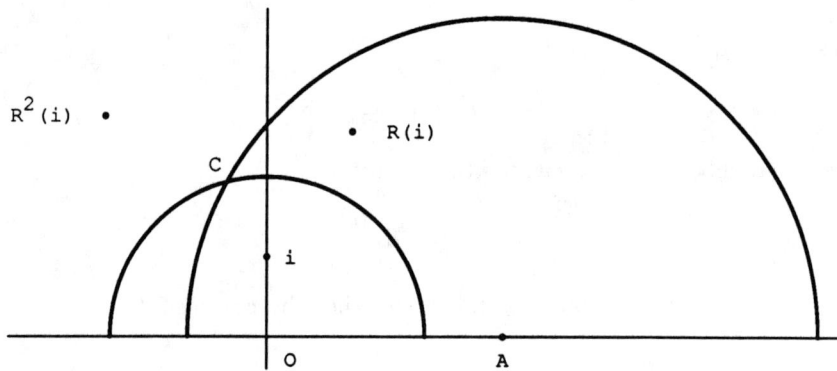

FIGURE 9.2b.

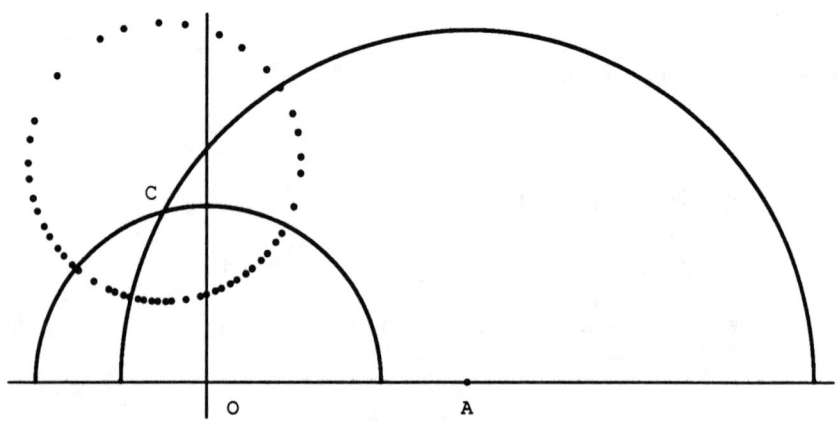

FIGURE 9.2c.

and so it maps the same point $1+i$ to the point
$$\frac{3(\overline{1+i})+7}{\overline{1+i}-3} = \frac{10-3i}{-2-i} = \frac{-17+16i}{5}.$$

The inversions $I_{O,2}$ and $I_{A,4}$ are hyperbolic reflections. Moreover, their axes intersect (Fig. 9.2a). Consequently, since Proposition 2.10 is absolute, the composition $R = I_{A,4} \circ I_{O,2}$ is a hyperbolic rotation. Let us examine this composition in greater detail. Keeping in mind that the conjugate of $\frac{4}{z}$ is $\frac{4}{\bar{z}}$ we compute

$$R(z) = I_{A,4} \circ I_{O,2}(z) = I_{A,4}\left(\frac{4}{\bar{z}}\right) = \frac{3\frac{4}{z}+7}{\frac{4}{z}-3} = \frac{7z+12}{-3z+4}.$$

Since this is a hyperbolic rotation, it should have a unique fixed point in the upper half-plane. Let us find this center directly without reference to Fig. 9.2. This is accomplished by solving the equation

$$R(z) = \frac{7z+12}{-3z+4} = z$$

which simplifies to the quadratic

$$z^2 + z + 4 = 0$$

which has roots

$$\frac{-1 \pm i\sqrt{15}}{2}$$

of which only

$$\frac{-1 + i\sqrt{15}}{2}$$

lies in the upper half-plane. Thus the composition $R = I_{A,4} \circ I_{O,2}$ is a hyperbolic rotation that is centered at the above point. The angle of the rotation can be found by means of Proposition 6.2. Accordingly, the angle between the axes of the hyperbolic reflection $I_{A,4}$ and $I_{O,2}$ is

$$\cos^{-1}\left(\frac{AC^2 + OC^2 - OA^2}{2 \cdot AC \cdot OC}\right) = \cos^{-1}\left(\frac{4^2 + 2^2 - 3^2}{2 \cdot 4 \cdot 2}\right) = \cos^{-1}\left(\frac{11}{16}\right)$$

and hence the angle of the hyperbolic rotation R is

$$2\cos^{-1}\left(\frac{11}{16}\right) = 93.13...°.$$

The rotational action of R is demonstrated in Fig. 9.2bc. In the first of these figures the points i, $R(i)$, and $R^2(i) = R(R(i))$ have been plotted. The second figure, 9.2c, contains a plot of the first 50 values of $R^n(i)$. There is a good reason for the circular shape of this plot. Since R is a hyperbolic rigid motion it follows that the hyperbolic distance from C to $R^k(i)$ equals the hyperbolic distance from $R(C) = C$ to $R(R^k(i)) = R^{k+1}(i)$. Hence, by mathematical induction, all the points $R^k(i)$ are at a constant hyperbolic distance from C. Consequently, this plot forms a hyperbolic circle about C. By Corollary 5.3, this accounts for the circular shape of the plot.

Example 9.11. Consider next the composition $T = I_{B,4} \circ I_{O,2}$ where $B = (-1, 0)$ (Fig. 9.3). Since the axes of the constituent reflections do not intersect, this composition, whose analytic expression is given by

$$T(z) = I_{B,4} \circ I_{O,2}(z) = \frac{4^2}{\frac{2^2}{z} + 1} - 1 = \frac{15z - 4}{z + 4},$$

is a hyperbolic counterpart of a Euclidean translation. The first few values of $T^n(1+i)$ and of $T^n(3i)$ (the latter represented by asterisks) are plotted in Fig. 9.3. It is clear that these converge, in the Euclidean sense, towards the same point on the x-axis. The significance of this common limit will be discussed in Section 5 of this chapter. Hyperbolically speaking, of course, both of these sequences of points diverge to hyperbolic infinity.

The foregoing discussion and examples demonstrate the utility of complex numbers in providing us with a very concrete and effective description of hyperbolic reflections and rotations. We remind the readers that by Theorem 5.6 every hyperbolic rigid motion is the composition of several hyperbolic reflections. Thus it is reasonable to expect that all the hyperbolic rigid motions should have similar explicit expressions. These expressions turn out to be surprisingly simple.

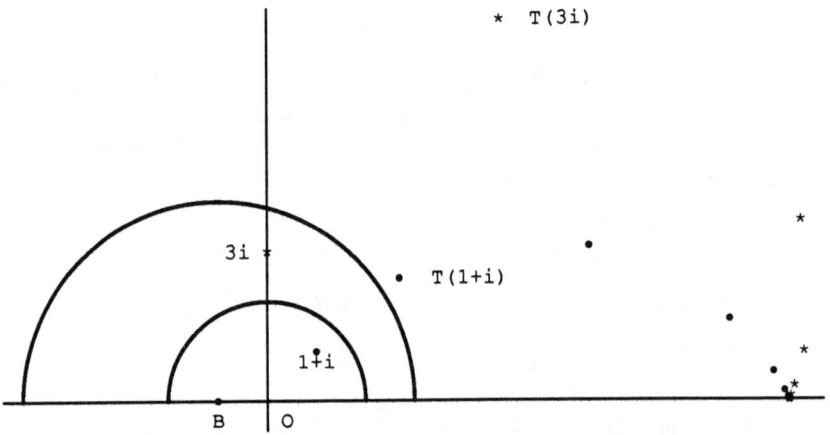

FIGURE 9.3.

COMPLEX NUMBERS AND RIGID MOTIONS   141

Theorem 9.12. *The rigid motions of the hyperbolic plane coincide with the complex functions that have the following forms:*

$$\text{i) } f(z) = \frac{\alpha z + \beta}{\gamma z + \delta} \quad \text{or} \quad \text{ii) } f(z) = \frac{\alpha(-\bar{z}) + \beta}{\gamma(-\bar{z}) + \delta}$$

where $\alpha, \beta, \gamma, \delta$ *are real numbers and* $\alpha\delta - \beta\gamma > 0$.

PROOF: It is clear that the horizontal translations have the form

$$\frac{1z + r}{0z + 1},$$

that the reflections in the straight geodesics have form

$$\frac{1(-\bar{z}) + r}{0(-\bar{z}) + 1},$$

that the reflections in the bowed geodesics have the form

$$\frac{k^2}{\bar{z} - a} + a = \frac{a\bar{z} + k^2 - a^2}{\bar{z} - a} = \frac{-a(-\bar{z}) + (k^2 - a^2)}{-(-\bar{z}) - a},$$

and that in all the cases $\alpha\delta - \beta\gamma > 0$.

Moreover, if f and g are two functions that have this format, so does their composition. For example, if

$$f(z) = \frac{\alpha z + \beta}{\gamma z + \delta} \quad \text{and} \quad g(z) = \frac{\alpha'(-\bar{z}) + \beta'}{\gamma'(-\bar{z}) + \delta'}$$

then

$$f \circ g(z) = \frac{\alpha \dfrac{\alpha'(-\bar{z}) + \beta'}{\gamma'(-\bar{z}) + \delta'} + \beta}{\gamma \dfrac{\alpha'(-\bar{z}) + \beta'}{\gamma'(-\bar{z}) + \delta'} + \delta}$$

$$= \frac{(\alpha\alpha' + \beta\gamma')(-\bar{z}) + (\alpha\beta' + \beta\delta')}{(\gamma\alpha' + \delta\gamma')(-\bar{z}) + (\gamma\beta' + \delta\delta')}$$

where the coefficients in parentheses are of course all real and

$$(\alpha\alpha' + \beta\gamma')(\gamma\beta' + \delta\delta') - (\alpha\beta' + \beta\delta')(\gamma\alpha' + \delta\gamma')$$
$$= \alpha\alpha'\delta\delta' + \beta\gamma'\beta'\gamma - \alpha\beta'\delta\gamma' - \beta\delta'\gamma\alpha'$$
$$= (\alpha\delta - \beta\gamma)(\alpha'\delta' - \beta'\gamma') > 0.$$

The verification of the remaining cases is left as Exercise 16. Since we know from Theorem 5.6 that every hyperbolic rigid motion is the composition of some hyperbolic reflections, it follows that they all do indeed have either form i or form ii.

Conversely, suppose $f(z)$ is of type i. Then, bearing in mind that α, β, γ, δ are reals, it is easily verified that

$$f(z) = -\left[\overline{\frac{(\alpha\delta - \beta\gamma)/\gamma^2}{\bar{z} - (-\delta/\gamma)} + \left(-\frac{\delta}{\gamma}\right)}\right] + \frac{\alpha - \delta}{\gamma}.$$

Thus $f(z)$ is the composition of the inversion $I_{(-\delta/\gamma, 0), \sqrt{\alpha\delta - \beta\gamma}/\gamma}$ with the reflection in the straight geodesic above the point $\left(\frac{\alpha-\delta}{2\gamma}, 0\right)$ both of which are hyperbolic rigid motions.

If, on the other hand, $f(z)$ has type ii, then

$$f(z) = \frac{(\alpha\delta - \beta\gamma)/\gamma^2}{\bar{z} - \delta/\gamma} + \frac{\delta}{\gamma} + \frac{\alpha - \delta}{\gamma}.$$

Since $\alpha\delta - \beta\gamma$ is positive, it follows that $f(z)$ is indeed the composition of an inversion in a bowed geodesic with a horizontal translation, both of which are hyperbolic rigid motions. Hence $f(z)$ itself is a hyperbolic rigid motion. Moreover, since every horizontal translation is the composition of two reflections in straight geodesics, it follows that every transformation of type ii is the composition of three hyperbolic reflections, unless, of course, $\alpha = \delta$, in which case the transformation is itself an inversion. q.e.d.

Example 9.13. The proof of Theorem 9.12 contains a recipe for expressing any hyperbolic rigid motion as the composition of hyperbolic reflections. Thus, the hyperbolic rigid motion given by

$$\frac{2z + 3}{z + 4}$$

is the composition consisting of the inversion

$$\frac{(8-3)/1}{\bar{z} - (-4)} + (-4) = I_{(-4, 0), \sqrt{5}}$$

followed by the reflection $-\bar{z} - 2$ whose axis is the vertical geodesic above $(-1, 0)$.

Transformations of the type

$$f(z) = \frac{az + b}{cz + d}$$

where a, b, c, d, are allowed to be any complex numbers as long as $ad - bc \neq 0$, are called *Moebius transformations*. As most of the transformations that will be discussed subsequently are Moebius transformations, it is important that the reader be comfortable with manipulating them. More specifically, we shall frequently make use of the fact that the composition of two such

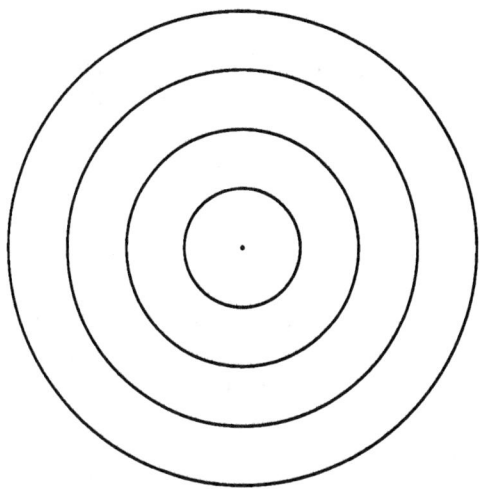

FIGURE 9.4.

transformations is given by the rule

$$\frac{az+b}{cz+d} \circ \frac{a'z+b'}{c'z+d'} = \frac{(aa'+bc')z+(ab'+bd')}{(ca'+dc')z+(cb'+dd')} \quad (1)$$

and consequently

$$\frac{az+b}{cz+d} \circ \frac{dz-b}{-cz+a} = \frac{dz-b}{-cz+a} \circ \frac{az+b}{cz+d}$$
$$= \frac{(ad-bc)z+0}{0z+(ad-bc)} = z \quad (2)$$

Hence the Moebius transformations are all invertible and

$$\text{if} \quad f(z) = \frac{az+b}{cz+d} \quad \text{then} \quad f^{-1}(z) = \frac{dz-b}{-cz+a}.$$

The mathematically experienced reader will no doubt notice the formal similarity between rules (1) and (2) on the one hand, and the algebra of 2×2 matrices on the other. The verification of these rules is left to the reader (Exercise 20).

9.3 EUCLIDEAN FLOW DIAGRAMS

The description of hyperbolic rigid motions given by Theorem 9.12 is algebraically quite complete and satisfactory. However, the readers are in all likelihood still very much lacking in an intuitive grasp of these transformations. Hyperbolic reflections are quite easy to visualize. They are essentially reflections in a mirror consisting of a bowed geodesic. But what about rotations? Moreover, we know that, strictly speaking, the hyperbolic plane is very lacking in hyperbolic translations, since the notion of parallelism is too ambiguous there. So what, if any, are the hyperbolic analogs of translations? Furthermore, once these questions have been answered, given a specific transformation of the type displayed in Theorem 9.12, how do we recognize whether it is a hyperbolic rotation, translation or whatever?

Before attempting to answer these questions, let us reconsider the Euclidean rigid motions. If we momentarily forget the mathematical definition of a rotation as a species of abstract functions and think of it as a physical operation that actually turns the plane about one of its points, then it is very tempting to associate a certain diagram with the rotation. This diagram (Figure 9.4) consists of concentric circles all centered at the pivot point of the rotation. For example, suppose we are to rotate the plane by 90° about the origin, and suppose that a light is attached to each integer point on both of the coordinate axes. If the aperture of a camera is held open throughout the execution of the rotation, the resulting photograph would then show that a collection of concentric circles.

The same diagram is associated with all rotations, except, of course, that the concentric circles must always be centered at the pivot of the rotation. We keep the angle of the rotation in our mind and think of the rotation as happening along its concentric circles. In other words, the rotation $R_{C,\alpha}$ is visualized by forming concentric circles about the point C and then imagining each point P of the plane to be tracing out an arc that subtends an angle α from the pivot C. We shall refer to these circles as the *flow lines* of $R_{C,\alpha}$ since we can think of the rotation as flowing along them.

A similar diagram can be associated with translations. Once again, think of the translation τ_{AB} as being executed by a continuous movement of the plane. In this case the flow lines consist of all the lines that are parallel to the line AB (Figure 9.5). Just as was the case for the angle of a rotation, the magnitude of the displacement associated with the translation is not manifest in the diagram. However, we can keep it too in mind as we imagine the translation as flowing along its flow lines. Arrowheads are attached to the flow lines of a translation in order to indicate the direction of the transformation. Such arrows cannot be used in the flow diagram of a rotation since, for example, every 180° counterclockwise rotation is also a 180° clockwise rotation. This phenomenon makes it impossible to unambiguously attach a direction to rotations.

Note that for both rotations and translations, the flow lines actually consist of fixed sets. In other words, if L is a *flow line* of the rotation (or translation) f, then L is a curve such that $f(L) = L$. As the rigorous definition of flow lines

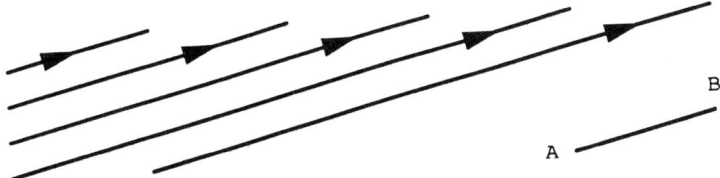

FIGURE 9.5.

requires concepts that go beyond the scope of this book, we use this property as the definition of flow lines. In other words, a *flow line* of the transformation f is a curve L such that $f(L) = L$.

A diagram consisting of some of the flow lines of a rigid motion will be called a *flow diagram*. Thus Figure 9.4 and 9.5 consist of flow diagrams of a rotation and a translation, respectively. The readers may well wonder about flow diagrams of reflections and glide-reflections. However, a moment's thought (not to say reflection) will convince them that such a diagram would necessarily be rather awkward. The reason for this is that the physical execution of a reflection involves revolving the plane by an angle of 180° through space. Thus a flow diagram of a reflection would have to be three dimensional, and so be too complicated to be helpful here. The same, of course, holds for glide-reflections.

9.4 HYPERBOLIC FLOW DIAGRAMS — ROTATIONS

We now turn to the flow diagrams of hyperbolic rigid motions in the hope that their description will make the reader more comfortable with the hyperbolic plane. Example 9.10 is quite typical of hyperbolic rotations. A pivot point C and a rotation angle α are specified. To each point P of the hyperbolic plane assign a point P' such that the hyperbolic angle PCP' equals a and $h(C, P) = h(C, P')$. The function f defined by $f(P) = P'$ is called a hyperbolic rotation. The same argument that was needed in Proposition 2.8 to establish the fact that every Euclidean rotation is a Euclidean rigid motion can be used here verbatim to show that these hyperbolic rotations are hyperbolic rigid motions.

Since $h(C, P) = h(C, P')$ for any P and $P' = f(P)$, it follows that f maps every hyperbolic circle into itself. Thus, as the reader will surely not be surprised to discover, the flow lines of f are concentric circles whose common hyperbolic center is C (Figure 9.6).

More examples of hyperbolic rotations are provided in section 7 of this chapter.

9.5 HYPERBOLIC FLOW DIAGRAMS — TRANSLATIONS

It has been repeatedly mentioned in this text that there are no such

146 THE POINCARÉ HALF-PLANE

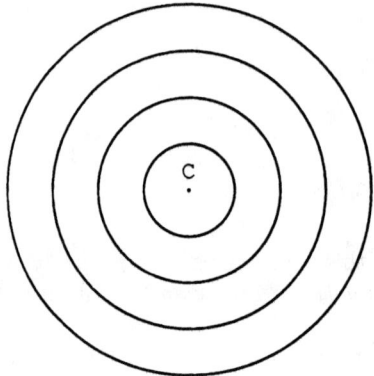

FIGURE 9.6a. A hyperbolic view of the flow lines of a hyperbolic rotation.

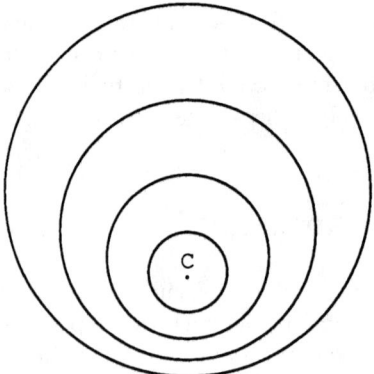

FIGURE 9.6b. A Euclidean view of the flow lines of a hyperbolic rotation.

transformations as hyperbolic translations. This assertion should be taken with a grain of salt. To see why, let us reexamine the rotations and translations of the Euclidean plane. We know that each of the rigid motions of the Euclidean plane is the composition of several reflections. Out of all these transformations, rotations and translations are characterized by the fact that they are expressible as the composition of two reflections. It would therefore not be unreasonable to focus attention on the hyperbolic rigid motions that are the composition of two hyperbolic reflections. As we know from Theorem 9.12 these are the hyperbolic rigid motions that are also Moebius transformations. Let us call them *Moebius rigid motions*. By analogy with the Euclidean plane, we define a *hyperbolic translation* to be a Moebius rigid motion that does not have fixed points. We now go on to search out such hyperbolic translations in a methodical manner.

In the wider context of the entire complex plane, the fixed points of the Moebius rigid motion

$$f(z) = \frac{\alpha z + \beta}{\gamma z + \delta} \tag{3}$$

are the solutions of

$$z = \frac{\alpha z + \beta}{\gamma z + \delta} \quad \text{or} \quad \gamma z^2 - (\alpha - \delta)z - \beta = 0,$$

and so, as long as $\gamma \neq 0$, they are the numbers

$$\frac{\alpha \pm \delta \sqrt{\Delta}}{2\gamma}, \quad \Delta = (\delta - \alpha)^2 + 4\beta\gamma \tag{4}$$

When $\gamma = 0$ the fixed point is

$$\frac{\beta}{\delta - \alpha} \quad \text{provided } \delta \neq \alpha.$$

If $\delta = \alpha$ when $\gamma = 0$, we have $f(z) = z + \frac{\beta}{\delta}$ and so f has no fixed points.

The last case to be described, namely where $f(z) = z + \frac{\beta}{\delta}$ is not new. Since β and δ are real numbers, this hyperbolic rigid motion also happens to be a Euclidean translation, and so its flow diagram consists of all the Euclidean straight lines that are parallel to the x-axis (Figure 9.7). The direction of the flow depends on the sign of β/δ. If this fraction is positive, the flow is directed to the right. If it is negative, the flow is directed to the left.

In the preceding case, where $\gamma = 0$ but $\delta \neq \alpha$, the function $f(z)$ has only one fixed point, namely $\left(\frac{\beta}{\delta-\alpha}, 0\right)$. Since this fixed point is on the x axis, it lies outside of the hyperbolic plane and so does not constitute a fixed point of the associated hyperbolic rigid motion. This case is illustrated by the function

$$D(z) = 2z$$

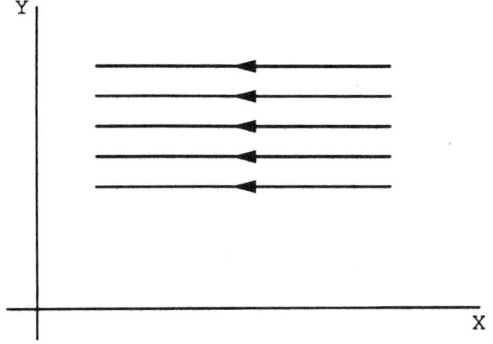

FIGURE 9.7.

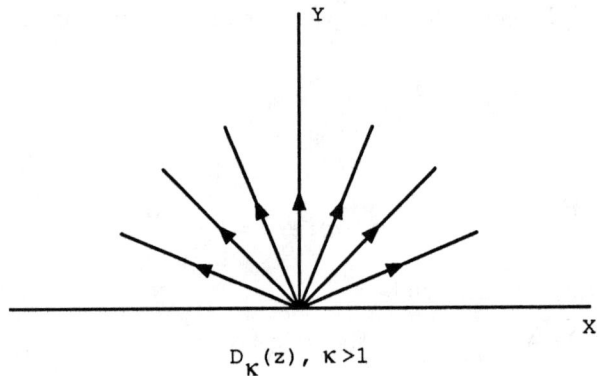

FIGURE 9.8a.

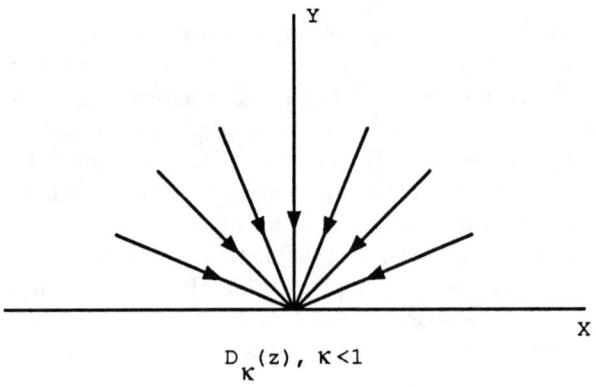

FIGURE 9.8b.

whose only fixed point is the origin $(0,0)$. The flow lines of $D(z)$ become evident when we write z in its polar form $re^{i\theta}$ since it is then clear that $D(z) = 2re^{i\theta}$ has the same argument as z and twice its modulus. Thus each flow line consists of a set of points all of which have the same argument, in other words, a ray from the origin. The flow diagram of $D(z) = 2z$ is depicted in Figure 9.8a. It is clear that the same figure depicts the flow diagram of $D_\kappa(z) = \kappa z$ for any positive κ. It also follows from the fact that $|D_\kappa(i)| = |\kappa i| = \kappa$ that the flow is directed outwards when $\kappa > 1$ and inwards when $\kappa < 1$.

We now turn to the examination of the fixed points of those Moebius rigid motions for which $\gamma \neq 0$. Of these only those fixed points that lie in the upper half-plane, namely those with a positive imaginary part, constitute fixed points

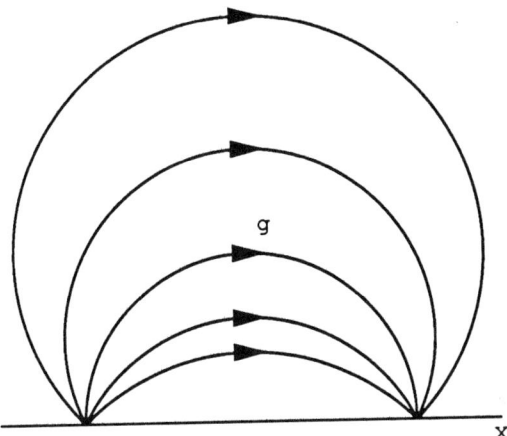

FIGURE 9.9.

of the associated hyperbolic rigid motion. Since $\alpha, \beta, \gamma, \delta$ are all real, it is the sign of Δ that determines the location of the fixed point relative to the x-axis.

When Δ is negative, the two potential fixed points have form $a \pm bi$ where b may be assumed to be positive. Of these two, $a - bi$ lies outside the upper half-plane and so does not constitute a fixed point of the hyperbolic rigid motion. On the other hand $a + bi$ does lie in the upper half-plane and so the rigid motion f of (3) has exactly one fixed point. Since we are interested in Moebius rigid motions without fixed points, we can clearly focus on the case where Δ is either positive or zero.

Rather than continue this discussion in its general form, let us now specialize by considering the Moebius rigid motion

$$T(z) = \frac{2z+1}{z+2}.$$

For this hyperbolic rigid motion $\Delta = (2-2)^2 + 4 = 4$, and so the two fixed points of the action of $T(z)$ on the whole plane are $(-1, 0)$ and $(1, 0)$, neither one of which lies in the upper half-plane. As it happens, we are already in possession of a sufficient amount of information to draw a flow diagram for the rigid motion of T. Since T is a rigid motion it maps geodesics into geodesics. Hence, if g denotes the bowed geodesic joining $(-1, 0)$ to $(1, 0)$ then $T(g)$ is another geodesic. However, since the action of T on the whole plane is continuous (except, of course, at $z = -2$), it follows that $T(g)$ is a geodesic that joins the point $T(-1, 0) = (-1, 0)$ to the point $T(1, 0) = (1, 0)$. In other words, $T(g) = g$, and so g is one of the flow lines of T.

Once we have this flow line, the others are easy to find. Observe that if P is any point of the hyperbolic plane, then, since T is a rigid motion, the distance d of the point P from g equals the distance of the point $T(P)$ from $T(g) = g$. Thus

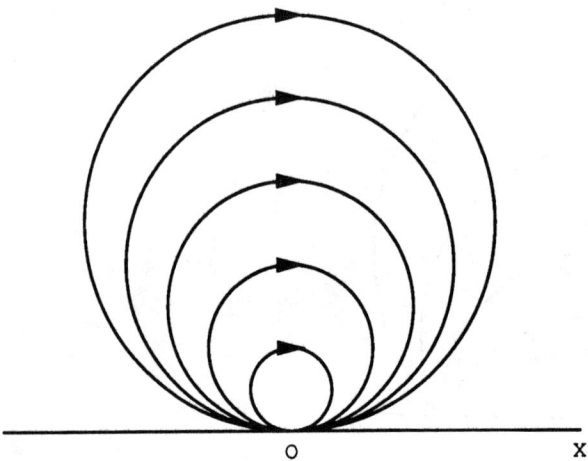

FIGURE 9.10.

P and $T(P)$ are at equal distances from g. Consequently, if m is the locus of all the points that are at the constant distance d from the geodesic g (and on one side of it) then $T(m) = m$ and so m is a flow line of T. We already know from Chapter 5 that m is necessarily the arc of a Euclidean circle that joins $(-1, 0)$ and $(1, 0)$, and so we have all the flow lines necessary for the flow diagram depicted in Figure 9.9. The direction of the flow is determined from the fact that

$$T(i) = \frac{2i+1}{i+2} = \frac{4+3i}{5}.$$

A very similar figure desribes the flow of the Moebius rigid transformation

$$T_{\alpha,\beta,\gamma}(z) = \frac{\alpha z + \beta}{\gamma z + \alpha}, \quad \frac{\beta}{\gamma} > 0.$$

For, by (2) above, the fixed points of $T_{\alpha,\beta,\gamma}$ in the Euclidean plane are

$$z = \pm\sqrt{\frac{\beta}{\gamma}}$$

and they play exactly the same role that $z = \pm 1$ had for T. The direction of the flow can be easily determined by examining the position of $T_{\alpha,\beta,\gamma}(i)$.

A somewhat different behavior is exhibited by the Moebius rigid transformation

$$M(z) = \frac{z}{z+1}.$$

An easy calculation shows that for this transformation $\Delta = 0$, and hence it has only one (albeit double) fixed point at $z = 0$. One can heuristically derive the flow diagram of $M(z)$ by replacing the fixed points $(\pm 1, 0)$ of $T(z)$ above with $(\pm \epsilon, 0)$ where ϵ is a very small real number. It is then very tempting to conclude that the flow lines of $M(z)$ consist of the Euclidean circles of the upper half-plane that are tangent to the x-axis at the origin, as depicted in Figure 9.10. A somewhat more rigorous derivation of the same conclusion now follows.

Let I denote the inversion $I_{0,1}$ i.e., the inversion that leaves the unit circle pointwise fixed. Then
$$I(z) = \bar{z}^{-1},$$
and so
$$(I \circ M \circ I)(z) = (I \circ M)(\bar{z}^{-1}) = I\left(\frac{\bar{z}^{-1}}{\bar{z}^{-1} + 1}\right)$$
$$= I\left(\frac{1}{1+\bar{z}}\right) = z + 1.$$

Therefore, as was noted above in Fig. 9.7, the flow lines of $I \circ M \circ I$ are Euclidean straight lines parallel to the x-axis. Note, however, that if m is any flow line of $I \circ M \circ I$ then $I(m)$ is a flow line of M, because of the following algebraic manipulations:
$$(I \circ M \circ I)(m) = m$$
so
$$(I \circ (I \circ M \circ I))(m) = I(m)$$
or
$$((I \circ I) \circ (M \circ I))(m) = I(m).$$
Since I^2 is the identity transformation it follows that
$$M(I(m)) = I(m)$$
and so $I(m)$ is indeed a flow line of M. Since $I(m)$ consists of a circle that is tangent to the x-axis at the origin, we see once again that Figure 9.10 does indeed describe the flow diagram of $M(z)$.

It is clear that the analysis of the flow diagram of $M(z)$ holds verbatim for any Moebius rigid motion which has a unique double fixed point at $z = 0$. Thus it holds for all Moebius rigid motions of the form
$$M_{\alpha,\gamma}(z) = \frac{\alpha z}{\gamma z + \alpha} \quad \alpha, \gamma \neq 0.$$

9.6 HYPERBOLIC FLOW DIAGRAMS — THE GENERAL CASE

In the previous section the reader was led through a variety of examples chosen to illustrate the various cases that arise in analyzing the flow structure of the general Moebius rigid motion. One more general lemma about the behavior of flow lines is necessary before we can bring this project to a close. The proof of this lemma is implicit in the above discussion of the flow lines of $M(z)$ but will nevertheless be repeated.

Proposition 9.14. *Let F and f be any functions on the complex plane, and let f have inverse f^{-1}. Then S is a fixed set of F if and only if $f(S)$ is a fixed set of $f \circ F \circ f^{-1}$.*

PROOF: Suppose S is a fixed set of F. In other words, suppose $F(S) = S$. Then,
$$f \circ F \circ f^{-1}(f(S)) = f \circ F(S) = f(S),$$
so $f(S)$ is indeed a fixed set of $f \circ F \circ f^{-1}$.

Conversely, suppose $f(S)$ is a fixed set of $f \circ F \circ f^{-1}$, i.e., suppose
$$f \circ F \circ f^{-1}(f(S)) = f(S).$$
Then
$$S = f^{-1}(f(S)) = f^{-1}(f \circ F \circ f^{-1}(f(S))) = f^{-1} \circ f \circ F \circ f^{-1} \circ f(S)$$
$$= F(S),$$
and so S is indeed a fixed set of F. q.e.d.

Since our interest lies mainly in the set of Moebius transformations we extract the following corollary.

Corollary 9.15. *Let f and F be two Moebius transformations. Then*
  a) *z is a fixed point of F if and only if $f(z)$ is a fixed point of $f \circ F \circ f^{-1}$;*
  b) *m is a flow line of F if and only if $f(m)$ is a flow line of $f \circ F \circ f^{-1}$.*
□

It is our goal to show that the flow diagrams of Figures 9.6–10 describe all the hyperbolic rigid motions.

Theorem 9.16. *The flow lines of all Moebius rigid motions are either Euclidean straight lines or arcs of circles.*

PROOF: Let $T = T(z)$ be the Moebius rigid motion
$$\frac{\alpha z + \beta}{\gamma z + \delta}.$$
We saw in the previous section that as long as $\gamma \neq 0$, the fixed points of T are given by
$$\frac{\alpha - \delta \pm \sqrt{\Delta}}{2\gamma}, \quad \Delta = (\delta - \alpha)^2 + 4\beta\gamma \tag{5}.$$

If $\gamma = 0$ and $\alpha = \delta$, than $T(z) = z + \frac{\beta}{\delta}$ is also a Euclidean translation, and so its flow lines are the horizontal Euclidean straight lines of Figure 9.7. On the other hand, if $\gamma = 0$ and $\alpha \neq \delta$, then

$$T(z) = \frac{\alpha z + \beta}{\delta}$$

and so it has the unique fixed point

$$a = \frac{\beta}{\delta - \alpha}$$

which lies on the x-axis. Let $f(z) = z - a$. Then $f^{-1}(z) = z + a$ and

$$f \circ T \circ f^{-1}(z) = \frac{\alpha(z + a) + \beta}{\delta} - a$$

$$= \frac{\alpha z + a(\alpha - \delta) + \beta}{\delta} = \frac{\alpha z - \beta + \beta}{\delta} = \frac{\alpha}{\delta} z.$$

Thus $f \circ T \circ f^{-1} = D_{\alpha/\delta}$ and so its flow lines are the Euclidean rays emanating from the origin. By Corollary 9.15, the flow lines of T are the Euclidean rays emanating from a. This takes care of the case $\gamma = 0$, and we assume henceforth that $\gamma \neq 0$.

If Δ is negative, then exactly one of these two values of (5), say

$$w = \frac{\alpha - \delta + \sqrt{\Delta}}{2\gamma}$$

lies in the hyperbolic plane. We shall now show that under these assumptions T is necessarily a hyperbolic rotation. Let w' be a point of the hyperbolic plane that is distinct from w and let R be a hyperbolic rotation that is pivoted at w and that maps w' to $T(w')$. Now R^{-1} is also a rotation of the hyperbolic plane and so, by the hyperbolic analog of Proposition 2.10 it follows that it is the composition of two inversions. By Theorem 9.12 we now conclude that R^{-1} is itself a Moebius rigid motion. The same theorem now guarantees that the composition $R^{-1} \circ T$ is also a Moebius rigid transformation. However,

$$R^{-1} \circ T(w) = R^{-1}(w) = w$$

and

$$R^{-1} \circ T(w') = R^{-1}(T(w')) = w'.$$

In other words, $R^{-1} \circ T$ is a Moebius rigid motion with two distinct fixed points in the hyperbolic plane. By the fixed point analysis at the beginning of this proof, $R^{-1} \circ T$ must be the identity map, and so $T = R$ and T is indeed a rotation. Consequently, in this case the flow lines of T are indeed hyperbolic, and therefore also Euclidean, circles.

Next, assume that Δ is positive. Thus the two values of (5) are both real.

Set
$$f(z) = z - \frac{\alpha - \delta}{2\gamma} \quad \text{so that} \quad f^{-1}(z) = z + \frac{\alpha - \delta}{2\gamma}.$$

It is easily verified that

$$f \circ T \circ f^{-1}(z) = \frac{\alpha\left(z + \frac{\alpha-\delta}{2\gamma}\right) + \beta}{\gamma\left(z + \frac{\alpha-\delta}{2\gamma}\right) + \delta} - \frac{\alpha - \delta}{2\gamma}$$

$$= \frac{2\gamma(\alpha + \delta)z + \Delta}{4\gamma^2 z + 2\gamma(\alpha + \delta)} = T_{2\gamma(\alpha+\delta),\Delta,4\gamma^2}$$

Thus, the flow lines of $f \circ T \circ f^{-1}$ are given by Figure 9.9, and by Corollary 9.15, the flow lines of T are the Euclidean circular arcs joining its two real fixed points.

It only remains to dispose of the case $\Delta = 0$. Here T has a double real fixed point at $a = (\alpha - \delta)/2\gamma$. Using the same f as above, we get

$$f \circ T \circ f^{-1}(z) = \frac{2\gamma(\alpha + \delta)z}{4\gamma^2 z + 2\gamma(\alpha + \delta)} = M_{2\gamma(\alpha+\delta),4\gamma^2},$$

and yet another application of Corollary 9.15 allows us to conclude that the flow lines of T consist of the Euclidean circles of the upper half-plane that are tangent to the x-axis at the real point $a = (\alpha - \delta)/2\gamma$. q.e.d.

While the statement of Theorem 9.16 specifies the nature of the flow lines of a Moebius rigid motion, it says nothing about their arrangement. This information can be easily obtained by examining the fixed points of the associated Moebius transformation. If there are two distinct real fixed points, the flow diagram is that of Figure 9.9. If there is a double (and necessarily real) fixed point, then the flow diagram is the one depicted in Figure 9.10. If there are two non-real fixed points, only one lies in the upper half-plane and so Figure 9.6 yields the flow diagram. If there is only one fixed point it is necessarily real and the flow lines are Euclidean rays that either emanate from or converge into it, as in Figure 9.8. Finally, if there are no fixed points at all, then the flow diagram consists of the horizontal lines of Figure 9.7.

The flow lines of all the Moebius rigid motions have an interesting property which is not immediately evident. Any two flow lines in a flow diagram are at a constant distance from each other. In the Euclidean case, flow lines consist of either a family of parallel Euclidean straight lines (Fig. 9.5) or a family of concentric circles (Fig. 9.4). It is clear that the constant distance property is possessed by both kinds of flow diagrams.

We shall now show, case by case, that the flow diagrams of the Moebius hyperbolic rigid motions also possess that property. Such is obviously the case for the flow diagram of $w = z + \alpha$, α real, which consists of a family of

Euclidean straight lines that are parallel to the x axis (Fig. 9.7). Consider next the flow diagram that consists of a family of circles that are all tangent to the x axis at a common point P. By Theorem 3.3 the inversion $I_{P,1}$ transforms these circles into Euclidean straight lines. Moreover, since these circles are all tangent to the x axis at P, and since the inversion $I_{P,1}$ fixes the x axis, it follows that the inversion transforms these circles into Euclidean straight lines that do not intersect the x axis. Thus, the inversion $I_{P,1}$ transforms the the flow diagram of circles tangent to the x axis at P into the diagram of Euclidean straight lines that are parallel to the x axis and that are already known to possess the constant distance property. Since $I_{P,1}$ is also a hyperbolic rigid motion, the circles of the given flow diagram must also possess the constant distance property.

We turn next to the flow diagram that consists of Euclidean rays emanating from a point P on the x axis (Fig. 9.8). It follows from the Euclidean properties of circles that every bowed geodesic that is centered at P is orthogonal to every one of these rays. Moreover, by Corollary 4.2, the arcs intercepted on two such circular geodesics by any two of the rays of the flow diagram have equal hyperbolic lengths. Hence any two such rays are at a constant distance from each other.

Every hyperbolic flow diagram that consists of circular arcs joining two distinct points on the x axis (Fig. 9.9) can be converted to one of the previous type by an inversion that is centered at one of these two points. Thus, these diagrams too possess the constant distance property.

We leave it to the readers to convince themselves that flow diagrams that consist of circles that have the same hyperbolic center (Fig. 9.6) also possess the constant distance property.

9.7 HYPERBOLIC RIGID MOTIONS – CONSTRUCTION

We now turn to the construction of specific hyperbolic rigid motions. It is clear that if $z_0 = a + bi$ is any point of the half-plane then the Moebius transformation

$$f(z) = \frac{z-a}{b}$$

is a hyperbolic **translation** that carries z_0 to i, and its inverse

$$f^{-1}(z) = bz + a$$

carries i to z_0. Hence, given any other point $z_1 = c + di$ of the half-plane, the composition

$$d\left(\frac{z-a}{b}\right) + c = \frac{dz + (bc - ad)}{0z + b}$$

is a hyperbolic rigid motion that carries z_0 onto z_1. Note that if $b = d$, then this is a horizontal Euclidean translation. If $b \neq d$ then the flow diagram of this transformation is that of Fig. 9.8.

Example 9.17. If $z_0 = 2 - 3i$ and $z_1 = 3 + 4i$ then
$$\frac{4z + (-3 \cdot 3 - 2 \cdot 4)}{-3} = \frac{4z - 17}{-3}$$
is a hyperbolic rigid motion that carries z_0 onto z_1.

Hyperbolic **rotations** are characterized by the fact that they have a single fixed point in the half-plane. Hence it is only necessary to find this center and to locate the image of one other point in order to have a good grasp on the rotation.

Example 9.18. The Moebius transformation
$$f(z) = -\frac{1}{z} = \frac{0z - 1}{1z + 0}$$
is the hyperbolic rotation by 180° about the point $z = i$. To see this observe that $f(i) = -1/i = i$, and hence this is a rotation that is pivoted at i. Since $f(2i) = -1/2i = i/2$ it follows that the angle of this rotation is 180°.

All the other hyperbolic rotations that are centered at i also possess a simple description.

Lemma 9.19. *If θ is any angle then the Moebius transformation*
$$f_\theta(z) = \frac{\cos\theta z + \sin\theta}{-\sin\theta z + \cos\theta}$$
induces a hyperbolic rotation of the half-plane by the angle 2θ about i.

PROOF: Since
$$\cos\theta\cos\theta - \sin\theta(-\sin\theta) = 1 > 0$$
it follows that f_θ does indeed induce a hyperbolic rigid motion of the half-plane. Moreover,
$$f_\theta(i) = \frac{\cos\theta\, i + \sin\theta}{-\sin\theta\, i + \cos\theta} = \frac{i(\cos\theta - \sin\theta\, i)}{-\sin\theta\, i + \cos\theta} = i$$
and hence f_θ is in fact a hyperbolic rotation pivoted at i.

When $\theta = n\pi + \pi/2$, $f_\theta(z) = -1/z$, which was already shown in the above example to describe the 180° hyperbolic rotation about i. For the rest of the values of θ, $\tan\theta$ is a finite real number and we let A be the point $(\tan\theta, 0)$ (Fig. 9.11). Since f_θ fixes i and $f_\theta(0) = \tan\theta$ it follows that the hyperbolic rotation f_θ transforms the half (straight) geodesic from i to O into the half (bowed) geodesic from i to A. If $C = (-c, 0)$ denotes the Euclidean center of this bowed geodesic, let α denote $\angle ACi$. Since the Euclidean length of Oi is 1 it follows that the Euclidean lengths of Ci (and therefore also of CA) is $\csc\alpha$. For the same reason $c = \cot\alpha$. Hence, upon decomposing CA into its parts CO and

COMPLEX NUMBERS AND RIGID MOTIONS   157

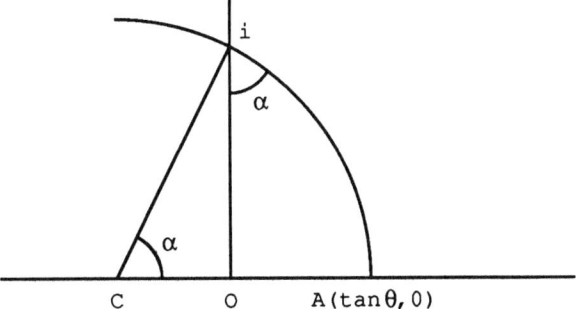

FIGURE 9.11.

OA we obtain
$$\csc \alpha = \cot \alpha + \tan \theta,$$
$$\csc \alpha - \cot \alpha = \tan \theta,$$
$$\frac{1 - \cos \alpha}{\sin \alpha} = \tan \theta,$$
$$\tan \frac{\alpha}{2} = \tan \theta,$$
or
$$\alpha = 2\theta.$$

It follows from Proposition 6.2 that the angle between the two geodesics iO and iA is 2θ and hence that is also the angle of the hyperbolic rotation f_θ.

q.e.d.

We now have sufficient tools to describe any hyperbolic rotation.

Proposition 9.20. *The hyperbolic rotation by angle 2θ that is centered at $a + bi$ is given by the composition*
$$(bz + a) \circ \frac{\cos \theta z + \sin \theta}{-\sin \theta z + \cos \theta} \circ \frac{z - a}{b}.$$

PROOF: Exercise 25.

□

Example 9.21. The $90°$ rotation about the point $1 + i$ is the composition
$$(z + 1) \circ \frac{z + 1}{-z + 1} \circ (z - 1) = (z + 1) \circ \frac{z - 1 + 1}{-(z - 1) + 1}$$
$$= (z + 1) \circ \frac{z}{-z + 2} = \frac{z}{-z + 2} + 1 = \frac{2}{-z + 2}.$$

9.8 EXERCISES

1. Let $t = 2 - i$, $w = 1 + i$, $z = -1 + i$.
    i) Express each of the numbers t, w, z, t^2, w^2, z^2 in the form $re^{i\theta}$;
    ii) Express each of the numbers $t + w, t - w, tw, t/w, w/t$ in the form $re^{i\theta}$;
    iii) Express each of the numbers $w + z, w - z, wz, w/z, z/w$ in the form $re^{i\theta}$.
2. Find the square roots of the following complex numbers:
    i) i ii) $1 + i$ iii) $2 - 3i$.
3. Express each of the following Euclidean rigid motions as a function of the form $f(z) = e^{i\alpha}z + c$:
    i) $\tau_{1+i, 2-i}$ ii) $R_{i, \pi}$ iii) $R_{3, \pi/2}$
    iv) $R_{1-i, \pi/2}$ v) $R_{-2, \pi/4}$ vi) $R_{-2i, \pi/3}$.
4. Let m, n denote the Euclidean straight lines joining the point i to the points $1 + i$, $1 + 2i$ respectively. Express each of the following Euclidean rigid motions as a function of the form $f(z) = e^{i\alpha}\bar{z} + c$:
    i) ρ_m ii) ρ_n iii) $\gamma_{i, 2i}$
    iv) $\gamma_{i, 1+2i}$ v) $\gamma_{1+i, 1+2i}$ vi) $\gamma_{1+i, i}$.
5. Identify the following Euclidean rigid motions:
    a) $f(z) = z + 3$ b) $f(z) = -z + 3$ c) $f(z) = -\bar{z} + 3$
    d) $f(z) = iz - 3$ e) $f(z) = -iz - 3$ f) $f(z) = -i\bar{z} - 3$
    g) $f(z) = \left(\frac{1}{2} + i\frac{\sqrt{3}}{2}\right)z - 2$ h) $f(z) = \left(\frac{1}{2} - i\frac{\sqrt{3}}{2}\right)z - 2$
    i) $f(z) = \left(\frac{1}{2} + i\frac{\sqrt{3}}{2}\right)\bar{z} - 2$ j) $f(z) = \left(-\frac{1}{2} + i\frac{\sqrt{3}}{2}\right)\bar{z} - 2$
6. Use the methods of this chapter to find the image of the point $(1, 1)$ when subjected to the following Euclidean rigid motions where $A = (-1, 1)$ and $B = (1, -2)$:
    a) ρ_{BA} b) τ_{BA} c) γ_{BA}
    d) $R_{A, \pi/2}$ e) $R_{B, \pi/3}$ f) $R_{A, 4\pi/3}$.
7. Prove (geometrically) that for any two complex numbers z and w, $|z + w| \le |z| + |w|$.
8. Let λ be a real number such that $0 < \lambda < 1$, and let z and w be two arbitrary complex numbers. Show that the complex number $(1 - \lambda)z + \lambda w$ is the one which divides the Euclidean line segment from z to w in the ratio $\lambda : (1 - \lambda)$.
9. Let t, w, z be three complex numbers. Show that the center of gravity of the triangle spanned by these three points is
$$\frac{t + w + z}{3}.$$
10. Show that the three complex numbers t, w, z span a Euclidean equilateral triangle if and only if either
$$t + \omega w + \omega^2 z = 0 \quad \text{or} \quad z + \omega w + \omega^2 t = 0,$$
where $\omega = e^{2\pi i/3}$.
11. Let ABC be an arbitrary Euclidean triangle. Suppose A' is the point such that $A'BC$ is an equilateral triangle and A and A' are on opposite sides of BC. Let B'

and C' be similarly defined. Prove that the centers of gravity of triangles $A'BC$, $AB'C$, ABC' form an equilateral triangle.

12. Show that Exercise 11 still holds when A' is chosen so that it and A are on the same side of BC, and B' and C' are chosen in the same manner.

13. Find both values of z such that z forms an equilateral triangle together with the numbers $1 + i$ and $2 - 3i$.

14. Find the image of the point i, $1 + i$, $-3 + 4i$ when subjected to the following hyperbolic rigid motions where $O = (0,0)$ and $A = (3,0)$:
   i) $I_{O,3}$ ii) $I_{A,2}$ iii) $I_{A,2} \circ I_{O,3}$
   iv) $f(z) = \dfrac{2z-1}{z+2}$ v) $f(z) = \dfrac{2(-\bar z)-1}{(-\bar z)+2}$.

15. Express the following compositions as Moebius transformations, where $O = (0,0)$ and $A = (3,0)$:
   i) $I_{A,2} \circ I_{O,3}$ ii) $I_{A,2} \circ \dfrac{2z-1}{z+2}$
   iii) $\dfrac{2(-\bar z)-1}{(-\bar z)+2} \circ I_{O,3}$ iv) $\dfrac{2(-\bar z)-1}{(-\bar z)+2} \circ \dfrac{2z-1}{z+2}$

16. Supply the missing details in the proof of Theorem 9.12 when
   a) f and g both have format i;
   b) f and g both have format ii.

17. Express each of the following hyperbolic rigid motions as the composition of two hyperbolic reflections:
   i) $\dfrac{2z}{5}$ ii) $\dfrac{z}{2z+3}$ iii) $\dfrac{-2z}{z-1}$
   iv) $\dfrac{z-1}{z+1}$ v) $\dfrac{2z+1}{-z+3}$ vi) $\dfrac{3z-4}{z+2}$.

18. Draw the flow diagrams for the following hyperbolic rigid motions:
   a) $z+3$ b) $z-2$ c) $5z$
   d) $\dfrac{2z}{5}$ e) $\dfrac{z}{2z+3}$ f) $\dfrac{-2z}{z-1}$
   g) $\dfrac{z-1}{z+1}$ h) $\dfrac{2z+1}{-z+3}$ i) $\dfrac{3z-4}{z+2}$.

19. Find the cartesian equations of those flow lines of the transformations of Exercise 18 v–ix that contain the point $2i$.

20. Verify rules (1) and (2) of Section 2.

21. Find a Moebius rigid motion one of whose fixed points is $1+i$.

22. Find a Moebius rigid motion whose fixed points are 5 and 7.

23. Find a Moebius rigid motion which has a double fixed point at -3.

24. Find a Moebius rigid motion which has a single fixed point at -3.

25. Prove Proposition 9.20.

26. Express the following hyperbolic rotations as Moebius transformations:
   i) center i and angle $2\pi/3$;
   ii) center $-1+i$ and angle π;
   iii) center $3i$ and angle $\pi/2$;
   iv) center $2-3i$ and angle $\pi/4$;
   v) center $\tfrac{3}{2}+\tfrac{\sqrt 3}{2}i$ and angle $2\pi/3$.

27. Prove that the hyperbolic rigid motion
$$\frac{\alpha z + \beta}{\gamma z + \delta}$$
is a 180° rotation of the upper half-plane if and only if $\alpha + \delta = 0$.

28. Prove that the hyperbolic rigid motion
$$\frac{\alpha(-\bar{z}) + \beta}{\gamma(-\bar{z}) + \delta}$$
is a hyperbolic reflection of the upper half-plane if and only if $\alpha = \delta$.

29. Prove that the hyperbolic rigid motion
$$\frac{\alpha(-\bar{z}) + \beta}{\gamma(-\bar{z}) + \delta}$$
is a hyperbolic reflection of the upper half-plane if and only if it fixes at least one point.

10
Absolute Geometry and the Angles of the Triangle

10.1 THE SUM OF THE ANGLES OF THE TRIANGLE

Perhaps the most glaring difference between the two geometries described so far in this book, namely Euclidean geometry and hyperbolic geometry, is the sum of the angles of a triangle in each. This sum always equals π in Euclidean geometry and is less than π for every hyperbolic triangle. It will now be shown that this is not a coincidence, and that such a behavior is to be expected in any geometry that is derived from absolute geometry – the geometry that is based on the first four of Euclid's postulates only, and that ignores the parallel postulate.

Theorem 10.1. *The sum of the three angles of a triangle cannot be greater than π.*

PROOF: Suppose, to the contrary, that there exists a triangle ABC the sum of whose angles is $\pi + \delta$ where δ is a positive angle. We shall show that this assumption results in the conclusion that there exists a triangle two of whose angles have a sum that exceeds π. Since this contradicts Proposition 17 of Euclid, we shall then have a proof of this theorem.

Let H be the midpoint of the side BC (Figure 10.1). Draw the line AH and extend it to a point D such that $AH = HD$. Join BD to create a new triangle BHD, which, by the SAS Theorem, is congruent to triangle CHA. Hence,

$$\angle BAC = \angle BAH + \angle HAC = \angle BAH + \angle HDB \qquad (1)$$

This equation has two consequences. First,

$$\angle DBA + \angle ADB + \angle BAD = (\angle CBA + \angle DBH) + \angle HDB + \angle BAH$$
$$= \angle CBA + \angle ACB + \angle BAC = \pi + \delta.$$

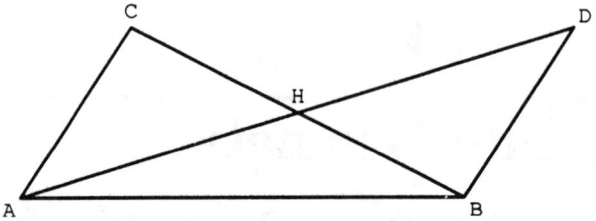

FIGURE 10.1.

Thus the sum of the angles of triangle ABD is also $\pi + \delta$. Moreover, it also follows from (1) that if $\alpha = \angle BAC$ then one of the angles $\angle HDB$, $\angle BAH$ has size at most $\frac{\alpha}{2}$. We may assume without loss of generality that $\angle BAH$ has size at most $\frac{\alpha}{2}$. If the above argument is now applied to triangle ABD instead of ABC, we will obtain a new triangle the sum of whose angles is again $\pi + \delta$, and one of whose angles is no greater than $\frac{\alpha}{4}$. By mathematical induction we may conclude that for every positive integer n there exists a triangle T_n the sum of whose angles is $\pi + \delta$ and one of whose angles is no greater than $\frac{\alpha}{2^n}$. Let m be a positive integer such that $\frac{\alpha}{2^m} < \delta$. Since the sum of the angles of the corresponding triangle T_m is $\pi + \delta$, and one of its angles is no greater than $\frac{\alpha}{2^m} < \delta$ it follows that the sum of the other two angles is greater than π, which is the required contradiction. q.e.d.

The *defect* of the triangle ABC with angles α, β, γ is the difference $\pi - (\alpha + \beta + \gamma)$. It follows from Theorem 10.1 that the defect of any triangle is always nonnegative. Such is of course the case in the two geometries we know; in Euclidean geometry every triangle has defect zero, whereas in hyperbolic geometry every triangle has positive defect. The defect of triangle ABC will be denoted by *defect(ABC)*.

Lemma 10.2. *Let ABC be a triangle and let D be a point in the interior of side BC. Then*

$$\text{defect}(ABC) = \text{defect}(ABD) + \text{defect}(ACD).$$

PROOF: The following easy argument refers to Figure 10.2.

$$\text{defect}(ABC) = \pi - (\angle CBA + \angle ACB + \angle BAC)$$
$$= \pi - (\angle CBA + \angle ACB + \angle DAC + \angle BAD)$$
$$= 2\pi - (\angle DBA + \angle ACD + \angle DAC + \angle BAD + \angle ADB + \angle CDA)$$
$$= 2\pi - [\pi - \text{defect}(ABD) + \pi - \text{defect}(ACD)]$$
$$= \text{defect}(ABD) + \text{defect}(ACD).$$

q.e.d.

ABSOLUTE GEOMETRY AND THE ANGLES OF THE TRIANGLE   163

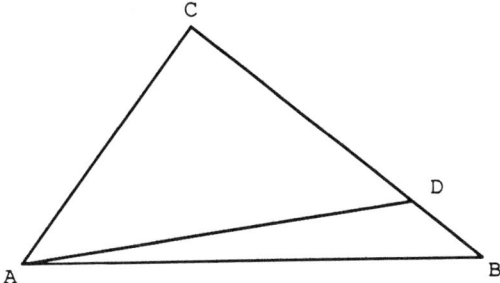

FIGURE 10.2.

The following theorem accounts for the uniform behavior of the sum of the angles of a triangle in both Euclidean and hyperbolic geometry.

Theorem 10.3. *If in any absolute geometry the sum of the angles of one triangle is π, then so is the sum of the angles of every triangle in that geometry.*

PROOF: Let the sum of the angles of triangle ABC be π; in other words, triangle ABC has defect zero. As argued in the proof of Theorem 6.6, we may assume without loss of generality that the foot D of the altitude from A to BC lies in the interior of BC (Figure 10.3). Since

$$0 = \text{defect}(ABC) = \text{defect}(ABD) + \text{defect}(ACD)$$

and since all triangles have nonnegative defect, it follows that the right triangle ABD has defect 0. In particular, this implies that

$$\angle DBA + \angle BAD = \frac{\pi}{2}.$$

Let E be the point on the other side of AB from D such that $AE = BD$ and $BE = AD$. Since the two triangles ABD and BAE have all their corresponding sides equal, it follows that they are congruent and so

$$\angle EAB = \angle DBA, \quad \angle ABE = \angle BAD, \quad \text{and} \quad \angle BEA = \angle ADB = \frac{\pi}{2}.$$

Consequently,

$$\angle DBE = \angle DBA + \angle ABE = \angle DBA + \angle BAD = \frac{\pi}{2}$$

and

$$\angle EAD = \angle EAB + \angle BAD = \angle DBA + \angle BAD = \frac{\pi}{2}.$$

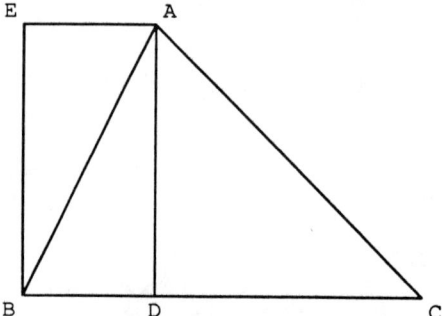

FIGURE 10.3.

Thus, figure $ADBE$ is a *rectangle* in the sense that its opposite sides are equal and all of its angles are right angles. However, when two such rectangles $ADBE$ and $ADB'E'$ are abutted along one of their equal edges (Fig. 10.4a) the result is again a rectangle $EE'B'B$ of the same nature, since their right angled corners at A and D will complement each other so that the lines EAE' and BDB' are both straight lines. Consequently, for any two positive integers m and n there is a rectangle with sides of lengths mAD and nBD, respectively. (Fig. 10.4b illustrates the case $m = 2$ and $n = 3$.) In other words, we have demonstrated the existence of rectangles with arbitrarily large sides.

We can now show that consequently every right triangle has defect zero. Let XYZ be a triangle with a right angle at Z, and let $PQRS$ be a rectangle

a)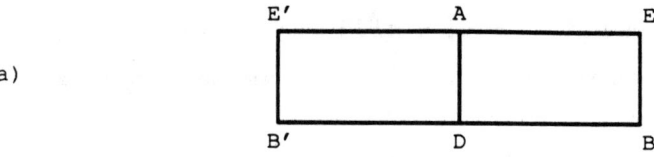

b)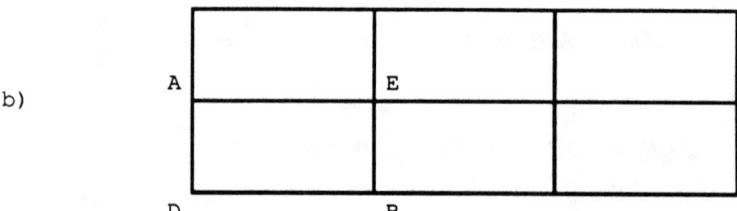

FIGURE 10.4.

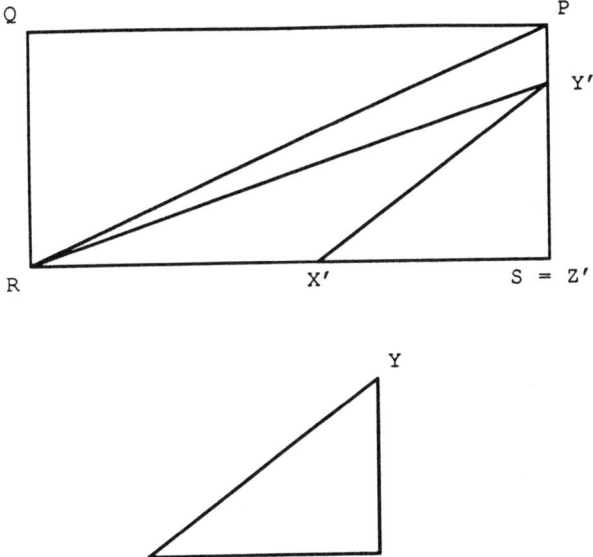

FIGURE 10.5.

such that $PS > YZ$ and $RS > XZ$ (Fig. 10.5). Let Y' and X' be points in the interiors of PS and RS, respectively, such that $Y'S = YZ$ and $X'S = XZ$. Since the angles at S and Z are both right angles, it follows that triangles XYZ and $X'Y'S$ are congruent and so they have the same defects.

On the other hand, $\triangle QPR$ and $\triangle SRP$ are congruent since their corresponding sides are all equal. Consequently,

$$\angle RSP + \angle SPR + \angle PRS$$
$$= \frac{1}{2}(\angle PQR + \angle RSP + \angle QRP + \angle SPR + \angle RPQ + \angle PRS)$$
$$= \frac{1}{2}(\angle PQR + \angle QRS + \angle RSP + \angle SPQ) = \frac{1}{2}(2\pi) = \pi.$$

In other words, $\triangle PRS$ has defect 0. Now draw the lines RY' and $X'Y'$. Since all defects are nonnegative it follows from Lemma 10.2 that

$$0 \le \text{defect}\,(XYZ) = \text{defect}\,(X'Y'S) \le \text{defect}\,(RY'S) \le$$
$$\text{defect}\,(PRS) = 0.$$

Thus, every right triangle has defect zero. Since, as was noted above, every triangle can be split into two right triangles by means of an interior altitude, it follows from Lemma 10.2 that under the hypothesis of this theorem every triangle has defect zero and hence the sum of the angles of every such triangle is π.

q.e.d.

Corollary 10.4. *If the sum of the angles of one triangle is less than π, then so is the sum of the angles of every triangle.* □

It follows from Corollary 10.4 and Theorem 1.1 that Euclid's Postulate 5 is equivalent to the assumption that *there exists at least one triangle the sum of whose angles is π*. The combined, and futile, efforts of many of the mathematicians of the eighteenth and nineteenth centuries to prove the dependence of Postulate 5 on the other postulates resulted in the formulation of many other statements that are also logically equivalent to it. Some of these are listed below.

There exist two noncongruent triangles that have the same angles.

There exist triangles of arbitrarily large areas.

Every triangle can be circumscribed in a circle.

There exist two parallel straight line p and q and a point P on p such that p is the only straight line containing P that is parallel to q.

10.2 EXERCISES

1. Prove that if every triangle can be circumscribed in a circle, then Postulate 5 holds.
2. Prove that Postulate 5 holds if and only if there exist two noncongruent triangles that have the same angles.
3. Prove that if Postulate 5 does not hold then there exist triangles the sum of whose angles is arbitrarily small.
4. Suppose p and q are two straight lines and P is a point on p such that p is the only straight line containing P that is parallel to q. Prove that Postulate 5 holds.
5. Let *PQRS* be a cyclic quadrilateral that contains the center of the circle in its interior. Show that if Postulate 5 does not hold then the sum of the opposite angles of PQRS is less than p.

11
Spherical Trigonometry and Elliptic Geometry

11.1 INTRODUCTION

The discovery of hyperbolic geometry was first announced by Nikolai Lobachevsky in an article published in 1829. This article was followed by Janos Bolyai's famous appendix which appeared in 1832. Though these two mathematicians worked independently their works have some amazing similarities. In contrast with the analytic approach used in chapters 4–8 of this book, both Bolyai and Lobachevsky obtained the hyperbolic plane in a synthetic manner. They assumed, in addition to Euclid's first four postulates, that given any line m and a point P not on m there are at least two lines through P that are parallel to m. They went on to derive many non-Euclidean theorems from these hypotheses in much the same style that Euclid derived the theorems of the ordinary plane from his postulates. What is very striking is that both Bolyai and Lobachevsky even went so far as to develop a trigonometry of this non-Euclidean geometry. If a contradiction had appeared at any time during their developments, their labors would have resulted in a proof of the dependence of Euclid's Postulate 5 on the other postulates. However, no such contradiction was obtained. That, of course, did not preclude the possibility that such a contradiction might be encountered if one persisted long enough. What convinced Bolyai and Lobachevsky that this would never be the case was the very striking similarity between their non-Euclidean trigonometric formulas and the formulas of spherical trigonometry. We shall now go on to describe the geometry of the sphere and develop its trigonometry to the point where the reader can appreciate this analogy.

Like Euclidean geometry, spherical geometry was invented by the Greeks. Its raison d'être was the facilitation of exact astronomical calculations. The best known and most influential of their books on this topic is without question the *Almagest* written by Ptolemy in the second century A.D. This work was later preserved and extended by the Indians and Arabs. The European world resumed its work in this and many other mathematical areas in the sixteenth century. The modern form of this discipline is due to Leonhard Euler whom we have already mentioned in the context of rigid motions.

11.2 GEODESICS ON THE SPHERE

Let $\mathbb{S}$ denote the surface of the sphere with radius 1. We shall assume in the sequel that a three dimensional Cartesian coordinate system has been placed with its origin O at the center of $\mathbb{S}$. If P is any point of $\mathbb{S}$ (Fig. 11.1) let Q denote the point on the xy plane directly below P, and let u and v denote the measures of $\angle ZOP$ and $\angle XOQ$, respectively. Then, according to the standard equations for transforming spherical to Cartesian coordinates, the point P has the Cartesian coordinates

$$P = (\sin u \cos v,\ \sin u \sin v,\ \cos u)$$

where $0 \leq u \leq \pi$ and $0 \leq v < 2\pi$. Note that the locus of all those points P for

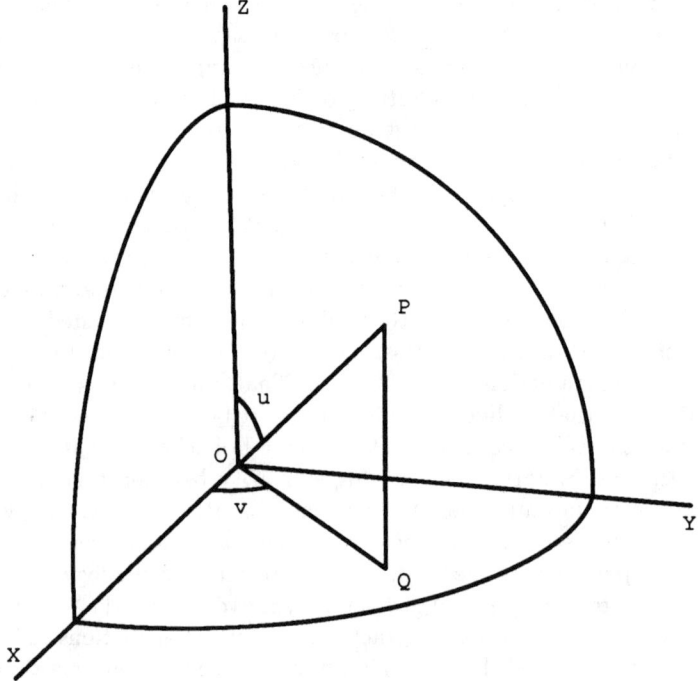

FIGURE 11.1.

which u is some constant c is a latitude, whereas the locus of those points P for which v is some constant d is a meridian.

Let P and Q be any two points on $\mathbb{S}$. By a *geodesic segment joining P and Q* we mean a shortest curve that joins P and Q and lies entirely on $\mathbb{S}$. We shall later see that while in most cases any two such points P and Q are joined by a unique geodesic segment, there do indeed exist points that are joined by more than one such segment.

A *great circle* of $\mathbb{S}$ is a circle, lying entirely on $\mathbb{S}$, whose center coincides with that of $\mathbb{S}$. In other words, a great circle of $\mathbb{S}$ is the intersection of $\mathbb{S}$ with a plane that passes through the origin O.

Theorem 11.1. *Every geodesic segment on $\mathbb{S}$ is an arc of some great circle of $\mathbb{S}$.*

PROOF: Let P and Q be two distinct points of $\mathbb{S}$. By rotating the sphere, if necessary, we may assume that the Cartesian coordinate system has been placed so that P is the north pole $(0,0,1)$ and Q has coordinates $(\sin A \cos B, \sin A \sin B, \cos A)$ for some appropriate angles A and B. If γ is any curve joining P and Q and lying entirely on $\mathbb{S}$, then γ can be paramterized as

$$P(t) = [\sin u(t) \cos v(t), \quad \sin u(t) \sin v(t), \quad \cos u(t)], \quad 0 \le t \le T,$$

where

$$P(0) = P \quad \text{and} \quad P(T) = Q.$$

Since the length of an arbitrary curve $[x(t), y(t), z(t)]$ from $t = a$ to $t = b$ is given by

$$\int_a^b \sqrt{\left(\frac{dx}{dt}\right)^2 + \left(\frac{dy}{dt}\right)^2 + \left(\frac{dz}{dt}\right)^2}\, dt, \tag{1}$$

it follows that the length of γ can be expressed in terms of the derivatives of its components. These are

$$\frac{dx}{dt} = \frac{d}{dt}[\sin u(t) \cos v(t)]$$

$$= \left[\cos u(t) \cos v(t) \frac{du}{dt} - \sin u(t) \sin v(t) \frac{dv}{dt}\right],$$

$$\frac{dy}{dt} = \frac{d}{dt}[\sin u(t) \sin v(t)]$$
$$= \left[\cos u(t) \sin v(t) \frac{du}{dt} + \sin u(t) \cos v(t) \frac{dv}{dt}\right].$$
$$\frac{dz}{dt} = \frac{d}{dt}[\cos u(t)] = -\sin u(t) \frac{du}{dt}.$$

An application of the Binomial Theorem, the cancellation of like terms and several applications of a well known trigonometric identity yield

$$\left(\frac{dx}{dt}\right)^2 + \left(\frac{dy}{dt}\right)^2 = \left[\cos^2 u(t) \cos^2 v(t) \left(\frac{du}{dt}\right)^2 + \sin^2 u(t) \sin^2 v(t) \left(\frac{dv}{dt}\right)^2\right.$$
$$\left. + \cos^2 u(t) \sin^2 v(t) \left(\frac{du}{dt}\right)^2 + \sin^2 u(t) \cos^2 v(t) \left(\frac{dv}{dt}\right)^2\right]$$
$$= \left[\cos^2 u(t) \left(\frac{du}{dt}\right)^2 + \sin^2 u(t) \left(\frac{dv}{dt}\right)^2\right],$$

and

$$\left(\frac{dx}{dt}\right)^2 + \left(\frac{dy}{dt}\right)^2 + \left(\frac{dz}{dt}\right)^2$$
$$= \left[\cos^2 u(t) \left(\frac{du}{dt}\right)^2 + \sin^2 u(t) \left(\frac{dv}{dt}\right)^2 + \sin^2 u(t) \left(\frac{du}{dt}\right)^2\right]$$
$$= \left[\left(\frac{du}{dt}\right)^2 + \sin^2 u(t) \left(\frac{dv}{dt}\right)^2\right].$$

Thus the length of C is

$$\int_0^T \sqrt{\left(\frac{du}{dt}\right)^2 + \sin^2 u(t) \left(\frac{dv}{dt}\right)^2} \, dt. \tag{2}$$

If the curve γ happens to be a meridian joining P to Q, then, as observed above, the parameter v is constant along γ and so $dv/dt = 0$ at every point of γ. Moreover, since γ is a geodesic segment, it cannot backtrack on itself and hence we also have $du/dt \geq 0$. Thus the above integral reduces to

$$\int_0^T \sqrt{\left(\frac{du}{dt}\right)^2} \, dt = \int_0^T \frac{du}{dt} \, dt = \int_0^A du = A, \tag{3}$$

a result that is of course consistent with what Euclidean geometry tells us.

If γ is an arbitrary curve joining P and Q then it follows from the basic properties of the Riemann integral that the integral of (2) is greater than or equal to the integral of (3). This proves that the arc of the meridian joining P and Q is indeed a geodesic segment of $\mathbb{S}$. Since the points P and Q were arbitrary it follows that every arc of a meridian is a geodesic segment.

It now only remains to demonstrate that all the geodesic segments are indeed such arcs. Suppose therefore that γ is indeed a geodesic segment joining P and Q. Consequently the integrals of (2) and (3) are equal. This can only happen provided that

$$\sin^2 u(t)\left(\frac{dv}{dt}\right)^2 = 0$$

at each point on γ. However, $\sin u(t) = 0$ only if $u(t) = 0$ or π, i.e., only if $P(t)$ is either the north or the south pole. At every other point of γ, we must have

$$\frac{dv}{dt} = 0 \tag{4}$$

Since γ is a geodesic segment it will clearly contain each pole at most once. Hence (4) holds at every interior point of the curve γ, and so v is constant along γ. This of course means that γ is indeed an arc of some meridian.

q.e.d.

It is implicit in the above proof that since the length of the radius of the sphere $\mathbb{S}$ has no units, the lengths of curves on the surface of $\mathbb{S}$ are measured in radians. Moreover, unless P and Q are the opposite ends of a diameter of $\mathbb{S}$, they are joined by exactly one geodesic segment. When P and Q are on the opposite ends of a diameter they are of course joined by an infinitude of such geodesic segments.

On the basis of Theorem 11.1 a case can made that arcs of great circles are the spherical analogs of Euclidean line segments. We now go on to define other spherical analogs of Euclidean notions. A *spherical angle* consists of two such arcs that share an endpoint which, in turn, is called the *vertex* of the angle. The *measure* of a spherical angle is the measure of the Euclidean angle formed by the planes that contain the arcs that form the spherical angle. This angle, in turn, is the angle that is formed by the intersections of these two planes with any plane that is perpendicular to their line of intersection. Alternately, this is the angle between the normals to these two planes. It is clear that congruent spherical angles have the same measure. Moreover, if AB, AC, and AD are three geodesic segments with respective tangent lines AT, AU, AV (Fig. 11.2), then these three tangent lines all lie in the plane tangent to the sphere $\mathbb{S}$ at the

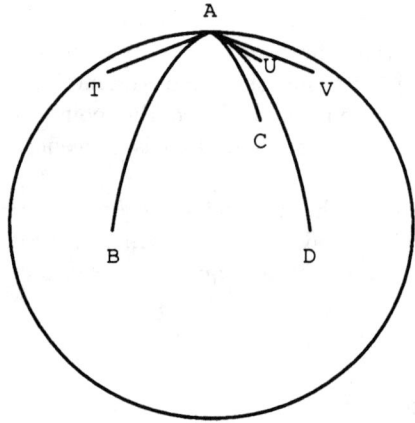

FIGURE 11.2.

point A. Consequently,

$$\angle BAD = \angle TAV = \angle TAU + \angle UAV$$
$$= \angle BAC + \angle CAD.$$

Hence, this manner of measuring spherical angles satisifies the requirements specified in Euclid's Common Notions.

A *spherical triangle* consists of the geodesic segments joining three points of $\mathbb{S}$ that do not all lie on the same great circle. In discussing these triangles it will be convenient to establish the following conventions. The letters A, B, C will denote the vertices of the triangle. and the lower case letters a, b, c, will denote the lengths of the geodesic segments opposite the vertices A, B, C, respectively.

11.3 SPHERICAL TRIGONOMETRY

Theorem 11.5 below consists of spherical analogs of the Law of Cosines and the Law of Sines. Their derivation is greatly facilitated by the use of some well known vector identities. If **a** and **b** are two three dimensional vectors, then (**a**, **b**) will denote their dot, or inner, product, and $\mathbf{a} \times \mathbf{b}$ denotes their cross product. The next two lemmas state, without proof, two fundamental properties of these vector products.

Lemma 11.2. *Let θ be the angle between the vectors **a** and **b**. Then*

$$(\mathbf{a}, \mathbf{b}) = |\mathbf{a}||\mathbf{b}| \cos \theta.$$

□

SPHERICAL TRIGONOMETRY AND ELLIPTIC GEOMETRY   173

Lemma 11.3(Lagrange). *Let a, b, c, d, be three-dimensional vectors. Then*

$$(a \times b, c \times d) = (a, c)(b, d) - (a, d)(b, c).$$

□

In deriving the formulas of spherical trigonometry we shall designate by a, b, c the vectors from the origin to the vertices A, B, C of the spherical triangle, and by α, β, γ the angles at A, B, C.

Lemma 11.4. *If ABC is a spherical triangle, then*

$$\cos \alpha = \left(\frac{a \times b}{|a \times b|}, \frac{a \times c}{|a \times c|} \right).$$

PROOF: The vector $\frac{a \times b}{|a \times b|}$ is the unit normal to the plane containing the geodesic segment AB, and the vector $\frac{a \times c}{|a \times c|}$ is the unit normal to the plane containing the geodesic segment AC. The lemma now follows from Lemma 11.2.

q.e.d.

The following theorem describes the spherical versions of the Laws of Sines and Cosines. As is the case in the hyperbolic plane, there are two versions of the latter law on the sphere.

Theorem 11.5. *Let ABC be a spherical triangle with sides a, b, c. Then*

i) $$\cos \alpha = \frac{\cos a - \cos b \cos c}{\sin b \sin c}$$

ii) $$\cos a = \frac{\cos \alpha + \cos \beta \cos \gamma}{\sin \beta \sin \gamma}$$

iii) $$\frac{\sin \alpha}{\sin a} = \frac{\sin \beta}{\sin b} = \frac{\sin \gamma}{\sin c}.$$

PROOF: It follows from Lemmas 11.2 and 11.3 that

$$(a \times b, a \times c) = (b, c) - (a, c)(a, b) = \cos a - \cos b \cos c,$$

$$|b \times c| = \sqrt{|b|^2 |c|^2 - (b, c)^2} = \sqrt{1 - \cos^2 a} = \sin a$$

and similarly $|a \times b| = \sin c$ and $|c \times a| = \sin b$.

It now follows from Lemma 11.4 that

$$\cos \alpha = \left(\frac{\mathbf{a} \times \mathbf{b}}{|\mathbf{a} \times \mathbf{b}|}, \frac{\mathbf{a} \times \mathbf{c}}{|\mathbf{a} \times \mathbf{c}|} \right)$$

$$= \frac{(\mathbf{a} \times \mathbf{b}, \mathbf{a} \times \mathbf{c})}{|\mathbf{a} \times \mathbf{b}||\mathbf{a} \times \mathbf{c}|} = \frac{\cos a - \cos b \cos c}{\sin b \sin c}.$$

This concludes the proof of part i. The proof of part ii is relegated to Exrecises 23, 24. We now turn to part iii. An application of i to triangle ABC yields:

$$\frac{\sin^2 \alpha}{\sin^2 a} = \frac{1 - \cos^2 \alpha}{\sin^2 a}$$

$$= \frac{\sin^2 b \sin^2 c - (\cos a - \cos b \cos c)^2}{\sin^2 a \sin^2 b \sin^2 c}$$

$$= \frac{(1 - \cos^2 b)(1 - \cos^2 c) - (\cos a - \cos b \cos c)^2}{\sin^2 a \sin^2 b \sin^2 c}$$

$$= \frac{1 - \cos^2 a - \cos^2 b - \cos^2 c + 2 \cos a \cos b \cos c}{\sin^2 a \sin^2 b \sin^2 c}.$$

This last expression, however, is symmetrical in a, b, and c, and hence it follows that

$$\frac{\sin^2 \alpha}{\sin^2 a} = \frac{\sin^2 \beta}{\sin^2 b} = \frac{\sin^2 \gamma}{\sin^2 c}.$$

Since all the quantities $\alpha, \beta, \gamma, a, b, c$ are angles between 0 and π, their sines are all positive and so it follows that

$$\frac{\sin \alpha}{\sin a} = \frac{\sin \beta}{\sin b} = \frac{\sin \gamma}{\sin c}.$$

q.e.d.

The similarity between the formulas of Theorems 8.5 and 11.5 is of course very striking. We use Lobachevsky's own words to show that this similarity is even deeper than we might at first suspect.

"After we have found equations [Theorem 8.5] which represent the dependence of the angles and sides of a triangle; when, finally, we have given general expressions for elements of lines, areas and volumes of solids, all else in the Geometry is a matter of analytics, where calculations must necessarily agree with each other, and we cannot discover anything new that is not included in these first equations from which must be taken all relations of geometric magnitudes, one to another. Thus if one now needs to assume that some contradiction will force us subsequently to refute the principles that we accepted in this geometry, then such contradiction can only hide in the very equations [of Theorem 8.5]. We note however, that these equations become equations [Theorem 11.5] of spherical Trigonometry as soon as, instead of the sides a, b, c we put $a\sqrt{-1}$, $b\sqrt{-1}$, $c\sqrt{-1}$; but in ordinary Geometry and in spherical Trigonometry there enter only ratios of lines: therefore ordinary Geometry, Trigonometry and the new Geometry will always agree among themselves."

Lobachevsky's argument for the consistency of his new geometry, while plausible, is not adequate and the readers are referred to Chapter 15 for a further discussion of this topic. In the meantime we note that the first of the spherical Laws of Cosines yields the following spherical analog of the Theorem of Pythagoras.

Corollary 11.6. *Let ABC be a spherical right triangle on the unit sphere $\mathbb{S}$ with legs a, b and hypotenuse c. Then*

$$\cos c = \cos a \cos b.$$

□

Thus, if a spherical right triangle has legs 1 and 2, then its hypotenuse has length

$$\cos^{-1}(\cos 1 \cos 2) = 1.79\ldots.$$

The readers are reminded here that the unit of length is the radius of the sphere $\mathbb{S}$.

11.4 SPHERICAL AREAS

Our next topic for this chapter is the problem of measuring the area of a spherical triangle. This will be facilitated by two definitions. Given a point A on the sphere $\mathbb{S}$, the *antipode of A*, denoted by A' is the other endpoint of the diameter of $\mathbb{S}$ that contains A. If A is any point of $\mathbb{S}$ and g and h are two geodesic segments joining A to A', they clearly divide the surface of the sphere $\mathbb{S}$ into two parts, each of which is called a *lune*. If we now denote the angle at the corner of the lune by the letter α, then we refer to that lune as a *lune of angle α*. It is clear that any two lunes of the same angle α can be made congruent by a series of rotations of the sphere $\mathbb{S}$. Consequently, every two such lunes have the same area. This, in turn, implies that the area of a lune is proportional to its angle. Since the lune of angle 2π radians has area 4π, we have the following lemma.

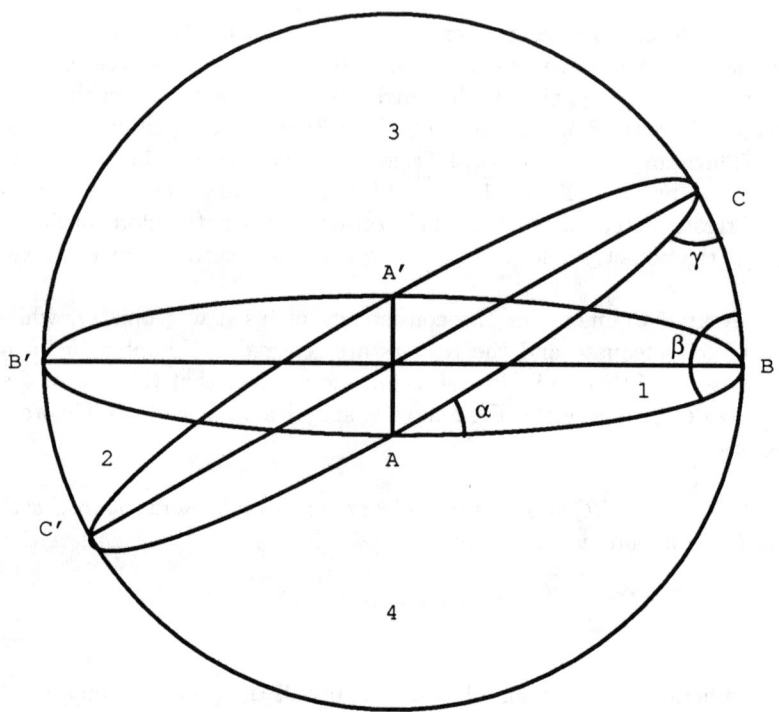

FIGURE 11.3.

Lemma 11.7. *The area of the lune of angle α radians is 2α.*

□

The following theorem is attributed to Thomas Harriot (1560–1621) and Albert Girard (1595–1632). The proof presented here is due to Euler.

Theorem 11.8. *The area of the spherical triangle ABC is $\alpha + \beta + \gamma - \pi$.*

PROOF: Let the spherical triangle ABC be given (Fig. 11.3). Join its vertices to their antipodes A', B', C' by means of diameters, and draw the great circles that contain the geodesic segments AB, BC, and CA. The hemisphere in front of the great circle $BCB'C'$ is thus divided into four spherical triangles ABC, $AB'C'$, $AB'C$, ABC', whose areas are denoted, respectively, by T_1, T_2, T_3, T_4.

From the construction it follows that the spherical triangle $A'BC$ is congruent to the triangle $AB'C'$ of area T_2. Hence, denoting the area of a lune of angle α by *lune* α, we have

$$T_1 + T_2 = \text{lune } \alpha.$$

Similarly,

$$T_1 + T_3 = \text{lune } \beta$$

and

$$T_1 + T_4 = \text{lune } \gamma.$$

Consequently,

$$2T_1 = \text{lune } \alpha + \text{lune } \beta + \text{lune } \gamma - (T_1 + T_2 + T_3 + T_4)$$

$$= 2\alpha + 2\beta + 2\gamma - 2\pi,$$

and the statement of the theorem now follows immediately.

q.e.d.

This very surprising similarity between the area formulas for spherical and hyperbolic triangles, above and beyond its intrinsic interest, also indicates that spherical and hyperbolic geometry should be susceptible to a unified treatment that would shed a light on their relationship. Such a treatment is indeed well known and will be described in Chapter 12.

11.5 A DIGRESSION INTO GEODESY (OPTIONAL)

Our main purpose in incorporating some formulas of spherical trigonometry into this book was to point out the similarity between them and the formulas of hyperbolic geometry. As was mentioned earlier, this similarity provided much needed psychological support to the pioneers of non-Euclidean geometry in that it indicated that their bizzare and unnatural creation might after all be just as valid as the better known spherical geometry. However, having gone so far, it might be appropriate to present some elementary applications of these spherical formulas to problems in geodesy.

Since the unit of the unit sphere was left unspecified, it follows that Theorem 11.1 actually holds for arbitrary spheres. Thus, the geodesic segments of any sphere are arcs of its great circles. A similar argument based on units makes it possible to obtain analogs of Theorems 11.5 and 11.8. Alternately, having obtained the formulas on the sphere of radius 1, one needs only to project the figures on a sphere of radius r onto a sphere of radius 1 that is concentric with the given sphere. This has the effect of leaving the angle measures unchanged whereas the lengths of the arcs are divided by r.

Theorem 11.9. *Let ABC be a spherical triangle with sides a, b, c on a sphere of radius r. Then*

i) $\quad \cos\alpha = \dfrac{\cos\dfrac{a}{r} - \cos\dfrac{b}{r}\cos\dfrac{c}{r}}{\sin\dfrac{b}{r}\sin\dfrac{c}{r}}$

ii) $\quad \cos\dfrac{a}{r} = \dfrac{\cos\alpha + \cos\beta\cos\gamma}{\sin\beta\sin\gamma}$

iii) $\quad \dfrac{\sin\alpha}{\sin\dfrac{a}{r}} = \dfrac{\sin\beta}{\sin\dfrac{b}{r}} = \dfrac{\sin\gamma}{\sin\dfrac{c}{r}}.$

□

Theorem 11.10. *The area of the spherical triangle ABC on a sphere of radius r is $r^2(\alpha + \beta + \gamma - \pi)$.*

□

Example 11.11. Assuming that the earth is a sphere of radius 5280 miles, find the length of the sides, the measure of the angles and the area of the spherical triangle with vertices $A(70°N, 10°E)$, $B(10°S, 100°E)$, and $C(50°S, 80°W,)$.

Let us temporarily take the radius of the earth as the unit of length. This makes it possible to ignore the 5280 miles until the final steps. The spherical coordinates (r, v, u) of the three vertices are $(1, 10°, 20°)$, $(1, 100°, 100°)$, and $(1, -80°, 140°)$, respectively. Hence their Cartesian coordinates are

$$(\sin 20° \cos 10°, \ \sin 20° \sin 10°, \ \cos 20°) = (.3368, .0594, .9397)$$

$$(\sin 100° \cos 100°, \ \sin^2 100°, \cos 100°) = (-.1710, .9698, -.1736)$$

$$(\sin 140° \cos(-80°), \ \sin 140° \sin(-80°), \ \cos 140°)$$

$$= (.1116, -.6330, -.7660).$$

By Lemma 11.2, the cosines of the angles between the radii OA, OB and OC are equal to the dot product of the corresponding position vectors. Thus

$$\angle AOB = \cos^{-1}(-.1631) = 1.7347 \text{ radians}$$

$$\angle BOC = \cos^{-1}(-.5000) = 2.0944 \text{ radians}$$

$$\angle COA = \cos^{-1}(-.7198) = 2.3743 \text{ radians}.$$

Since the length of the arc of a circle is the product of its radius by the radian measure of its central angle, it follows that the lengths of the sides of the spherical triangle ABC are

$$9,159 \text{ miles}, \quad 11,058 \text{ miles}, \quad \text{and} \quad 12,536 \text{ miles}.$$

The angles of this triangle are derived by means of the appropriate Law of Cosines. Thus

$$\cos A = \frac{\cos 2.0944 - \cos 1.7346 \cos 2.3743}{\sin 1.7346 \sin 2.3743}$$

$$= -.9013.$$

So

$$A = 2.6941 \text{ radians} = 154°.$$

Similarly,

$$B = 2.7874 \text{ radians} = 160°$$
$$C = 2.6261 \text{ radians} = 150°.$$

Finally, the area of this spherical triangle is given by Theorem 11.9 as

$$5280^2 (2.6941 + 2.7874 + 2.6261 - 3.1416)$$

$$= 1.3844 \times 10^8 \text{ square miles}.$$

The readers may find the conclusion of the following example somewhat surprising.

Example 11.12. A plane is to fly nonstop from Moscow (56°N, 38°E) to Melbourne (38°S, 145°E). In what direction of the compass should it set out from Moscow?

It is, of course, advantageous for the plane to fly along the great circle joining Moscow and Melbourne. To locate this great circle we incorporate it into a spherical triangle. Let A denote the location of Moscow, B that of Melbourne, and let C be the point on the equator directly south of Moscow (0°N, 38°E). Using the techniques of Example 11.11, the angle A of the spherical triangle ABC is computed to be 78.5°. Hence the plane should set out in the direction of 11.5° south of due east.

11.6 ELLIPTIC GEOMETRY

There are many similarities between the geometry of the sphere and the two other geometries, to wit Euclidean and hyperbolic geometry, but there are some important differences too. Great circles are clearly the spherical analogs of the Euclidean straight lines. However, two distinct Euclidean straight lines intersect in at most a single point, whereas two distinct great circles **always** intersect and the intersection always consists of a **pair** of antipodal points.

It is possible to overcome the second of these anomalies by yet another act of the imagination, one that is similar to that which was used to motivate the creation of hyperbolic geometry. Imagine a spherical flatland which consists of the surface of the sphere $\mathbb{S}$. Imagine further that each of the inhabitants of this land consists of two parts which always occupy antipodal positions on $\mathbb{S}$, but that the two parts always regard themselves as a single entity and are in fact unaware of this split. Thus any displacement of such an inhabitant consists of a simultaneous displacement of both its halves such that the halves are always at diametrically opposite points of the sphere. Such an inhabitant will necessarily regard a pair of antipodal points on the sphere as a single location. From this creature's point of view, every two great circles intersect in a single location which looks to us like two points. Of course, this inhabitant will experience a great circle as half of what we observe as a great circle. Moreover, when this inhabitant traces out a triangle, we will see two triangles being traced out, each the antipodal image of the other. The geometry of the spherical flatland is called elliptic geometry. It resembles Euclidean and hyperbolic geometry in that every two elliptic points can be joined by a unique elliptic line. However, there are no parallel elliptic lines. Thus, elliptic geometry is also non-Euclidean in the sense that Playfair's postulate does not hold in it. The trigonometric formulas of Theorem 11.5 do hold for elliptic geometry, and so it is indeed, just like its hyperbolic counterpart, a very rich geometry. Rather than display its properties in detail, we have assigned the development of some of its contents to the exercises below.

11.7 EXERCISES

1. Let ABC be a right spherical triangle with right angle at C. Prove the following formulas:

    i) $\sin a = \sin \alpha \sin c$ ii) $\tan a = \tan \alpha \sin b$

    iii) $\tan a = \cos \beta \tan c$ iv) $\cos c = \cos b \cos a$

    v) $\cos \alpha = \sin \beta \cos a$ vi) $\sin b = \sin \beta \sin c$

    vii) $\tan b = \tan \beta \sin a$ vii) $\tan b = \cos \alpha \tan c$

    ix) $\cos c = \cot \alpha \cot \beta$ x) $\cos \beta = \sin \alpha \cos b$.

2. Prove Lobachevsky's assertion that the formulas of Theorem 8.5 can be obtained from the formulas of Theorem 11.5 by replacing a, b, c, with $a\sqrt{-1}$, $b\sqrt{-1}$, $c\sqrt{-1}$. (Hint: use Euler's formula $\exp(ix) = \cos x + i \sin x$.)

3. Discuss Euclid's Postulates 1–5 in the context of elliptic geometry.

4. Discuss Euclid's Propositions 1–5 in the context of elliptic geometry.
5. Discuss Euclid's Propositions 6–10 in the context of elliptic geometry.
6. Discuss Euclid's Propositions 11–15 in the context of elliptic geometry.
7. Discuss Euclid's Propositions 16–20 in the context of elliptic geometry.
8. Discuss Euclid's Propositions 20–25 in the context of elliptic geometry.
9. Discuss Euclid's Propositions 26–28 in the context of elliptic geometry.
10. Prove that every elliptic circle is also a Euclidean circle with the same center.
11. Find the relation between the elliptic radius and the Euclidean radius of the same circle.
12. Express the circumference and area of an elliptic circle in terms of its elliptic radius. Comment on the elliptic π.
13. What is an elliptic reflection?
14. What is an elliptic translation?
15. What is an elliptic rotation?
16. Is the difference between the hyperbolic and spherical expressions for the area of a triangle consistent with Lobachevsky's claim of how hyperbolic trigonometric formulas can be obtained from their spherical counterparts?
17. Find a formula for the area of a polygon in elliptic geometry.
18. What angles can an elliptic equilateral triangle have?
19. What angles can the elliptic regular n-gon have?
20. Find the lengths of the sides, the measures of the angles, and the area of the spherical triangle whose vertices are the cities of Kalamazoo (42°N, 85°W), Kathmandu (28°N, 85°E), and Timbuktu (17°N, 3°W).
21. A plane is to fly nonstop from London (51.5°N, 0°W) to Lima (12°S, 77°W). Determine its initial compass direction.
22. Let $\mathbf{a}, \mathbf{b}$, and $\mathbf{c}$ be three vectors in R^3. Prove the following identities:
    i) $\mathbf{a} \times (\mathbf{b} \times \mathbf{c}) = (\mathbf{a}, \mathbf{c})\mathbf{b} - (\mathbf{a}, \mathbf{b})\mathbf{c}$;
    ii) $(\mathbf{a} \times \mathbf{b}) \times \mathbf{c} = (\mathbf{a}, \mathbf{c})\mathbf{b} - (\mathbf{b}, \mathbf{c})\mathbf{a}$;
    iii) $(\mathbf{a} \times \mathbf{b}, \mathbf{c}) = (\mathbf{b} \times \mathbf{c}, \mathbf{a}) = (\mathbf{c} \times \mathbf{a}, \mathbf{b})$.
23. Let $\mathbf{a}, \mathbf{b}, \mathbf{c}, A, B, C, a, b, c, \alpha, \beta, \gamma$ be as in Theorem 11.5. Let $\mathbf{a}^*$ be the unit vector in the direction of $\mathbf{b} \times \mathbf{c}$, and let A^* be its terminal point on the unit sphere $\mathbb{S}$. If $\mathbf{b}^*, B^*, \mathbf{c}^*, C^*$ are similarly defined, then $\triangle A^* B^* C^*$ is called the polar triangle of $\triangle ABC$. If $\alpha^*, \beta^*, \gamma^*$ are the angles of $\triangle A^* B^* C^*$ and a^*, b^*, c^* are its sides, prove that
    i) $\alpha^* = \pi - a, \quad \beta^* = \pi - b, \quad \gamma^* = \pi - c$;
    ii) $\alpha = \pi - a^*, \quad \beta = \pi - b^*, \quad \gamma = \pi - c^*$.
24. Use Exercise 23 to prove part ii of Theorem 11.5.
25. Use the spherical Law of Sines to formulate and prove a spherical version of the Theorem of Menelaus. (See Exercises 1.20 and 8.6).
26. Use the spherical Law of Sines to formulate and prove a spherical version of the Theorem of Ceva. (See Exercises 1.21 and 8.7).
27. Prove that the medians of a spherical triangle are concurrent.

12

Differential Geometry and Gaussian Curvature

12.1 DIFFERENTIAL GEOMETRY

Having developed spherical geometry in great detail it was natural for the mathematicians of the eighteenth century to turn their attention to the geometry of other surfaces. In contrast with the plane and the sphere, most such surfaces are non-homogeneous in the sense that they are differently shaped, or curved, at different points. Differential geometry is the branch of mathematics that is concerned with the quantification of shape, both for curves and for surfaces. While several prior mathematicians including Lagrange and Euler made substantial contributions to the topic, it was Gauss himself who gave this topic the direction it has today. He accomplished this by making the vague of notion of the curvature of a surface precise and then proving interesting theorems about it. We shall describe Gauss's definition and theorems because of their pivotal role in modern geometry and because of their strong connection with the various theorems regarding the areas of triangles in the hyperbolic, Euclidean, and spherical geometries.

For our purposes here a surface is the end result of a smooth distortion of the plane or a portion thereof. By a smooth distortion we mean that the resulting surface will contain no sharp corners or edges; in other words, the tangent plane to the surface should be well defined at every one of its points. In seeking to quantify the notion of curvature we must adhere to the general guideline that the numerical curvature of the surface should in some sense be proportional to its distortion. Moreover, we have some preconceptions regarding some specific surfaces to which any reasonable definition of curvature must conform. For example, the curvature of the plane should be 0, since it is the end result of no distortion at all. In addition, the sphere should have the same curvature at all of its points since its distortion is the same at all of its points. The same, of course must hold for the circular cylinder.

184 THE POINCARÉ HALF-PLANE

FIGURE 12.1a.

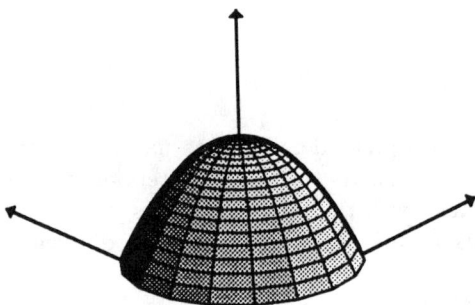

FIGURE 12.1b.

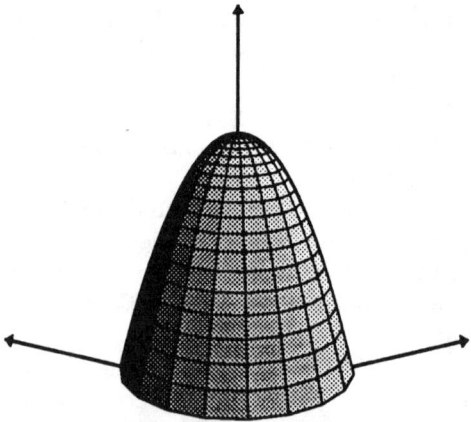

FIGURE 12.1c.

The amount of distortion that must be applied to the plane in order to obtain a given surface was pictured by Gauss by means of a device that has come to be known as the *Gauss map*. Let $\mathbb{F}$ denote a surface, let P denote a general point on it, and let Φ be a portion of the surface that surrounds P and in which the surface is fairly uniformly curved. The author apologizes for the vagueness of these terms, but believes that this is the best that can be done under the circumstances. Let $\overrightarrow{PP'}$ be the unit vector that is perpendicular to the surface $\mathbb{F}$ – the *unit normal*. If the surface is greatly curved in Φ, then a small displacement in the position of P will result in a disproportionately large change in the direction of the normal $\overrightarrow{PP'}$, whereas if Φ is nearly flat, the same small displacement in the position of P will result in very little change in the direction of $\overrightarrow{PP'}$. This is illustrated in Fig. 12.1 wherein the surfaces of a, b, and c are obtained from a *circle* by the application of successively greater amounts of distortion. It is clear in the picture that the greater the distortion, the more the various normal vectors on the surface diverge in direction. In fact, if the surface $\mathbb{F}$ happens to be a plane, then the direction of the normal $\overrightarrow{PP'}$ remains unchanged regardless of the position of P. On the other hand, if the surface is a sphere, then diametrically opposite points will have normal vectors that have **opposite** directions, regardless of the smallness of the radius of the sphere. Thus, this change in the direction of the normal is a reasonable measure of the distortion.

To obtain a concrete representation of the amount of variation in the direction of the normal $\overrightarrow{PP'}$ as P varies over all the points of the portion Φ of the surface $\mathbb{F}$ Gauss next anchors the vector $\overrightarrow{PP'}$ at the origin. That is, he chooses a unit vector $\overrightarrow{OP^*}$ emanating from the origin that is equal and parallel to the original $\overrightarrow{PP'}$ (Fig. 12.2). The point P^* traces out a portion Φ^* of the unit sphere $\mathbb{S}$ as P varies over the original Φ. For example, if $\mathbb{F}$ happens to be a piece of the xy plane then $P^*$is constantly $(0,0,1)$. If $\mathbb{F}$ is the yz plane, then P^* is constantly $(1,0,0)$. It should now be clear to the reader that if $\mathbb{F}$ is any plane surface, then Φ^* consists of a single point. On the other hand, if $\mathbb{F}$ is a sphere, then P^* traces out the whole of the unit sphere as P varies over all of $\mathbb{F}$. If $\mathbb{F}$ is the paraboloid of revolution whose equation is $z = x^2 + y^2$, then P^* traces out a hemisphere on the unit sphere.

Gauss defined the *total curvature* of the portion Φ of $\mathbb{F}$ as

$$\text{area}\,(\Phi^*).$$

If the surface $\mathbb{F}$ is bounded (i.e., has finite extent), then its total curvature can be obtained by subdividing it into many sufficiently small portions Φ and adding up their total curvatures.

It is clear that this definition satisfies the requirement that the curvature of a plane surface be zero, since in that case each Φ^* consists of a single point, and so its area is zero. There are some other surfaces for which the total curvature is always zero. Such, for example, are the circular cylinder and all portions thereof. To see this picture the cylinder as standing on its end so that its cross

186 THE POINCARÉ HALF-PLANE

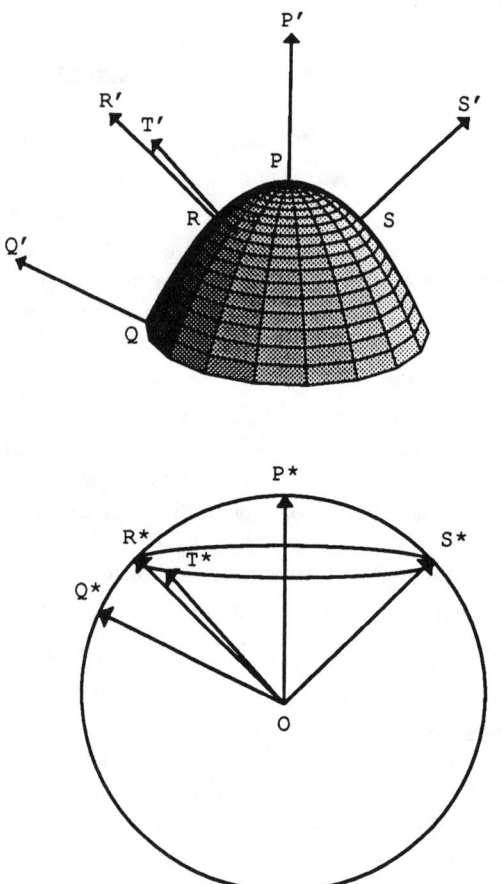

FIGURE 12.2.

sections by horizontal planes are circles. Then the normal vector $\overrightarrow{PP'}$ is everywhere parallel to the xy plane and so the locus of the point P^* is a circle that lies in the xy plane and is centered at the origin. Since the area of this circle and all of its arcs are zero, it follows that every portion of the circular cylinder has total curvature 0. A similar argument can be used to demonstrate that all cylinders and all cones also have total curvature zero. These are all examples of *developable* surfaces which are defined as those surfaces which can be obtained by bending planar regions without distorting the distances on them. In other words, they are the surfaces that can be formed from a sheet of paper. It is one of the strengths of this notion of total curvature that it can be used to characterize developable surfaces. Loosely speaking, a surface is developable if and only if every portion of it has total curvature zero.

DIFFERENTIAL GEOMETRY AND GAUSSIAN CURVATURE 187

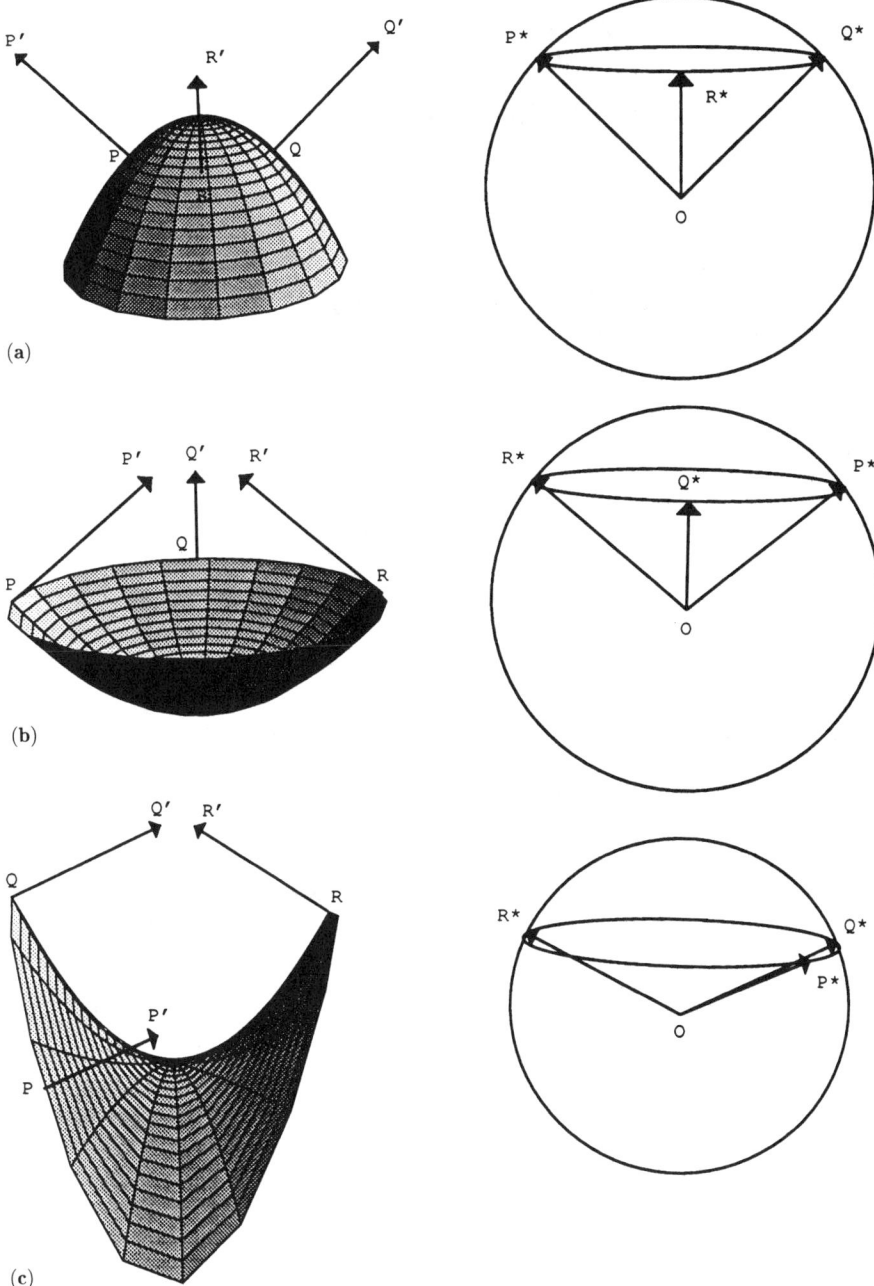

FIGURE 12.3. (a) The Gauss map on a convex surface. (b) The Gauss map on a concave surface. (c) The Gauss map near a saddle point.

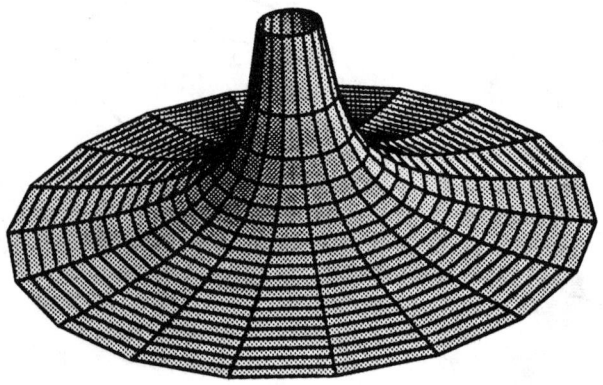

FIGURE 12.4.

Next, let us turn to the unit sphere $\mathbb{S}$ and let it play the role of the surface $\mathbb{F}$. Since the radius of the sphere is perpendicular to its tangent plane, this has the very helpful consequence that $\overrightarrow{OP} = \overrightarrow{PP'} = \overrightarrow{OP^*}$. In other words, $P = P^*$! Thus, for every portion Φ of the unit sphere, $\Phi = \Phi^*$ and hence area(Φ) = area(Φ^*), and so $\mathbb{F}$ has total curvature area($\mathbb{F}$).

At this point it is necessary to focus on a difficulty that was glossed over in the preceding discussion. There are circumstances wherein it is necessary to define the total curvature to be negative, i.e., -area(Φ^*) rather than area(Φ^*). Such is the case when P happens to be a saddle point of $\mathbb{F}$, i.e., a point where the tangent plane cuts through Φ rather than lies on one side of it. Fig. 12.3 makes it clear that the Gauss map behaves differently in the vicinity of such a saddle point than it does at a concave or convex portion of the sphere. Specifically, when Φ is saddle-shaped P^* will trace out a *clockwise* circle on $\mathbb{S}$ as P traces out a *counterclockwise* circle on Φ. Conversely, when Φ is either convex of concave P^* traces out a *counterclockwise* circle as P moves along a counterclockwise circle of Φ. Such saddle points are in no way exceptional. For example, the surface obtained by revolving the *hyperbola* $y = 1/x$ about the y axis consists exclusively of such points (Fig. 12.4).

We are now in position to state, without proof, Gauss's generalization of Theorem 11.8 to arbitrary surfaces. On such surfaces geodesic triangles and their angles are to be understood as the obvious generalization of their counterparts on the unit sphere. Namely, a *geodesic segment* joining two points on a surface $\mathbb{F}$ is the shortest of all the curves that join these points and lie in $\mathbb{F}$. A *geodesic triangle* on $\mathbb{F}$ consists of three geodesic segments each of which joins two vertices of the triangle. Because of the generality of the context it is necessary to add the proviso that the constituent geodesics of such a triangle intersect each other only at the vertices of the triangle.

Theorem 12.1 (Gauss). *On an arbitrary surface F, let Φ be the interior of a geodesic triangle with vertex angles α, β, γ. Then the total curvature of Φ is equal to*

$$\alpha + \beta + \gamma - \pi.$$

□

Since the total curvature of any region on the unit sphere is equal to its area, Gauss's theorem does indeed specialize to Theorem 11.8 on the surface of the unit sphere.

Gauss went on to define the curvature of a surface $\mathbb{F}$ at a point P to be the ratio

$$K(P) = \frac{\text{total curvature}(\Phi)}{\text{area}(\Phi)}$$

for infinitesimally small regions Φ on $\mathbb{F}$ that contain P. In other words, K(P) can be estimated by taking a small region Φ of $\mathbb{F}$ that contains P and dividing its total curvature by its area. This quantity is now known as the *Gaussian curvature* of the surface at the point P. He then found an explicit formula for the local curvature of the surface $z = f(x, y)$, namely,

$$K = \frac{\left(\frac{\partial^2 z}{\partial x^2}\right)\left(\frac{\partial^2 z}{\partial y^2}\right) - \left(\frac{\partial^2 z}{\partial x \partial y}\right)^2}{\left\{1 + \left(\frac{\partial z}{\partial x}\right)^2 + \left(\frac{\partial z}{\partial y}\right)^2\right\}^2}. \tag{1}$$

As the justification of this formula is rather involved and also relies on some subsequent material, its sketch is postponed to a later section.

It should be noted here that if a surface has negative Gaussian curvature at each of its points then, according to Gauss's theorem, the sum of the angles of its geodesic triangles will be less than π, just as is the case in the hyperbolic plane. Thus, such surfaces could serve, in principle, as models for hyperbolic geometry. The aforementioned surface obtained by revolving the hyperbola $y = 1/x$ ($y > 0$) about the y axis is of course such a surface. However, its Gaussian curvature is not constant – in fact, it is clear that points far from the axis of revolution have a curvature that is almost zero, whereas points that are close to the y axis have a curvature that is negatively very large (Exercise 4). Consequently, such a surface cannot serve as a model for hyperbolic geometry, since such a geometry, by virtue of Euclid's Postulate 4 must be homogeneous in the sense that given any two points of the geometry, there exists a rigid motion that carries one onto the other. Surfaces with constant negative Gaussian curvature were constructed by Minding in 1840, but he was unaware of their relationship to the hyperbolic plane. It is interesting to note that Minding went so far as to develop the trigonometry of the geodesic triangles on these surfaces and obtained the same formulas that Lobachevsky had derived for the hyperbolic plane. He too noted that his formulas could be obtained from those of spherical trigonometry by replacing the radius 1 of the sphere S with the imaginary number $i = \sqrt{-1}$. It remained for Beltrami to point out in 1868 that the geometry of these surfaces, which he called *pseudospheres*, provided a concrete realization of Lobachevsky's non-Euclidean geometry.

12.2   A REVIEW OF LENGTHS AND AREAS ON SURFACES

In order to describe some more of Gauss's contributions to modern geometry, as well as some further generalizations of the theorems about the areas of geodesic triangles, it necessary to pause for a review of some elementary facts about the calculus of surfaces. These are contained in the later chapters of most calculus texts.

We begin with the arclength formula for curves in a three dimensional Cartesian coordinate system. If γ is a curve which is parametrized as

$$\gamma(t) = [x(t),\ y(t), z(t)], \quad a \le t \le b,$$

then its length is

$$\int_a^b \sqrt{\left(\frac{dx}{dt}\right)^2 + \left(\frac{dy}{dt}\right)^2 + \left(\frac{dz}{dt}\right)^2}\, dt. \tag{2}$$

All the curves we will be discussing in this chapter will be lying on some surface $\mathbb{F}$ which is the graph of a function $z = f(x, y)$. Thus it is possible to specify any curve on the surface by specifying its projection in the xy plane. Effectively this means that any curve γ on the surface of $z = f(x, y)$ can be parametrized as

$$\gamma(t) = [x(t),\ y(t),\ f(x(t), y(t))], \quad a \le t \le b,$$

where

$$[x(t),\ y(t)]$$

is the parametrization of its projection onto the xy plane.

We now go on to incorporate this parametrization of γ on $\mathbb{F}$ into the above arclength formula. The chain rule implies that

$$\frac{dz}{dt} = \frac{\partial z}{\partial x}\frac{dx}{dt} + \frac{\partial z}{\partial y}\frac{dy}{dt}$$

and hence, for a curve γ that lies on the graph of $z = f(x, y)$ the arclength formula (2) can be rewritten as

$$\int_a^b \sqrt{\left[1 + \left(\frac{\partial z}{\partial x}\right)^2\right]\left(\frac{dx}{dt}\right)^2 + 2\frac{\partial z}{\partial x}\frac{\partial z}{\partial y}\frac{dx}{dt}\frac{dy}{dt} + \left[1 + \left(\frac{\partial z}{\partial y}\right)^2\right]\left(\frac{dy}{dt}\right)^2}\, dt$$

or in the abbreviated form of

$$\int_\gamma \sqrt{\left[1 + \left(\frac{\partial z}{\partial x}\right)^2\right] dx^2 + 2\frac{\partial z}{\partial x}\frac{\partial z}{\partial y} dx dy + \left[1 + \left(\frac{\partial z}{\partial y}\right)^2\right] dy^2} \qquad (3)$$

Example 12.2. Find the length of the curve γ on the paraboloid of revolution $z = x^2 + y^2$ that lies directly over the line $y = x$ in the xy plane and joins the origin to the point $(1,1,2)$.

Here

$$\frac{\partial z}{\partial x} = 2x, \quad \frac{\partial z}{\partial y} = 2y, \quad \text{and} \quad y = x$$

along γ. Hence the integral of (3) reduces to

$$\int_0^1 \sqrt{2 + 16x^2}\, dx = 2.56\ldots$$

Next we turn our attention to the surface $\mathbb{F}$ itself (Fig. 12.5). If this surface has equation $z = f(x, y)$ then its cross section C with a plane containing

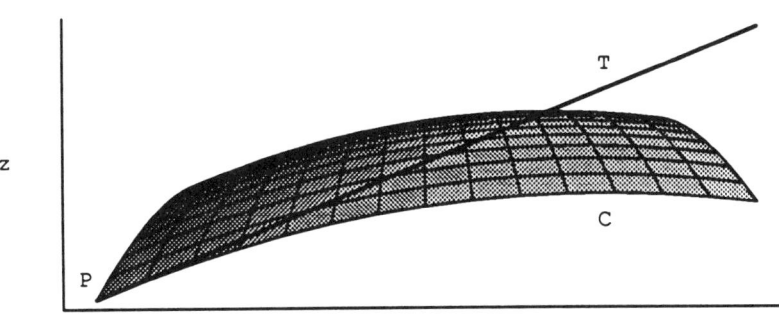

FIGURE 12.5.

$P = (a, b, c)$ and parallel to the xz plane can be parametrized as

$$C(x) = [x, b, f(x, b)]$$

which has the derivative vector

$$\frac{dC(x)}{dx} = \left(1, 0, \frac{\partial f}{\partial x}\right).$$

Hence the tangent line S to C at P can be parametrized as

$$S(x) = (a, b, c) + (x - a)\left(1, 0, \frac{\partial f}{\partial x}\right) = \left[x, b, c + (x - a)\frac{\partial f}{\partial x}\right].$$

Similarly, the cross section D of the surface with a plane containing P and parallel to the yz plane can be parametrized as

$$D(y) = [a, y, f(a, y)]$$

which has the derivative vectro

$$\frac{dD(y)}{dy} = \left(0, 1, \frac{\partial f}{\partial y}\right)$$

and hence the tangent line T to D at P can be parametrized as

$$T(y) = (a, b, c) + (y - b)\left(0, 1, \frac{\partial f}{\partial y}\right) = \left[a, y, c + (y - b)\frac{\partial f}{\partial y}\right].$$

The reader should keep in mind that all the above derivatives and partial derivatives are to be evaluated at $x = a$ and $y = b$. The only reason this is not explicitly specified in the above equations is that it would make them too difficult to read.

Example 12.3. Consider the graph of $z = x^2 + y^3$ at the point $(1,1,2)$. Here,

$$\frac{\partial z}{\partial x} = 2x = 2 \quad \text{and} \quad \frac{\partial z}{\partial y} = 3y^2 = 3.$$

Consequently, the tangent lines have the vector parametric equations

$$S(x) = [x, 1, 2 + (x - 1)2] = (x, 1, 2x),$$
$$T(y) = [1, y, 2 + (y - 1)3] = (1, y, 3y - 1).$$

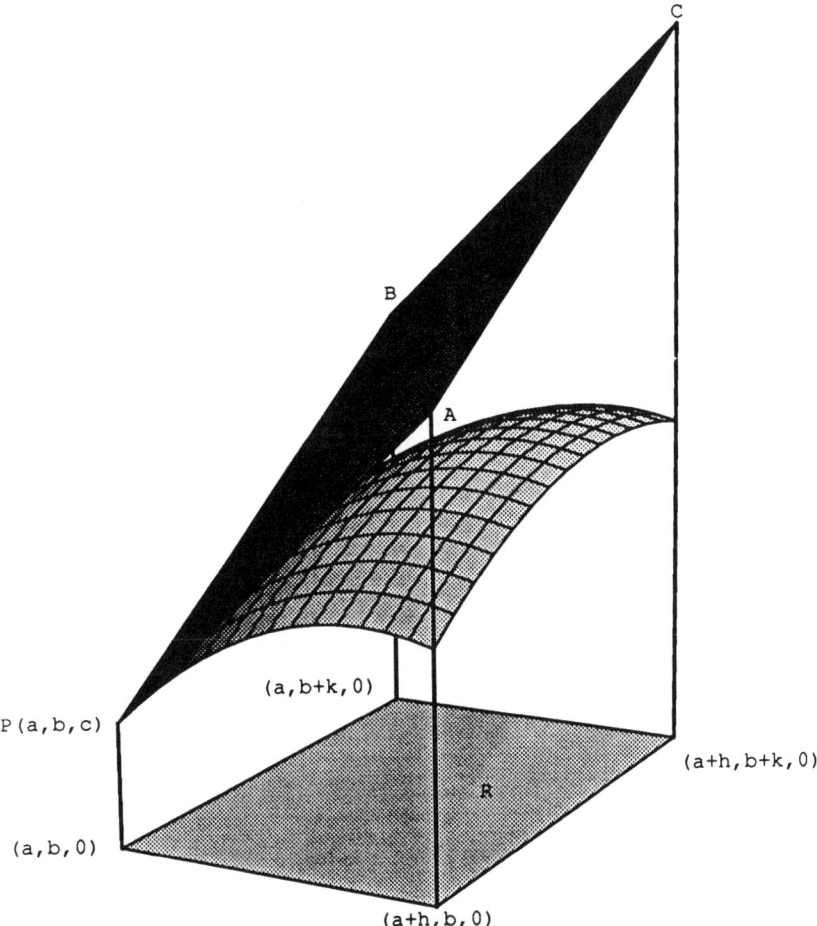

FIGURE 12.6.

The portion Φ of the surface $\mathbb{F}$ that lies directly above the rectangle R with vertices $(a, b, 0)$, $(a + h, b, 0)$, $(a, b + k, 0)$, $(a + h, b + k, 0)$ is approximated by the parallelogram $PACB$ which is the portion of the tangent plane to $\mathbb{F}$ at P that lies above the same rectangle (Figure 12.6). This parallelogram has vertices whose coordinates can be obtained by approporiate substitutions in the parametric equations of the tangent lines S and T above. These substitutions yield

$$A = \left(a + h, b, c + h\frac{\partial f}{\partial x}\right) \quad \text{and} \quad B = \left(a, b + k, c + k\frac{\partial f}{\partial y}\right)$$

as the vertices of the parallelogram that are adjacent to P. Hence the area of the parallelogram $PACB$ is the length of the cross product

$$|\overrightarrow{PA} \times \overrightarrow{PB}| = \left|\left(h, 0, h\frac{\partial f}{\partial x}\right) \times \left(0, k, k\frac{\partial f}{\partial y}\right)\right|$$

$$= \left|\left(-\frac{\partial f}{\partial x}, -\frac{\partial f}{\partial y}, 1\right)\right| hk = \sqrt{1 + \left(\frac{\partial f}{\partial x}\right)^2 + \left(\frac{\partial f}{\partial y}\right)^2}\, hk$$

Consequently, we have the expression

$$\iint_R \sqrt{1 + \left(\frac{\partial z}{\partial x}\right)^2 + \left(\frac{\partial z}{\partial y}\right)^2}\, dxdy \qquad (4)$$

for the area of the portion of the surface of $\mathbb{F}$ that lies directly above the region R in the xy plane.

Example 12.4. Find the area of the portion of the paraboloid $z = x^2 + y^2$ that lies above the unit disk $\triangle$ in the xy plane.

Since the relevant partial derivatives are $2x$ and $2y$ the area is

$$\iint_\triangle \sqrt{1 + 4x^2 + 4y^2}\, dxdy = \int_0^{2\pi} \int_0^1 \sqrt{1 + 4r^2}\, rdrdq$$

$$\int_0^{2\pi} \frac{1}{12}(\sqrt{1+4r^2})^3\Big]_0^1 d\theta = \frac{1}{12}\int_0^{2\pi} (5\sqrt{5} - 1)\, d\theta$$

$$= \frac{(5\sqrt{5} - 1)\pi}{6}.$$

The integrand

$$\sqrt{1 + \left(\frac{\partial z}{\partial x}\right)^2 + \left(\frac{\partial z}{\partial y}\right)^2}\, dxdy$$

of (4) is called the *area element* and is denoted by dA. It should be thought of as the area of an infinitesimal portion of the surface that lies over an infinitesimal rectangle in the xy plane with sides dx and dy.

Occasionally it is necessary to sum the values of a function that is defined at every point of a surface. For example, if $\delta(x, y)$ denotes the density of the surface at the point $[x, y, f(x, y)]$, then the sum of the of the values of δ over a portion Φ of the surface yields the mass of Φ. Now, to continue this analogy, the mass of the infinitesimal portion above the rectangle in the xy plane with

DIFFERENTIAL GEOMETRY AND GAUSSIAN CURVATURE 193

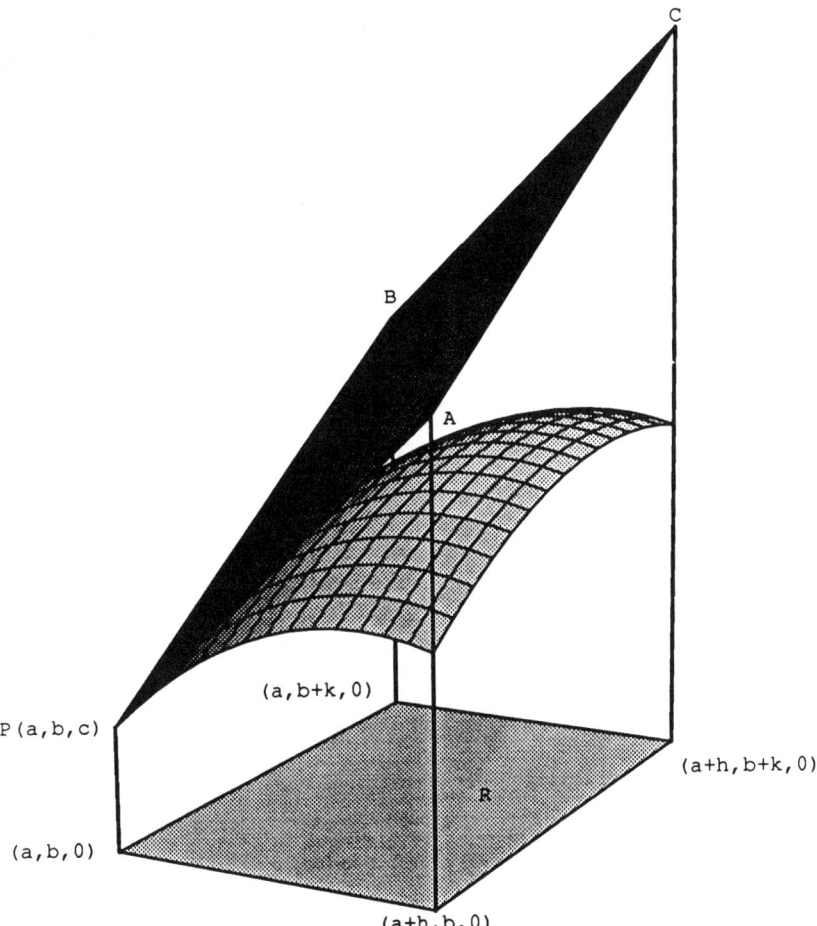

FIGURE 12.6.

The portion Φ of the surface $\mathbb{F}$ that lies directly above the rectangle R with vertices $(a, b, 0)$, $(a + h, b, 0)$, $(a, b + k, 0)$, $(a + h, b + k, 0)$ is approximated by the parallelogram $PACB$ which is the portion of the tangent plane to $\mathbb{F}$ at P that lies above the same rectangle (Figure 12.6). This parallelogram has vertices whose coordinates can be obtained by approporiate substitutions in the parametric equations of the tangent lines S and T above. These substitutions yield

$$A = \left(a + h, b, c + h\frac{\partial f}{\partial x}\right) \quad \text{and} \quad B = \left(a, b + k, c + k\frac{\partial f}{\partial y}\right)$$

as the vertices of the parallelogram that are adjacent to P. Hence the area of the parallelogram $PACB$ is the length of the cross product

$$|\vec{PA} \times \vec{PB}| = \left| \left(h, 0, h\frac{\partial f}{\partial x} \right) \times \left(0, k, k\frac{\partial f}{\partial y} \right) \right|$$

$$= \left| \left(-\frac{\partial f}{\partial x}, -\frac{\partial f}{\partial y}, 1 \right) \right| hk = \sqrt{1 + \left(\frac{\partial f}{\partial x}\right)^2 + \left(\frac{\partial f}{\partial y}\right)^2} \, hk$$

Consequently, we have the expression

$$\iint_R \sqrt{1 + \left(\frac{\partial z}{\partial x}\right)^2 + \left(\frac{\partial z}{\partial y}\right)^2} \, dxdy \qquad (4)$$

for the area of the portion of the surface of $\mathbb{F}$ that lies directly above the region R in the xy plane.

Example 12.4. Find the area of the portion of the paraboloid $z = x^2 + y^2$ that lies above the unit disk $\triangle$ in the xy plane.

Since the relevant partial derivatives are $2x$ and $2y$ the area is

$$\iint_\triangle \sqrt{1 + 4x^2 + 4y^2} \, dxdy = \int_0^{2\pi} \int_0^1 \sqrt{1 + 4r^2} \, r \, dr \, dq$$

$$\int_0^{2\pi} \frac{1}{12} (\sqrt{1+4r^2})^3 \Big|_0^1 d\theta = \frac{1}{12} \int_0^{2\pi} (5\sqrt{5} - 1) \, d\theta$$

$$= \frac{(5\sqrt{5} - 1)\pi}{6}.$$

The integrand

$$\sqrt{1 + \left(\frac{\partial z}{\partial x}\right)^2 + \left(\frac{\partial z}{\partial y}\right)^2} \, dxdy$$

of (4) is called the *area element* and is denoted by dA. It should be thought of as the area of an infinitesimal portion of the surface that lies over an infinitesimal rectangle in the xy plane with sides dx and dy.

Occasionally it is necessary to sum the values of a function that is defined at every point of a surface. For example, if $\delta(x, y)$ denotes the density of the surface at the point $[x, y, f(x, y)]$, then the sum of the of the values of δ over a portion Φ of the surface yields the mass of Φ. Now, to continue this analogy, the mass of the infinitesimal portion above the rectangle in the xy plane with sides dx and dy is the product of its (almost uniform) density with its area, i.e.,

$$\delta(x,y)\sqrt{1+\left(\frac{\partial z}{\partial x}\right)^2+\left(\frac{\partial z}{\partial y}\right)^2}\,dxdy,$$

and so the mass of the portion of the surface $\mathbb{F}$ that lies over the rectangle R in the xy plane is given by the integral

$$\iint_R \delta(x,y)\sqrt{1+\left(\frac{\partial z}{\partial x}\right)^2+\left(\frac{\partial z}{\partial y}\right)^2}\,dxdy \qquad (5)$$

Example 12.5. The plate that consists of the portion of the plane $z = 3x + 2y + 1$ above the square $0 \leq x, y \leq 1$ has density function $\delta(x,y) = x + y$. Find the total mass of the plate.

Applying the formula of (5) the mass of the plate is

$$\int_0^1 \int_0^1 (x+y)\sqrt{1+9+4}\,dxdy = \sqrt{14}.$$

The notions of density and mass were used above only for the purpose of making the notion of addition over a surface more concrete, but the discussion was completely general. Even when δ is an arbitrary function, the *sum of δ over a portion Φ of the surface $\mathbb{F}$* is given by the integral of (5), where R is the projection of Φ onto the xy plane. It is customary to abbreviate the integral of (5) as

$$\iint_\Phi \delta\,dA \qquad (6)$$

It is important that the readers realize that formula (6) is merely a convenient abbreviation. When it comes to nitty gritty calculations, one must revert to the integral of (5).

Example 12.6. Evaluate

$$\iint_\Phi (x^2 - y)\,dA$$

where Φ is the portion of the plane $z = 3x + 2y + 1$ that lies over the unit square $0 \leq x, y \leq 1$ in the xy plane.

Reverting to integral (5) we obtain

$$\int_0^1 \int_0^1 (x^2 - y)\sqrt{1+9+4}\,dxdy = -\frac{\sqrt{14}}{6}.$$

Notice that when δ is identically 1, the integral (6) simply yields the area of Φ.

196   THE POINCARÉ HALF-PLANE

This completes our review of calculus on surfaces. Its main purpose was to remind the reader how to add values that are defined at each point of a surface. We will be particularly interested in two cases. When this value is the constant 1, the summation process yields the area of the surface. When this summed value is the curvature at each point, the result will be seen to be the total curvature.

12.3 GAUSS'S FORMULA FOR THE CURVATURE AT A POINT

This section contains only a proof of the aforementioned formula derived by Gauss for curvature at a point.

Theorem 12.7. *If the surface $\mathbb{F}$ is the graph of the function $z = f(x, y)$ then*

$$K = \frac{\left(\frac{\partial^2 z}{\partial x^2}\right)\left(\frac{\partial^2 z}{\partial y^2}\right) - \left(\frac{\partial^2 z}{\partial x \partial y}\right)^2}{\left[1 + \left(\frac{\partial z}{\partial x}\right)^2 + \left(\frac{\partial z}{\partial y}\right)^2\right]^2}.$$

PROOF: Let P be any point on the given surface $\mathbb{F}$, let Φ be an infinitesimal portion of $\mathbb{F}$ surrounding P, and let Φ^* be the image of Φ under the Gauss map. Then

$$K = \frac{\text{area}(\Phi^*)}{\text{area}(\Phi)} = \frac{dA(\text{on sphere})}{dA(\text{on }\mathbb{F})}.$$

However, the unit sphere has equation $Z = \sqrt{1 - X^2 - Y^2}$ and hence

$$dA \text{ (on sphere)} = \sqrt{1 + \left(\frac{\partial Z}{\partial X}\right)^2 + \left(\frac{\partial Z}{\partial Y}\right)^2}\, dXdY$$

$$= \sqrt{1 + \frac{X^2 + Y^2}{1 - X^2 - Y^2}}\, dXdY = \frac{dXdY}{Z}.$$

Now think of the point (X, Y, Z) on the sphere as being the endpoint of the unit normal vector at the general point (x, y, z) on the surface $\mathbb{F}$. In other words,

$$(X, Y, Z) = \frac{\left(-\frac{\partial z}{\partial x}, -\frac{\partial z}{\partial y}, 1\right)}{\sqrt{1 + \left(\frac{\partial z}{\partial x}\right)^2 + \left(\frac{\partial z}{\partial y}\right)^2}}.$$

Hence,

$$\frac{\partial X}{\partial x} = \frac{-\dfrac{\partial^2 z}{\partial x^2}\left[1+\left(\dfrac{\partial z}{\partial y}\right)^2\right] + \dfrac{\partial z}{\partial x}\dfrac{\partial z}{\partial y}\dfrac{\partial^2 z}{\partial x\partial y}}{\left[1+\left(\dfrac{\partial z}{\partial x}\right)^2+\left(\dfrac{\partial z}{\partial y}\right)^2\right]^{3/2}}$$

$$\frac{\partial X}{\partial y} = \frac{-\dfrac{\partial^2 z}{\partial x\partial y}\left[1+\left(\dfrac{\partial z}{\partial y}\right)^2\right] + \dfrac{\partial z}{\partial x}\dfrac{\partial z}{\partial y}\dfrac{\partial^2 z}{\partial y^2}}{\left[1+\left(\dfrac{\partial z}{\partial x}\right)^2+\left(\dfrac{\partial z}{\partial y}\right)^2\right]^{3/2}}$$

$$\frac{\partial Y}{\partial x} = \frac{-\dfrac{\partial^2 z}{\partial x\partial y}\left[1+\left(\dfrac{\partial z}{\partial x}\right)^2\right] + \dfrac{\partial z}{\partial x}\dfrac{\partial z}{\partial y}\dfrac{\partial^2 z}{\partial x^2}}{\left[1+\left(\dfrac{\partial z}{\partial x}\right)^2+\left(\dfrac{\partial z}{\partial y}\right)^2\right]^{3/2}}$$

$$\frac{\partial Y}{\partial y} = \frac{-\dfrac{\partial^2 z}{\partial y^2}\left[1+\left(\dfrac{\partial z}{\partial x}\right)^2\right] + \dfrac{\partial z}{\partial x}\dfrac{\partial z}{\partial y}\dfrac{\partial^2 z}{\partial x\partial y}}{\left[1+\left(\dfrac{\partial z}{\partial x}\right)^2+\left(\dfrac{\partial z}{\partial y}\right)^2\right]^{3/2}}.$$

Amazingly, the Jacobian

$$\frac{\partial(X,Y)}{\partial(x,y)} = \frac{\partial X}{\partial x}\frac{\partial Y}{\partial y} - \frac{\partial X}{\partial y}\frac{\partial Y}{\partial x}$$

simplifies to the expression

$$\frac{\dfrac{\partial^2 z}{\partial x^2}\dfrac{\partial^2 z}{\partial y^2} - \left(\dfrac{\partial^2 z}{\partial x\partial y}\right)^2}{\left[1+\left(\dfrac{\partial z}{\partial x}\right)^2+\left(\dfrac{\partial z}{\partial y}\right)^2\right]^2}.$$

Hence,

$$K = \frac{dA \text{ (on sphere)}}{dA \text{ (on } \mathbb{F})}$$

$$= \frac{\dfrac{dXdY}{Z}}{\sqrt{1+\left(\dfrac{\partial z}{\partial x}\right)^2+\left(\dfrac{\partial z}{\partial y}\right)^2}\,dxdy}$$

$$= \frac{\dfrac{1}{Z}\dfrac{\partial(X,Y)}{\partial(x,y)}\,dxdy}{\sqrt{1+\left(\dfrac{\partial z}{\partial x}\right)^2+\left(\dfrac{\partial z}{\partial y}\right)^2}\,dxdy}$$

$$= \frac{\left(\dfrac{\partial^2 z}{\partial x^2}\right)\left(\dfrac{\partial^2 z}{\partial y^2}\right)-\left(\dfrac{\partial^2 z}{\partial x\partial y}\right)^2}{\left[1+\left(\dfrac{\partial z}{\partial x}\right)^2+\left(\dfrac{\partial z}{\partial y}\right)^2\right]^2}.$$

q.e.d.

12.4 RIEMANNIAN GEOMETRY REVISITED

In order to describe the next major step in the evolution of differential geometry it is necessary to ask our readers for yet another act of the imagination. Suppose an observer named Polaris lives very far on the positive z axis of a three dimensional Cartesian coordinate system and that he is examining the behavior of some two dimensional inhabitants of the surface of the unit sphere through a telescope. Let us further suppose that the inhabitants of the northern hemisphere ($z > 0$) never wander off into the southern portion and that moreover, Polaris is unaware of the curved nature of the hemisphere he is observing. In other words, Polaris believes he is observing two dimensional creatures that are restricted to the interior of the unit disk. It might seem to him at first that these creatures tend to meander. Certainly he would observe that when leaving point A to reach point B, these creatures, even when they are in a hurry, do not take the shortest route. In fact, rather than follow the straight line from A to B they move in a circular arc that bends towards the circumference of their world. We, of course, know better. We know that these creatures live on the surface of a sphere and hence their geodesics are arcs of great circles. Polaris, because of his limitations, will see these great circles as what we know are only the projections of these spherical arcs onto the xy plane.

If Polaris is predisposed to believe in the common sense of these creatures, he might assume that the physical nature of their universe is such as to make them move faster when they are closer to the circumference of their universe. By analogy with the infinitely cold x-axis that borders the Poincaré upper half-plane, Polaris might suppose that the unit disk he is observing has an infinitely hot circumference which causes its inhabitants to expand as they move outwards. In other words, he may assume that there is a

Riemann metric that governs the notion of distance in the unit disk he believes they inhabit. In fact, this Riemann metric is easily calculated, and we shall do so momentarily. At this point let us state without justification that it is the metric

$$\frac{(1-y^2)dx^2 + 2xy\,dx\,dy + (1-x^2)dy^2}{1-x^2-y^2} \qquad (7)$$

which was examined in some detail in Exercises 20, 21 of Chapter 4.

If Polaris were to turn his telescope upon any other surface, his blindness to the topography of the surface would make him think he is observing other inhabitants of the xy plane whose sense of distance is governed by yet another Riemann metric. The Riemann metric associated with an arbitrary surface is quite easily derived. If the surface is the graph of the function $z = f(x, y)$, then, because of formula (3) the Riemann metric it induces on the xy plane is

$$\left[1 + \left(\frac{\partial z}{\partial x}\right)^2\right] dx^2 + 2 \frac{\partial z}{\partial x} \frac{\partial z}{\partial y} dx\,dy + \left[1 + \left(\frac{\partial z}{\partial y}\right)^2\right] dy^2 \qquad (8)$$

It is easily seen that the constraints that E, F, and $EG - F^2$ all be positive hold for every metric that is obtained in this way.

Example 12.8. The upper half of the unit sphere has equation

$$z = \sqrt{1 - x^2 - y^2}.$$

Hence

$$\frac{\partial z}{\partial x} = \frac{-x}{\sqrt{1-x^2-y^2}} \quad \text{and} \quad \frac{\partial z}{\partial y} = \frac{-y}{\sqrt{1-x^2-y^2}}$$

from which the metric of (7) above is easily derived.

Let us call all metrics of the type exhibited in (8) *Polaris* metrics. In other words, a Polaris metric is one which is induced on a region of the plane by a surface that lies above it. It follows immediately from the definition that every surface $z = f(x, y)$ has a Polaris metric associated with it, namely, the one given by (8). This brings up the natural question of whether all Riemann metrics are in fact Polaris metrics. The answer is negative. For example, Hilbert proved a theorem that implies, amongst other consequences, that the Poincaré metric is not a Polaris metric.

One of Gauss's deepest and most surprising observations in his investigation of curved surfaces is that the curvature of a surface can be expressed in terms of its Polaris metric. The reason this is surprising is that in focusing on the Polaris metric we have chosen to ignore the shape of the surface and are instead concentrating on the indirect manifestation of this shape as a distortion of distances. It is clear that this process entails a considerable loss of information about the surface. Nevertheless, it follows from Gauss's Theorem below that the Polaris metric contains enough information for us to be able to recover the curvature of the surface from it. The proof of this theorem lies beyong the scope of this text and can be found in Gauss's monograph or in Millman and Parker's book.

Theorem 12.9 (Gauss). *If the surface with equation $z = f(x, y)$ has the Polaris metric $Edx^2 + 2Fdxdy + Gdy^2$ and Gaussian curvature K, then*

$$4(EG - F^2)^2 K = E\left[\frac{\partial E}{\partial y}\frac{\partial G}{\partial y} - 2\frac{\partial F}{\partial x}\frac{\partial G}{\partial y} + \left(\frac{\partial G}{\partial x}\right)^2\right]$$

$$+ F\left[\frac{\partial E}{\partial x}\frac{\partial G}{\partial y} - \frac{\partial E}{\partial y}\frac{\partial G}{\partial x} - 2\frac{\partial E}{\partial y}\frac{\partial F}{\partial y} + 4\frac{\partial F}{\partial x}\frac{\partial F}{\partial y} - 2\frac{\partial F}{\partial x}\frac{\partial G}{\partial x}\right]$$

$$+ G\left[\frac{\partial E}{\partial x}\frac{\partial G}{\partial x} - 2\frac{\partial E}{\partial x}\frac{\partial F}{\partial y} + \left(\frac{\partial E}{\partial y}\right)^2\right]$$

$$- 2(EG - F^2)\left[\frac{\partial^2 E}{\partial y^2} - 2\frac{\partial^2 F}{\partial x \partial y} + \frac{\partial^2 G}{\partial x^2}\right].$$

□

Example 12.10. Consider the surface whose equation is $z = x^2 - y^2$. For this surface

$$\frac{\partial z}{\partial x} = 2x, \quad \frac{\partial z}{\partial y} = -2y, \quad \frac{\partial^2 z}{\partial x^2} = 2, \quad \frac{\partial^2 z}{\partial x \partial y} = 0, \quad \frac{\partial^2 z}{\partial y^2} = -2.$$

Consequently, by (1) above, the Gaussian curvature of this surface is given by

$$K = \frac{-4}{(1 + 4x^2 + 4y^2)^2}.$$

On the other hand, according to (8),

$$E = 1 + 4x^2 \qquad F = -4xy \qquad G = 1 + 4y^2$$

$$\frac{\partial E}{\partial x} = 8x \qquad \frac{\partial F}{\partial x} = -4y \qquad \frac{\partial G}{\partial x} = 0$$

$$\frac{\partial E}{\partial y} = 0 \qquad \frac{\partial F}{\partial y} = -4x \qquad \frac{\partial G}{\partial y} = 8y$$

$$\frac{\partial^2 E}{\partial y^2} = 0 \qquad \frac{\partial^2 F}{\partial x \partial y} = -4 \qquad \frac{\partial^2 G}{\partial x^2} = 0.$$

Hence, by Theorem 12.9,

$$4[(1+4x^2)(1+4y^2) - (-4xy)^2]K = (1+4x^2)[0 - 2(-4y)(8y) + 0]$$
$$+ (-4xy)[(8x)(8y) - 0 - 0 + 4(-4y)(-4x) - 0]$$
$$+ (1+4y^2)[0 - 2(8x)(-4x) + 0]$$
$$- 2(1+4x^2+4y^2)[0 - 2(-4) - 0]$$
$$= (1+4x^2)(64y^2) - (4xy)(128xy) + (1+4y^2)(64x^2)$$
$$- 16(1+4x^2+4y^2) = -16,$$

which yields the same value for K as was obtained above.

It is clear that the new expression for the Gaussian curvature K in terms of the coefficients E, F, and G of the Polaris metric is much more cumbersome than equation (1) above. However, it has the advantage that it is also applicable to non-Polaris metrics, i.e., Riemann metrics that are not induced by a surface lying over the xy plane. Let us now extend the notion of Gaussian curvature to an arbitrary Riemann metric. I.e., the *curvature* K of an arbitrary Riemann metric $E dx^2 + 2F dx dy + G dy^2$ is defined by the equation

$$4(EG - F^2)^2 K = E\left[\frac{\partial E}{\partial y}\frac{\partial G}{\partial y} - 2\frac{\partial F}{\partial x}\frac{\partial G}{\partial y} + \left(\frac{\partial G}{\partial x}\right)^2\right]$$
$$+ F\left[\frac{\partial E}{\partial x}\frac{\partial G}{\partial y} - \frac{\partial E}{\partial y}\frac{\partial G}{\partial x} - 2\frac{\partial E}{\partial y}\frac{\partial F}{\partial y} + 4\frac{\partial F}{\partial x}\frac{\partial F}{\partial y} - 2\frac{\partial F}{\partial x}\frac{\partial G}{\partial x}\right]$$
$$+ G\left[\frac{\partial E}{\partial x}\frac{\partial G}{\partial x} - 2\frac{\partial E}{\partial x}\frac{\partial F}{\partial y} + \left(\frac{\partial E}{\partial y}\right)^2\right] - 2(EG - F^2)\left[\frac{\partial^2 E}{\partial y^2} - 2\frac{\partial^2 F}{\partial x \partial y} + \frac{\partial^2 G}{\partial x^2}\right]$$

Example 12.11. Consider the Poincaré metric

$$\frac{dx^2 + dy^2}{y^2}.$$

Here,

$$E = \frac{1}{y^2} \qquad F = 0 \qquad G = \frac{1}{y^2}$$

$$\frac{\partial E}{\partial x} = 0 \qquad \frac{\partial F}{\partial x} = 0 \qquad \frac{\partial G}{\partial x} = 0$$

$$\frac{\partial E}{\partial y} = \frac{-2}{y^3} \qquad \frac{\partial F}{\partial y} = 0 \qquad \frac{\partial G}{\partial y} = \frac{-2}{y^3}$$

$$\frac{\partial^2 E}{\partial y^2} = \frac{6}{y^4} \qquad \frac{\partial^2 F}{\partial x \partial y} = 0 \qquad \frac{\partial^2 G}{\partial x^2} = 0.$$

So
$$4\left(\frac{1}{y^4}-0\right)^2 K = \frac{1}{y^2}\left(\frac{4}{y^6}\right) + \frac{1}{y^2}\left(\frac{4}{y^6}\right) - 2\frac{1}{y^4}\left(\frac{6}{y^4}\right)$$
or
$$\frac{4K}{y^8} = \frac{-4}{y^8},$$
from which it follows that for the Poincaré metric the constant curvature $K=-1$.

The surprise towards which we are leading is that the curvature of an arbitrary Riemann metric is also very meaningful, even though it does not describe the curvature of a surface. It makes possible a theorem which holds for arbitrary Riemann metrics and which is a natural generalization of Theorem 7.1 according to which the area of a hyperbolic triangle is equal to its defect. In order to state this new version of Gauss's Theorem about geodesic triangles, we need to discuss the notion of area relative to an arbitrary Riemann metric.

For an arbitrary Polaris metric
$$\left[1+\left(\frac{\partial z}{\partial x}\right)^2\right]dx^2 + 2\frac{\partial z}{\partial x}\frac{\partial z}{\partial y}dxdy + \left[1+\left(\frac{\partial z}{\partial y}\right)^2\right]dy^2 \tag{9}$$
associated with a surface $\mathbb{F}$, the area it induces on a region R in its domain should, of course, coincide with the area on the portion Φ of the surface $\mathbb{F}$ that lies directly above R. Thus, it should be
$$\iint_R \sqrt{1+\left(\frac{\partial z}{\partial x}\right)^2 + \left(\frac{\partial z}{\partial y}\right)^2}\,dxdy \tag{10}$$
So we must find a way to generalize this expression to an arbitrary Riemann metric
$$Edx^2 + 2Fdxdy + Gdy^2.$$
Note that the integrand of (10) can be expressed in terms of the coefficients of the Polaris metric (9) as follows
$$1+\left(\frac{\partial z}{\partial x}\right)^2+\left(\frac{\partial z}{\partial y}\right)^2$$
$$=\left[1+\left(\frac{\partial z}{\partial x}\right)^2\right]\left[1+\left(\frac{\partial z}{\partial y}\right)^2\right] - \left[\frac{\partial z}{\partial x}\frac{\partial z}{\partial y}\right]^2$$
$$= EG - F^2,$$

an expression that we have encountered several times above. In particular, we know that the same expression must be positive for every Riemann metric. Hence it makes sense to define the *area of an arbitrary region R in the domain of an arbitrary Riemann metric* $Edx^2 + 2Fdxdy + Gdy^2$ as the integral

$$\iint_R \sqrt{EG - F^2}\,dxdy \tag{11}$$

This definition is not very well motivated. Nevertheless it has the advantages that it coincides with the notion of areas associated with Polaris metrics, is well defined and positive over any two dimensional region of the domain of the metric, and is consistent with the requirement that the area of the whole should equal to the area of its parts. To this we add the observation that this definition of area also agrees with the well justified notion of hyperbolic area introduced in Chapter 7. To see this note that for the Poincaré metric

$$\frac{dx^2 + dy^2}{y^2}$$

we have

$$E = G = \frac{1}{y^2} \quad \text{and} \quad F = 0,$$

and hence

$$\iint_R \sqrt{EG - F^2}\,dxdy = \iint_R \frac{dxdy}{y^2}$$

which is indeed the hyperbolic area of R.

Example 12.12. Compute the area of the square $2 \leq x, y \leq 3$ relative to the metric $ydx^2 + 2xdxdy + x^2dy^2$ of Example 4.9.
Since $EG - F^2 = x^2y - x^2 = x^2(y-1)$, this area is

$$\int_2^3 \int_2^3 x\sqrt{y-1}\,dxdy = \frac{5}{3}[2\sqrt{2} - 1].$$

If we redefine $dA = \sqrt{EG - F^2}\,dxdy$ then we are finally ready to restate Gauss's Theorem 12.1 for arbitrary Riemann metrics.

Theorem 12.13. *Let K be the curvature of the Riemann metric $Edx^2 + 2Fdxdy + Gdy^2$, and let $\triangle$ be a geodesic triangle relative to this metric with angles α, β, γ. Then*

$$\iint_\triangle K dA = \alpha + \beta + \gamma - \pi.$$

□

To see that this is indeed a generalization of Theorem 7.1, recall that the curvature K of the Poincaré metric was shown to be a constant of -1, and hence for any hyperbolic triangle $\triangle$

$$\pi - (\alpha + \beta + \gamma) = -\iint_\triangle K dA = -\iint_\triangle (-1) dA$$
$$= \iint_\triangle \frac{dxdy}{y^2} = ha(\triangle).$$

Gauss's Theorem has been further generalized in many directions. One can first restate it, without much difficulty, for geodesic polygons with an arbitrary number of sides. It can also be generalized to regions whose boundary consists of curves that are not necessarily geodesic segments. In this form it is known as the Gauss Bonnet Theorem and constitutes the cornerstone of differential geometry. To demonstrate the significance of this theorem we shall show how it can be used to sometimes resolve the question of whether two given points can lie on several geodesics of a given Riemann metric.

Example 12.14. We shall show that the Riemann metric

$$\frac{dx^2 + dy^2}{y}$$

on the upper half-plane is such that any two of its geodesics intersect in at most one point. For this geodesic $E = G = 1/y$ and $F = 0$. An easy calculation then shows that

$$K = -\frac{1}{2y} < 0.$$

Hence if $\triangle$ is any geodesic triangle in the geometry defined by this metric

$$\iint_\triangle K dA < 0.$$

If α, β, γ are the angles of $\triangle$, then, by Theorem 12.13,

$$\alpha + \beta + \gamma = \pi + \iint_\triangle K dA < \pi.$$

Suppose now that some two geodesics do interesect in two points D and E, and let F be any point on one of the geodesics between D and E. Then the points D, E, F form a geodesic triangle with angles $\delta \geq 0$, $\epsilon \geq 0$, and π at D, E, F respectively. Since the sum of these angles is at least π we have a contradiction. Hence, any two geodesics can intersect in at most one point.

12.5 EXERCISES

1. Use either a geometrical argument, or Gauss's formula (5) for K to show that the total curvature of a region Φ on a sphere of radius r is
$$\frac{\text{area}(\Phi)}{r^2}.$$

2. Use Gauss's formula of Theorem 12.7 to show that the plane has Gaussian curvature 0 at every point.
3. Find the Gaussian curvature of the surface $z = x^2 + y^2$ at every one of its points.
4. Find the Gaussian curvature of the surface $z = 1/\sqrt{x^2 + y^2}$ at every one of its points. Use this expression to estimate the value of the Gaussian curvature when z is very small and when z is very large.
5. Find the length of the curve γ that lies on the graph of $z = x^2 - y^2$ directly over the line segment that joins the origin to the point $(2, 1, 0)$.
6. Find the area of the portion of the graph of $z = 3 + x^2 - y^2$ that lies above the unit disk in the xy plane.
7. Suppose the portion of the graph of $z = 1 - x^2 - y^2$ that lies over the unit disk in the xy plane has density $\delta(x, y) = \sqrt{x^2 + y^2}$. Find its total mass.
8. Let Φ be the portion of the graph of $z = x^2 + y^2$ that lies over the unit disk in the xy plane. Evaluate the following integrals.

   a) $\int_\Phi \int x \, dA$ b) $\int_\Phi \int xy \, dA.$

9. What is the Polaris metric of the surfaces that have the following equations?

   a) $z = x + y$ b) $z = \dfrac{1}{xy}.$

10. Show that the Riemann metric
$$\frac{dx^2 + dy^2}{[1 + \frac{\alpha}{4}(x^2 + y^2)]^2}$$
has constant Gaussian curvature α.
11. Find the curvature of the Riemann metric
$$\frac{dx^2 + dy^2}{y^r}$$
for arbitrary non zero r.
12. Compute the area of the square with vertices $(1, 1)$, $(1, 2)$, $(2, 2)$, and $(2, 1)$ relative to the metric of Exercise 11. Discuss what happens as r approaches either zero or infinity.
13. Discuss the question of whether the geodesics of the metrics of Exercise 11 can intersect in more than one point.

13

The Cross Ratio and the Unit Disk Model

13.1 INTRODUCTION

There is another model of hyperbolic geometry that was developed nearly simultaneously with the Poincaré upper half-plane. Since this model is obtained by imposing a Riemann metric on the interior of the unit circle, it is called the *unit disk model*. In some sense that cannot be clarified here this model actually predates Poincaré's construction, although, as we shall see, they are in fact logically equivalent. This equivalence notwithstanding, both models have been kept alive in the literature since each has its own advantages. For example, the derivation of the geodesics is more easily accomplished in the half-plane model, whereas the proof that all hyperbolic circles are also Euclidean circles is almost immediate in the unit disk model.

Loosely speaking, the hyperbolic geometry of the unit disk is constructed by squeezing the Poincaré half-plane into the disk's interior. Somewhat more formally, a transformation U will be described which maps the upper half-plane onto the interior of the unit disk. The length of any curve γ inside the unit disk is then defined as the hyperbolic length of $U^{-1}(\gamma)$ in the half-plane, and the measure of any angle α can be similarly defined as the hyperbolic (and Euclidean) measure of $U^{-1}(\alpha)$. Fortunately, the transformation U turns out to be conformal in the sense that the measure of $U^{-1}(\alpha)$ equals the Euclidean measure of α. This, of course, greatly simplifies matters. It also explains why it is necessary at this point to digress into the subject of conformal transformations.

13.2 CONFORMAL TRANSFORMATIONS

We remind the reader that transformations that preserve the measures of angles are said to be *conformal*. It is clear that every rigid motion must preserve angles and is hence conformal. There are, however, conformal transformations which fail to be rigid. Such, for instance, are the inversions. It will be shown in this section that there is indeed a wealth of transformations of the complex plane which can be expressed in terms of the four elementary arithmetical operations and which are in fact conformal. One of these transformations will be used in the following section to transfer the hyperbolic geometry of the half-plane onto the unit disk.

We already saw in Theorem 9.7 that every transformation of the type

$$T(z) = e^{i\alpha}z + c$$

is a Euclidean rigid motion and hence must be conformal. We also know from section 9.2 that for any real $k \neq 0$ the transformation

$$T(z) = \frac{k^2}{\bar{z}}$$

is an inversion, and hence, by Theorem 3.4 it too is conformal.

Theorem 13.1. *Every Moebius transformation is conformal.*

PROOF: Let $T(z)$ be the general Moebius transformation

$$\frac{az + b}{cz + d}$$

where a, b, c, d are arbitrary *complex* numbers. Then $T(z)$ can be written in the form

$$-\left[\frac{(\bar{a}\bar{d} - \bar{b}\bar{c})/\bar{c}^2}{\bar{z} - (-\bar{d}/\bar{c})} + \left(-\frac{\bar{d}}{\bar{c}}\right)\right] + \frac{a - d}{c}$$

that is of course very reminiscent of the device used in the proof of Theorem 9.12. Thus, $T(z)$ is the composition of several functions each of which has one of the following two forms

$$f(z) = kz + f, \quad (k, f \text{ complex})$$

$$g(z) = \frac{1}{\bar{z}}$$

$$h(z) = -\bar{z}.$$

The transformations $g(z)$ and $h(z)$, being an inversion and a Euclidean reflection respectively, are already known to be conformal. If $k = re^{i\theta}$ then the transformation $f(z)$ can be further factored into the functions

$$f_1(z) = e^{i\theta}z, \quad f_2(z) = rz, \quad \text{and} \quad f_3(z) = z + f.$$

The first and the third, being a Euclidean rotation and a Euclidean translation, respectively, are clearly conformal. The second function is the composition of the two inversions

$$I(z) = \frac{(\sqrt{r})^2}{\bar{z}} \quad \text{and} \quad J(z) = \frac{1}{\bar{z}}$$

and so it itself must also be conformal. Thus, every Moebius transformation is the composition of conformal maps and so it too must be conformal.

q.e.d.

We mention in passing that the converse of this theorem is also true. In other words, every conformal (and orientation preserving) transformation of the plane is necessarily a Moebius transformation.

13.3 THE CROSS RATIO

Because of the special role that circles (and semicircles) play in these considerations, it is necessary to digress into a description of circles by means of complex numbers. Like most other mathematical digressions, this one too must begin with its own definitions. Given four distinct points z_1, z_2, z_3, z_4, their cross ratio (z_1, z_2, z_3, z_4) is defined as the number

$$\frac{\dfrac{z_1 - z_3}{z_1 - z_2}}{\dfrac{z_4 - z_3}{z_4 - z_2}}.$$

The next lemma gives a geometrical interpretation of the numerator and the denominator of the cross ratio separately.

Lemma 13.2. *If z_1, z_2, z_3 are any three distinct complex numbers then*

$$\angle z_1 z_3 z_2 = \arg\left(\frac{z_2 - z_3}{z_1 - z_3}\right).$$

PROOF: Set $w_i = z_i - z_3$ for $i = 1, 2, 3$. Since the transformation $w = z - z_3$ is a Euclidean rigid motion, it follows that

$$\angle z_1 z_3 z_2 = \angle w_1 w_3 w_2.$$

Moreover, since $w_3 = 0$, it follows from Proposition 9.2 that

$$\angle w_1 w_3 w_2 = \arg\left(\frac{w_2}{w_1}\right) = \arg\left(\frac{z_2 - z_3}{z_1 - z_3}\right),$$

from which the lemma follows immediately.

q.e.d

A set of points is said to be *cocyclic* if there is either a circle or a Euclidean straight line that contains all of them. It is of course well known that any three points are cocyclic. The cross ratio provides us with a very simple calculational device for determining whether any given four points are cocyclic.

Proposition 13.3. *The four points z_1, z_2, z_3, z_4 are cocyclic if and only if their cross ratio (z_1, z_2, z_3, z_4) is real.*

PROOF: Suppose the four given points lie on a circle (or a straight line) in the given cyclic order. It then follows from Proposition 27 of Book III of Euclid's Elements that

$$\angle z_2 z_1 z_3 = \angle z_2 z_4 z_3.$$

We already know from Lemma 13.2 that

$$\angle z_2 z_1 z_3 = \arg\left(\frac{z_3 - z_1}{z_2 - z_1}\right) \quad \text{and} \quad \angle z_2 z_4 z_3 = \arg\left(\frac{z_3 - z_4}{z_2 - z_4}\right).$$

Hence,

$$\arg\left(\frac{z_3 - z_1}{z_2 - z_1}\right) = \arg\left(\frac{z_3 - z_4}{z_2 - z_4}\right)$$

If these ratios are written in polar form as

$$\frac{z_3 - z_1}{z_2 - z_1} = re^{i\theta} \quad \text{and} \quad \frac{z_3 - z_4}{z_2 - z_4} = Re^{i\theta}$$

then the cross ratio

$$\frac{\dfrac{z_1 - z_3}{z_1 - z_2}}{\dfrac{z_4 - z_3}{z_4 - z_2}} = \frac{\dfrac{z_3 - z_1}{z_2 - z_1}}{\dfrac{z_3 - z_4}{z_2 - z_4}} = \frac{r}{R}$$

and is real. Thus, this direction of the proof is complete when the points occur on the circle (or straight line) in the given order. The validity of the theorem when the points occur in an arbitrary order follows from Exercise 1.

The proof of the converse is left as Exercise 17.

q.e.d.

There is a very strong and simple connection between Moebius transformations and the cross ratio.

Proposition 13.4. *If T is a Moebius transformation and z_1, z_2, z_3, z_4 are any four points, then*

$$(T(z_1), T(z_2), T(z_3), T(z_4)) = (z_1, z_2, z_3, z_4).$$

In other words, Moebius transformations preserve the cross ratio.

PROOF: Suppose

$$T(z) = \frac{az + b}{cz + d}$$

and set

$$w_i = T(z_i) = \frac{az_i + b}{cz_i + d} \quad \text{for} \quad i = 1, 2, 3, 4.$$

The proof is then completed by verifying that

$$\frac{\dfrac{w_1 - w_3}{w_1 - w_2}}{\dfrac{w_4 - w_3}{w_4 - w_2}} = \frac{\dfrac{z_1 - z_3}{z_1 - z_2}}{\dfrac{z_4 - z_3}{z_4 - z_2}}.$$

The details of this verification are left as Exercise 18. □

Corollary 13.5. *If m is either a Euclidean straight line or a circle and T is any Moebius transformation, then T(m) is either a Euclidean straight line or a circle.*

PROOF: Let $T(z_1)$, $T(z_2)$, $T(z_3)$, $T(z_4)$ be any four points on $T(m)$. By the previous corollary,

$$(T(z_1), T(z_2), T(z_3), T(z_4)) = (z_1, z_2, z_3, z_4),$$

and by Proposition 13.3, the second of these cross ratios is real. Consequently $(T(z_1), T(z_2), T(z_3), T(z_4))$ is also real and so the four points $T(z_1)$, $T(z_2)$, $T(z_3)$, $T(z_4)$ are cocyclic. Thus $T(m)$ is either a Euclidean straight line or a circle.

q.e.d.

The cross ratio provides us with a new description of circles. Given any three points z_1, z_2, z_3, the circle (or straight line) containing them consists of the set of all points z such that the cross ratio (z_1, z_2, z_3, z) is real. In other words, it is the locus of all points z such that

$$\lambda \frac{z_3 - z_1}{z_2 - z_1} = \frac{z_3 - z}{z_2 - z} \quad \text{for some real } \lambda.$$

Example 13.6. According to the above equation, the circle containing the points $0, 1, i$ is given by

$$\lambda \frac{i - 0}{1 - 0} = \frac{i - z}{1 - z}.$$

When this is solved for z we get

$$z = \frac{1}{2}\left[i + 1 - \frac{(i-1)(i\lambda + 1)}{i\lambda - 1}\right].$$

Notice that $\lambda = -1$ yields the the point $1 + i$, which is clearly on this circle.

Having read through this example, it should be clear to the reader that this way of describing circles is of more theoretical than practical importance. If a specific circle is to be described, the cross ratio generally yields a very awkward description of it.

The cross ratio has yet another, much deeper, connection with the Poincaré half-plane. The following proposition indicates that the geometry of the half-plane can be developed by using the cross ratio rather than the Poincaré metric of Chapter 4 as a starting point.

Proposition 13.7. *Let z and w be two points of the upper half-plane that are joined by a bowed geodesic with endpoints z^* and w^* on the x axis. If these points are labelled so that their clockwise order on the geodesic is z^*, z, w, w^* then*

$$h(z, w) = \ln(z^*, z, w, w^*).$$

$\square$

The proof of this proposition is relegated to Exercise 19. In its stead we provide an example.

Example 13.8. Let $z = -1 + i$ and $w = 1 + i$. By Proposition 4.1

$$h(z, w) = \ln \frac{\csc \frac{3\pi}{4} - \cot \frac{3\pi}{4}}{\csc \frac{\pi}{4} - \cot \frac{\pi}{4}} = \ln \frac{\sqrt{2} + 1}{\sqrt{2} - 1}.$$

On the other hand, $z^* = -\sqrt{2}$ and $w^* = \sqrt{2}$ and so

$$\ln(-\sqrt{2}, -1 + i,\ 1 + i,\ \sqrt{2}) = \ln \frac{(\sqrt{2} + 1 + i)(\sqrt{2} + 1 - i)}{(\sqrt{2} - 1 + i)(\sqrt{2} - 1 - i)}$$

$$= \ln \frac{\sqrt{2} + 1}{\sqrt{2} - 1}.$$

13.4 THE UNIT DISK MODEL AND ITS FLOW DIAGRAMS

Consider the Moebius transformation

$$U(z) = \frac{iz + 1}{z + i}.$$

Clearly $U(0) = -i$, $U(1) = 1$, and $U(-1) = -1$. Since $0, 1, -1$ are all on the x-axis and $-i, 1, -1$ are all on the unit circle, it follows from Corollary 13.5 that U transforms the x-axis into the unit circle. The continuity of U then implies that U transforms the upper half-plane into either the interior of this disk or its exterior. The fact that $U(i) = 0$ then shows that the former is the case, namely, U transforms the upper half-plane into the interior of the unit disk. The reader who is dissatisfied with this argument can replace it with Exercise 5 wherein it is asserted that whenever z is in the upper half-plane, $U(z)$ has modulus less than 1 and so it lies in the interior of the unit disk.

The Moebius transformation

$$V(z) = \frac{iz - 1}{-z + i}$$

is the inverse of U (Exercise 6), in other words, $U \circ V = V \circ U = Id$. This means that U and V establish a one-to-one correspondence between the points of the upper half-plane and the points of the interior of the unit disk. Moreover, since both U and V are Moebius transformations, it follows from Corollary 13.5 that U transforms the geodesic segments of the half-plane into circular arcs or straight line segments inside the unit disk. In fact, we can be even more specific. The geodesics of the half-plane are orthogonal to the x-axis, and so, since U is conformal, these geodesics are mapped into diameters and circular arcs that are orthogonal to the unit circle (Fig. 13.1).

The hyperbolic distance on the half-plane can now be used to define a new distance function on the unit disk as follows. If P and Q are any two points of the interior of the unit disk set

$$j(P, Q) = h(V(P), V(Q)).$$

Although we chose not to include the details here, it is easily verified that the distance function $j(P, Q)$ imposes a hyperbolic geometry on the unit disk. Its

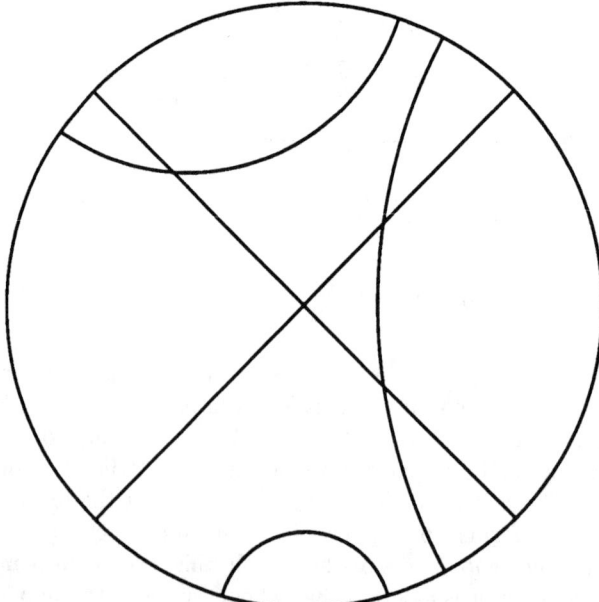

FIGURE 13.1. Some geodesics of the unit disk.

geodesics are simply the images, under the transformation U, of the geodesics of the half-plane. This hyperbolic geometry of the unit disk is in fact identical with that of the half-plane in the sense that if the inhabitants of one geometry were to communicate with those of the other by telephone, they would not be able to tell their worlds apart. It is only to us that their habitats look different. The rigid motions of the unit disk are readily obtained. In fact, if f is any hyperbolic rigid motion of the half-plane, then

$$f^* = U \circ f \circ V \tag{1}$$

is a hyperbolic rigid motion of the unit disk. For, if P and Q are any two point of the unit disk, then

$$\begin{aligned} j(f^*(P), f^*(Q)) &= j(U \circ f \circ V(P), \; U \circ f \circ V(Q)) \\ &= h(V(U \circ f \circ V(P)), \; V(U \circ f \circ V(Q))) \\ &= h(f \circ V(P), \; f \circ V(Q)) = h(V(P), \; V(Q)) = j(P, Q). \end{aligned}$$

The following proposition gives an even more explicit description of the rigid motions of the unit disk.

Proposition 13.9. *The rigid motions of the unit disk are the transformations of the forms*

i) $\quad g(z) = \dfrac{az + \bar{c}}{cz + \bar{a}} \qquad\qquad$ ii) $\quad g(z) = \dfrac{-a\bar{z} + \bar{c}}{-c\bar{z} + \bar{a}}$

where a and c are any two complex numbers such that $|a| > |c|$.

PROOF: We shall content ourselves with showing that the Moebius transformations of type i are in fact rigid motions of the unit disk model and that conversely, every Moebius rigid motion of the unit disk model is indeed of type i.

Let

$$f(z) = \frac{\alpha z + \beta}{\gamma z + \delta}$$

be any Moebius rigid motion of the upper half-plane (i.e., $\alpha, \beta, \gamma, \delta$ are real and $\alpha\delta - \beta\gamma > 0$). It induces on the unit disk model the Moebius transformation

$$\begin{aligned} f^*(z) = U(z) \circ f(z) \circ V(z) &= \frac{iz + 1}{z + i} \circ \frac{\alpha z + \beta}{\gamma z + \delta} \circ \frac{iz - 1}{-z + i} \\ &= \frac{az + \bar{c}}{cz + \bar{a}} \end{aligned}$$

where, by two applications of Exercise 9.20,

$$a = -(\alpha + \delta) + i(\gamma - \beta) \quad \text{and} \quad c = -(\beta + \gamma) + i(\alpha - \delta).$$

Moreover,

$$\begin{aligned}|a|^2 - |c|^2 &= [(\alpha + \delta)^2 + (\gamma - \beta)^2] - [(\beta + \gamma)^2 + (\alpha - \delta)^2] \\ &= 4(\alpha\delta - \beta\gamma) > 0.\end{aligned}$$

Thus $|a| > |c|$ and so f^* does indeed have the required format.

Conversely, suppose

$$g(z) = \frac{az + \bar{c}}{cz + \bar{a}}$$

has the format of i. Then set

$$f(z) = V(z) \circ g(z) \circ U(z) = \frac{iz-1}{-z+i} \circ \frac{az+\bar{c}}{cz+\bar{a}} \circ \frac{iz+1}{z+i}$$

$$= \frac{\alpha z + \beta}{\gamma z + \delta}$$

where

$$\alpha = -(a + \bar{a}) - i(c - \bar{c})$$
$$\beta = -(c + \bar{c}) + i(a - \bar{a})$$
$$\gamma = -(c + \bar{c}) - i(a - \bar{a})$$
$$\delta = -(a + \bar{a}) + i(c - \bar{c}).$$

Since for any complex number w, the sum $w + \bar{w}$ is real and the difference $w - \bar{w}$ is purely imaginary it follows that the quantities $\alpha, \beta, \gamma, \delta$ are all real. Moreover,

$$\begin{aligned}\alpha\delta - \beta\gamma &= [(a + \bar{a})^2 + (c - \bar{c})^2] - [(c + \bar{c})^2 + (a - \bar{a})^2] \\ &= 4(a\bar{a} - c\bar{c}) = 4(|a|^2 - |c|^2) > 0.\end{aligned}$$

Thus $f(z)$ is indeed a Moebius rigid motion of the upper half-plane. Since

$$f^* = U \circ f \circ V = U \circ (V \circ g \circ U) \circ V = (U \circ V) \circ g \circ (U \circ V)$$
$$= g,$$

it follows that g does indeed arise from a Moebius rigid motion of the half-plane in the usual way.

q.e.d

Before going on to some examples, the reader must be warned about a subtle point that can be the source of some confusion. The transformation

$$f(z) = \frac{2z - i}{iz + 2}$$

is, by the preceding proposition, a Moebius rigid motion of the unit disk model. However,

$$\frac{2z - i}{iz + 2} = \frac{i(2z - i)}{i(iz + 2)} = \frac{2iz + 1}{-z + 2i}$$

and this last expression does *not* have the format specified by this proposition. Nevertheless

$$f(z) = \frac{2iz + 1}{-z + 2i}$$

is a bona fide rigid motion of the unit disk model; it just happens to be wearing a disguise.

Example 13.10. The rigid motion $f(z) = 2z$ of the half-plane was discussed in detail in Chapter 9. In accordance with (1) above, the corresponding rigid motion of the unit disk model is

$$f^* = \frac{iz + 1}{z + i} \circ \frac{2z + 0}{0z + 1} \circ \frac{iz - 1}{-z + i} = \frac{3z + i}{-iz + 3}.$$

The reader can easily verify that

$$f^*(1) = \frac{4 + 3i}{5} \quad \text{and} \quad f^*(-1) = \frac{-4 + 3i}{5}$$

both have modulus 1, and so remain on the unit circle, and $f^*(0) = \frac{i}{3}$ which is in the interior of the unit circle.

The rigid motions of the unit disk also have flow diagrams. It follows from Corollary 9.15 that if f is any hyperbolic rotation or translation of the half-plane, and if m is any of its flow lines, then $U(m)$ is a flow line of the unit disk rigid motion $f^* = U \circ f \circ V$. Similarly, if z is a fixed point of f, then $U(z)$ is a fixed point of f^*. Since U is conformal and maps Euclidean straight lines and

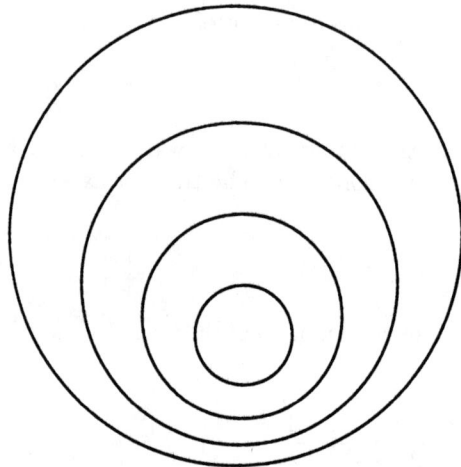

FIGURE 13.2.

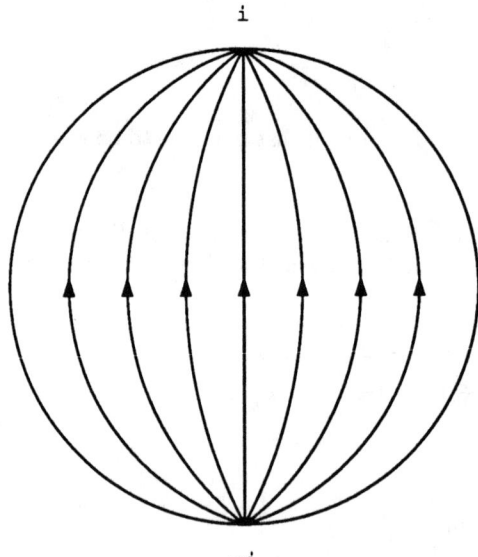

FIGURE 13.3.

circles into Euclidean straight lines and circles, it follows that the flow diagrams of Fig. 9.6, 9.8, 9.10 are transformed into the flow diagrams of Figs. 13.2, 13.3, 13.5, respectively. The general procedure for finding the flow diagram of a rigid motion of the unit disk model is the same as that for the half-plane. By Proposition 13.9 every Moebius rigid motion of the unit disk model has the format

$$g(z) = \frac{az + \bar{c}}{cz + \bar{a}} \quad \text{with} \quad |a| > |c|.$$

The fixed points of this Moebius transformation are the solutions of the equation

$$z = \frac{az + \bar{c}}{cz + \bar{a}}$$

or

$$cz^2 + (\bar{a} - a)z - \bar{c} = 0.$$

Now the product of the roots of this quadratic is $-\frac{\bar{c}}{c}$ which has modulus 1. Hence, by Proposition 9.2, the product of the moduli of the roots is 1. This means that either both of the roots lie on the unit circle, or else one is in its interior and the other is in its exterior. In the latter case the flow diagram consists of circles that are hyperbolically centered at the interior fixed point. In the first case the flow lines consist of circular arcs joining the two fixed points if they are distinct, or of circles tangent to the unit circle if there is only one (double) fixed point.

Example 13.11. Consider the transformation

$$f^* = \frac{3z + i}{-iz + 3}$$

which was found in Example 13.10 to be the rigid motion of the unit disk model that is induced by the rigid motion $f(z) = 2z$ of the half-plane. This $f(z)$ has as its flow diagram a set of Euclidean rays emanating from the origin (Fig. 9.6a). On the other hand, the fixed points of f^* are i and $-i$ and so f^* has the arcs of Fig. 13.3 as its flow diagram. The direction of the flow is determined by the fact that $f^*(0) = \frac{i}{3}$.

Example 13.12. Consider the transformation

$$f^*(z) = \frac{2z + 1}{z + 2}.$$

It has the format posited in Proposition 13.19 and so it yields a rigid

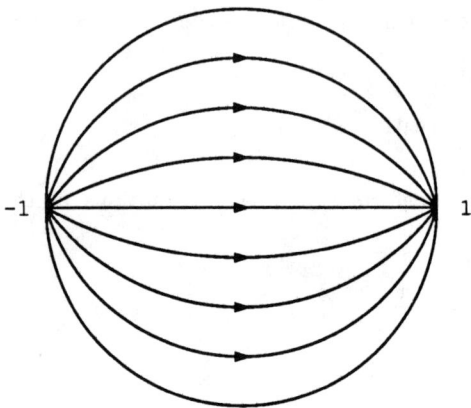

FIGURE 13.4.

transformation of the unit disk model. Since its fixed points are $z = 1, -1$, its flow diagram is given by Fig. 13.4.

Example 13.13. To find the flow diagram of the rigid motion of the unit disk model given by the Moebius transformation

$$T(z) = \frac{(2+i)z + (1+i)}{(1-i)z + (2-i)}$$

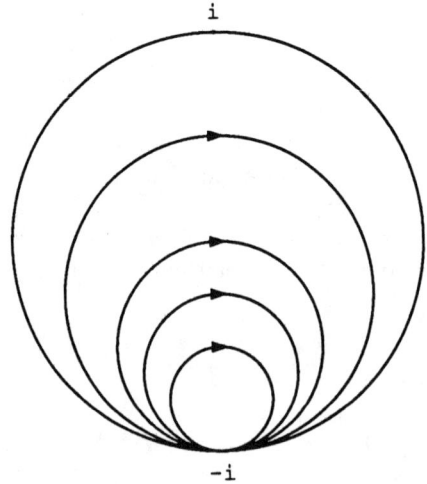

FIGURE 13.5.

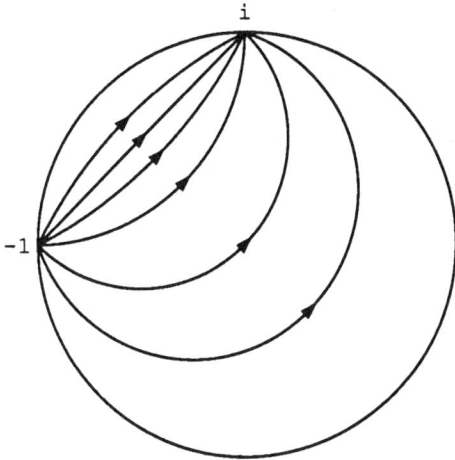

FIGURE 13.6.

we first find the fixed points of z. These are given as the solutions of

$$z = \frac{(2+i)z + (1+i)}{(1-i)z + (2-i)}$$

which is simplified to the quadratic equation

$$(1-i)z^2 - 2iz - (1+i) = 0$$

whose solutions are

$$\frac{2i \pm \sqrt{(-2i)^2 + 4(1-i)(1+i)}}{2(1-i)} = \frac{2i \pm 2}{2(1-i)}$$

which, in turn, are -1 and i. Fig. 13.6 depicts the flow whose direction is dictated by the fact that

$$T(0) = \frac{1+i}{2-i} = \frac{1+3i}{5}$$

13.5 EXPLICIT RIGID MOTIONS OF THE UNIT DISK MODEL

That the Euclidean plane is homogeneous in the sense that given any two points P and Q there is a rigid motion that carries P to Q, is so obvious a fact that it is hard to convince students that this statement calls for a proof. The analogous statement for the half-plane was implicitly proved when it was demonstrated in Chapter 5 that every two right angles are congruent. The following proposition asserts that the geometry of the unit disk model is also homogeneous in the sense that given any point in it, there is a Moebius rigid motion that carries that point to 0. Since such a transformation must map the unit circle into itself, this is by no means an obvious fact.

Proposition 13.14. *If b is any point in the interior of the unit disk, then the transformation*

$$T(z) = \frac{z - b}{\bar{b}z + 1}$$

is a translation of the unit disk model that carries b to 0.

PROOF: Since b is in the interior of the unit disk it follows that $|b| < 1$ and so, by Proposition 13.9, T is indeed a rigid motion of the unit disk model. The fact that $T(b) = 0$ says that T carries b to 0. Finally, to find the fixed points of T it is necessary to solve the equation

$$z = \frac{z - b}{-\bar{b}z + 1}.$$

This equation reduces to

$$z^2 = \frac{b}{\bar{b}}$$

whose solutions lie on the unit circle because

$$\left|\frac{b}{\bar{b}}\right| = \frac{|b|}{|b|} = 1.$$

Hence T has no fixed points inside the unit disk and so it constitutes a translation of the unit disk model.

q.e.d.

We already know that the function $R_{O,\theta}(z) = e^{i\theta}z$ is a Euclidean rotation about the origin. Because of the radial symmetry of the unit disk about the origin, the reader will probably not be surprised to learn that this transformation also happens to be a rigid motion of the unit disk model.

The reason for this is that

$$R_{O,\theta} = \frac{e^{i\theta/2}z + 0}{0z + e^{-i\theta/2}}$$

and since $e^{i\theta/2}$ and $e^{-i\theta/2}$ are conjugate complex numbers, the rigidity of $R_{O,\theta}$ follows from Proposition 13.9. This fact can be used to describe rotations of the unit disk model that are pivoted at arbitrary points. In fact, since

$$S(z) = \frac{z+b}{\bar{b}z+1}$$

is the inverse of the unit disk translation

$$T(z) = \frac{z-b}{-\bar{b}z+1}$$

that carries b to 0, it follows that

$$S \circ R_{O,\theta} \circ T(b) = S \circ R_{O,\theta}(0) = S(0) = b.$$

In other words, the composition $S \circ R_{O,\theta} \circ T$ has a fixed point at b, and so it is a hyperbolic rotation of the unit disk model pivoted at b with angle θ.

Example 13.15. The 90° counterclockwise rotation of the unit disk model about the point $z = \frac{1}{2}$ is given by

$$\frac{z+\frac{1}{2}}{\frac{1}{2}z+1} \circ \frac{e^{i\theta}z+0}{0z+1} \circ \frac{z-\frac{1}{2}}{-\frac{1}{2}z+1} = \frac{2z+1}{z+2} \circ \frac{iz+0}{0z+1} \circ \frac{2z-1}{-z+2}$$
$$= \frac{(4i-1)z + 2(1-i)}{2(i-1)z + (4-i)}.$$

Finally, we turn to the reflections of the unit disk model. These are of course the unit disk versions of the hyperbolic reflections of upper half-plane. Thus they are either Euclidean reflections in diameters of the unit disk or else they are inversions that fix every point of some geodesic. The Euclidean reflections were already demonstrated in Chapter 9 to have form $f(z) = e^{2i\theta}\bar{z}$ and so we turn to the inversions. Suppose the Euclidean center of the geodesic AB of Fig. 13.7 is $w = re^{i\theta}$. Let I_w denote the inversion that fixes every point of the geodesic AB. Since the geodesic AB is orthogonal to the unit circle, it follows that I_w transforms the unit disk onto itself and so it is indeed a reflection of the unit disk model. We obtain an analytic expression for this inversion by rotating this configuration about the origin till w falls on $R(r,0)$

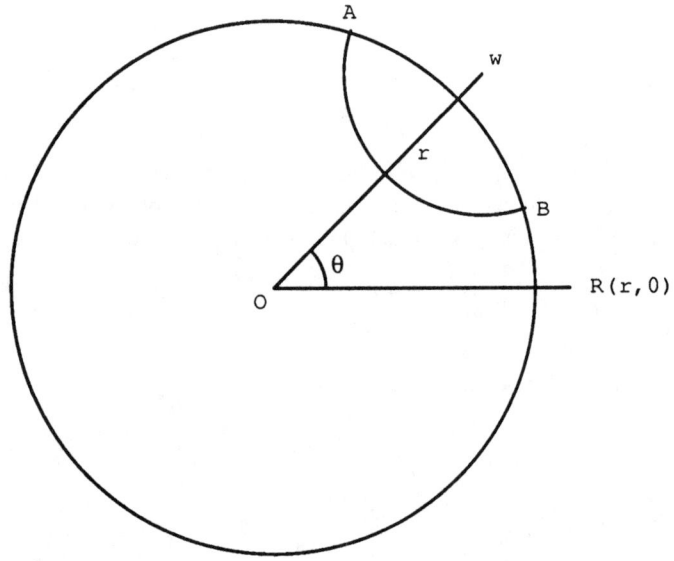

FIGURE 13.7.

on the positive x-axis, performing the inversion there, and then returning w to its original position. Since this is the rotation $R_{O,\theta}(z) = e^{i\theta}z$ and since AB has radius $\sqrt{r^2 - 1}$ so that

$$I_R(z) = \frac{r^2 - 1}{\bar{z} - r} + r = \frac{r\bar{z} - 1}{\bar{z} - r},$$

it follows that

$$I_w(z) = R_{O,\theta} \circ I_R \circ R_{O,-\theta}(z) = e^{i\theta} \cdot \overline{\frac{rze^{-i\theta} - 1}{ze^{-i\theta} - r}}$$

$$= \frac{1}{e^{-i\theta}} \cdot \frac{re^{i\theta}\bar{z} - 1}{\bar{z}e^{i\theta} - r} = \frac{re^{i\theta}\bar{z} - 1}{\bar{z} - re^{-i\theta}} = \frac{w(-\bar{z}) + 1}{(-\bar{z}) + \bar{w}}.$$

Note that I_w does have format ii of Proposition 13.9 since the required condition $|w| > 1$ is guaranteed by the fact that w is outside the unit circle.

Example 13.16. The hyperbolic reflection of the unit disk whose axis is the geodesic centered at $-1 + i$ has the analytic expression

$$\frac{(-1+i)(-\bar{z}) + 1}{(-\bar{z}) + (-1-i)}.$$

and it transforms the point $i/2$ to the point

$$\frac{(-1+i)(i/2)+1}{(i/2)+(-1-i)} = \frac{-1+3i}{5}.$$

13.6 THE RIEMANN METRIC OF THE UNIT DISK MODEL

It is, of course, possible to define this hyperbolic geometry on the unit disk directly, without recourse to the half-plane. We shall now derive the Riemann metric that endows the interior of the unit disk with the structure of hyperbolic geometry. Let $z = x + iy$ denote the general point in the interior of the unit disk and let $w = u + iv$ denote the corresponding point in the upper half-plane. Then

$$w = V(z) = \frac{iz-1}{-z+i},$$

or

$$u + iv = \frac{i(x+iy)-1}{-(x+iy)+i} = \frac{(y+1)-ix}{x+i(y-1)}$$

$$= \frac{2x + i(1-x^2-y^2)}{x^2+(y-1)^2}.$$

Hence

$$u = \frac{2x}{x^2+(y-1)^2}, \quad v = \frac{1-x^2-y^2}{x^2+(y-1)^2} \tag{2}$$

By the chain rule,

$$du = \frac{\partial u}{\partial x}dx + \frac{\partial u}{\partial y}dy, \quad dv = \frac{\partial v}{\partial x}dx + \frac{\partial v}{\partial y}dy.$$

(We remind the reader here that by the convention set in Chapter 4, we regard the differentials du, dv, dx, and dy as abbreviations for the derivatives du/dt, dv/dt, dx/dt, and dy/dt, where t is any unspecified parameter.)

A straightforward calculation yields the partial derivatives

$$\frac{\partial u}{\partial x} = \frac{\partial v}{\partial y} = \frac{-2x^2+2(y-1)^2}{[x^2+(y-1)^2]^2}$$

$$\frac{\partial u}{\partial y} = -\frac{\partial v}{\partial x} = \frac{-4x(y-1)}{[x^2+(y-1)^2]^2}.$$

Consequently

$$du^2 + dv^2 = \left(\left(\frac{\partial u}{\partial x}\right)^2 + \left(\frac{\partial v}{\partial x}\right)^2\right) dx^2 + 2\left(\frac{\partial u \partial u}{\partial x \partial y} + \frac{\partial v \partial v}{\partial x \partial y}\right) dx\, dy$$

$$+ \left(\left(\frac{\partial u}{\partial y}\right)^2 + \left(\frac{\partial v}{\partial y}\right)^2\right) dy^2$$

$$= \frac{4[x^2 - (y-1)^2]^2 + 16x^2(y-1)^2}{[x^2 + (y-1)^2]^4} [dx^2 + dy^2]$$

$$= \frac{4[x^2 + (y-1)^2]^2}{[x^2 + (y-1)^2]^4} [dx^2 + dy^2] = \frac{4[dx^2 + dy^2]}{[x^2 + (y-1)^2]^2}.$$

So if $\gamma(t) = [x(t), y(t)]$, $a \leq t \leq b$, is any parametrized curve in the unit disk, then, by definition, the length of γ in the unit disk model, equals the hyperbolic length of $\gamma^* = V(\gamma)$. The curve γ^* can be parametrized as $[u(t), v(t)]$, $a \leq t \leq b$ where u and v are given by (2) above, and so its hyperbolic length (relative to the Poincaré metric) is

$$\int_{\gamma^*} \frac{\sqrt{du^2 + dv^2}}{v} = \int_\gamma \frac{\frac{2\sqrt{dx^2 + dy^2}}{x^2 + (y-1)^2}}{\frac{1 - x^2 - y^2}{x^2 + (y-1)^2}} = \int_\gamma \frac{2\sqrt{dx^2 + dy^2}}{1 - x^2 - y^2}.$$

Hence the length of the arbitrary curve γ in the unit disk model is

$$\int_\gamma \frac{2\sqrt{dx^2 + dy^2}}{1 - x^2 - y^2}.$$

Thus, we have proved the following statement.

Proposition 13.17. *The Riemann metric that yields the hyperbolic geometry on the unit disk is*

$$\frac{4(dx^2 + dy^2)}{(1 - x^2 - y^2)^2}.$$

□

Starting with this metric one can develop the hyperbolic geometry of the unit disk in much the same way as was done for the upper half-plane in Chapters 4–8 of this book. The calculations are somewhat more complicated in the context of the unit disk. The main reason for this higher complexity is that the geodesics of the unit disk have their centers in inconvenient places, namely, at points that are in the exterior of the unit disk.

THE CROSS RATIO AND THE UNIT DISK MODEL 227

There is an alternate description of the hyperbolic distances on the unit disk that corresponds to Proposition 13.7 and whose validity follows from the fact that Moebius transformations preserve cross ratios (Proposition 13.4).

Proposition 13.18. *Let z and w be two points of the unit disk that are joined by a geodesic whose endpoints are z^* and w^* on the unit circle. If these points are labelled so that their clockwise order on the geodesic is z^*, z, w, z^*, then the unit disk distance between z and w is*

$$\ln\,(z^*, z, w, w^*).$$

$\square$

As a fitting conclusion for this section the author offers a new proof of one of the most surprising theorems of this branch of mathematics, namely the one

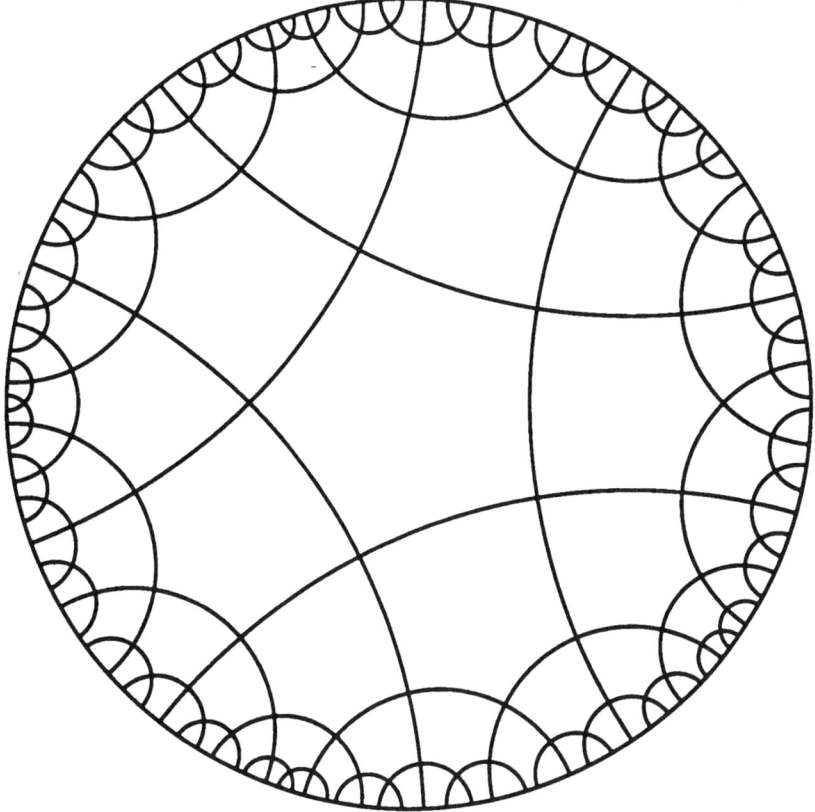

FIGURE 13.8.

FIGURE 13.9.
© 1960 M.C. Escher/Cordon Art-Baarn-Holland.

that states that every hyperbolic circle of the upper half-plane is also a Euclidean circle. Since all the hyperbolic circles of a given hyperbolic radius are hyperbolically congruent, it suffices to prove this fact for those circles that are hyperbolically centered at $z = i$. Let C_r be such a circle. Because of the way distances are defined in the unit disk model, the Moebius transformation

$$U(z) = \frac{iz + 1}{z + i}$$

transforms C_r into a hyperbolic circle D_r of the unit disk model, of the same radius, that is centered at $z = U(i) = 0$. Since the transformation $R_{O,\theta}$ is both a Euclidean and a hyperbolic rigid transformation, D_r is also a Euclidean circle. But the Moebius transformation U maps Euclidean circles into Euclidean circles, and so C_r is also a Euclidean circle.

13.7 REGULAR TESSELATIONS OF THE UNIT DISK MODEL

In Chapter 6 we constructed a tiling of the Poincaré half-plane with regular right angled pentagons. Since the conformal transformation $U(z)$ maps geodesics of the half-plane onto geodesics of the unit disk model, it follows that it also transforms this tesselation of the half-plane into a tesselation of the unit

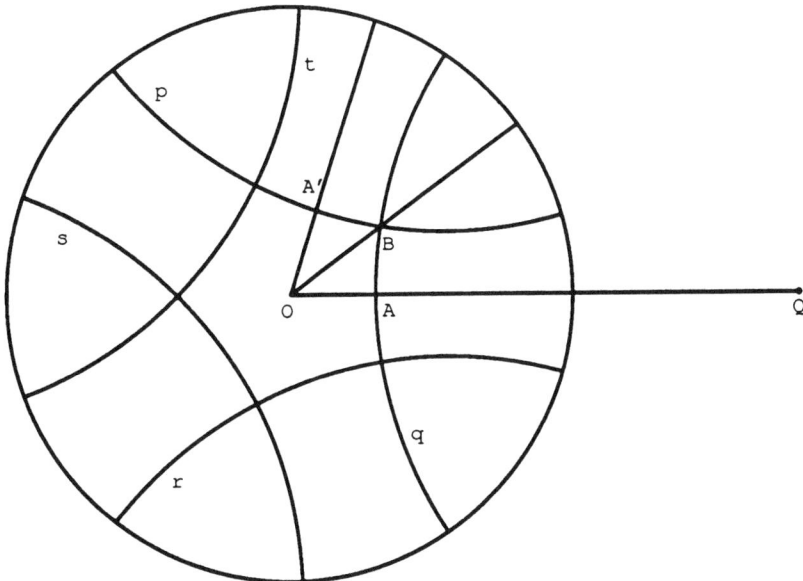

FIGURE 13.10.

disk with regular right angled pentagons. This tesselation consists of the arcs in the interior of the circle in Fig. 13.8. Interestingly enough, this configuration drew the interest of mathematicians even before its relationship to hyperbolic geometry was discovered by Poincaré. It was first constructed in 1872 by H.A. Schwarz in order to explain the behavior of certain solutions of Gauss's well known differential equation

$$\frac{d^2y}{dx^2} + \frac{\gamma - (\alpha + \beta + 1)x}{x(1-x)}\frac{dy}{dx} - \frac{\alpha\beta y}{x(1-x)} = 0.$$

We pause to dwell on this figure for several reasons. It is, of course, interesting in and of itself, and its symmetries as well as the aggregation of its arcs near the bounding circle (hyperbolic infinity) are visually fascinating. So much so that the artist M. Escher used its hexagonal analog as the framework for several of his well known prints, one of which is displayed in Fig. 13.9. The readers might be interested in a description of the nitty-gritty details involved in the construction of this figure. One approach, the classical one, is outlined H.S.M. Coxeter's article in the periodical *Leonardo*. It involves the very clever use of an auxiliary Euclideanly rectilineal configuration called *the scaffolding*. However, the proliferation of computers has made this approach unnecessarily complicated. In its stead we describe an easily implemented procedure in Fig. 13.10.

To locate the point Q which is the Euclidean center of the circle q that underlies one of the sides of the innermost hyperbolic pentagon we first locate the point A which is the intersection of this geodesic with the ray OQ. Once A is known, Q can be easily located since it is the midpoint of A and $I(A)$ where I is the inversion $I_{O,1}$ whose fixed circle is the circumference of the unit disk model. Let OB be a radius that is inclined at the angle $\pi/5$ to the ray OQ. Since the innermost right angled hyperbolic pentagon of Fig. 13.8 consists of ten congruent copies of the hyperbolic triangle OAB, it follows that the hyperbolic angle OBA measures $\pi/4$. Thus, the hyperbolic triangle OAB has angles $\pi/2$, $\pi/4$, $\pi/5$ and hence, by one of the hyperbolic Laws of Cosines the line segment OQ has hyperbolic length

$$\cosh^{-1}\left[\frac{\cos\frac{\pi}{5}\cos\frac{\pi}{2}+\cos\frac{\pi}{4}}{\sin\frac{\pi}{5}\sin\frac{\pi}{2}}\right]$$

$$\approx \ln(1.203 + \sqrt{1.203^2 - 1}) \approx \ln 1.872.$$

By Exercise 12, the Euclidean length of the segment OA is

$$\frac{1.872 - 1}{1.872 + 1} \approx .304.$$

Consequently the Euclidean distance of the point Q from O is

$$\frac{1}{2}\left(.304 + \frac{1}{.304}\right) \approx 1.797.$$

Thus we have one side q of the innermost pentagon. The Euclidean reflection of the circle q in the radius OB yields another side p, and the reflection of this p in the radius OA yields a third side r. If A' is the reflection of A in the radius OB, then t is the reflection of q in the radius OA', and s is the reflection of t in the line OA. This gives us all the sides of the innermost pentagon. The rest of the pentagons are obtained by first inverting this pentagon in each of its sides and then repeating this procedure for each of the new pentagons and so on.

13.8 EXERCISES

1. Prove that if $(z_1, z_2, z_3, z_4) = \lambda$ and p, q, r, s constitute a permutation of 1, 2, 3, 4, then (z_p, z_q, z_r, z_s) has one of the values

$$\lambda, \frac{1}{\lambda}, 1-\lambda, \frac{1}{1-\lambda}, \frac{\lambda-1}{\lambda}, \frac{\lambda}{\lambda-1}.$$

2. Prove that if T is any Moebius rigid motion and z is any complex number, then the cross ratio $(z, T(z), T^2(z), T^3(z))$ is real.
3. Draw the oriented flow diagrams of the following rigid motions of the unit disk model:

   a) $\dfrac{2z+1}{z+2}$ b) $\dfrac{2z+(1+i)}{(1-i)z+2}$ c) $\dfrac{3z+2i}{-2iz+3}$

   d) $\dfrac{3z-i}{iz+3}$ e) $\dfrac{(2-i)z+(1-i)}{(1+i)z+(2+i)}$ f) $-z$

   g) $\dfrac{-5z+4}{4z+5}$.

4. Compute the following cross ratios:
   a) $(i, 1+2i, 2+3i, 3+4i)$ b) $(1+i, 1-i, -1+i, -1-i)$
   c) $(0, 1, i, 1-i)$.
5. Show that if $z = x + iy$ with $y > 0$, then $|U(z)| < 1$, where
$$U(z) = \frac{iz+1}{z+i}.$$

6. Verify that the transformations
$$U(z) = \frac{iz+1}{z+i} \quad \text{and} \quad V(z) = \frac{iz-1}{-z+i}$$

   are inverses of each other.
7. Express the counterclockwise rotation of the unit disk model by the angle θ about the point $z = \tfrac{i}{3}$ as a Moebius transformation in the format specified in Proposition 13.8, where θ is the angle
   a) π b) $\pi/2$ c) $3\pi/2$ d) $\pi/3$.
8. Explain why $T(z) = iz$ does yield a rigid transformation of the unit disk model even though it does not have the format posited in Proposition 13.9.
9. Explain why $T(z) = 2z$ definitely does not yield a rigid transformation of the unit disk model.
10. Show that given any three points z_1, z_2, z_3, and another triple w_1, w_2, w_3, there is a Moebius transformation $T(z)$ such that $T(z_i) = w_i$ for $i = 1, 2, 3$.
11. Show that given any three points z_1, z_2, z_3, and another triple w_1, w_2, w_3, there is at most one Moebius transformation $T(z)$ such that $T(z_i) = w_i$ for $i = 1, 2, 3$.
12. Show that in the unit disk model the hyperbolic distance h from the origin to the point z is
$$\ln \frac{1+|z|}{1-|z|},$$

and so conversely,
$$|z| = \frac{e^h - 1}{e^h + 1}.$$

13. Prove directly that the Moebius transformation
$$g(z) = \frac{az + \bar{c}}{cz + \bar{a}}$$
where a and c are any two complex numbers such that $|a| > |c|$ is a rigid motion relative to the Riemann metric
$$\frac{4(dx^2 + dy^2)}{(1 - x^2 - y^2)^2}$$
of the unit disk.

14. Prove directly that the geodesics of the Riemann metric
$$\frac{4(dx^2 + dy^2)}{(1 - x^2 - y^2)^2}$$
on the interior of the unit disk are the diameters of, and the circular arcs perpendicular to, the bounding circle.

15. Give a complex theoretic description of the circle that contains the points $-1, 0, 1 + i$.

16. Construct a tesselation of the unit disk model by right angled hexagons.

17. Complete the proof of Proposition 13.3.

18. Complete the proof of Proposition 13.4.

19. Prove Proposition 13.7.

20. Prove that the Euclidean rigid motions preserve the cross ratio.

21. Prove that inversions preserve the cross ratio.

14
The Beltrami–Klein Model

14.1 INTRODUCTION

The Beltrami–Klein model for hyperbolic geometry was created in the same year (1868) as the two previously discussed models for hyperbolic geometry. Since these two other models have turned out to play a much more important role in the evolution of modern mathematics the author chose to emphasize them more than this one. In 1871 Felix Klein reinterpreted this model within the context of projective geometry and this accounts for this model's alias of *the projective model*. Because of this book's point of view, the properties of this model will be developed using inversive geometry instead.

The hyperbolic inhabitants of the hyperbolic unit disk experience their geodesics in the same way that we observe ours, namely as straight lines, whereas we see them as curved arcs. The Beltrami–Klein model, on the other hand, has the advantage that its geodesics are **Euclidean** chords of the unit circle. Thus both we and the inhabitants of the Beltrami–Klein universe would agree about the straightness of the same lines. However, it turns out that our measurements of angles would differ from theirs. A model whose measurement of angles agrees with their Euclidean measurement is said to be *conformal*. Thus the half-plane and the unit disk models are conformal, whereas the Beltrami–Klein model turns out to be nonconformal. It stands to reason that every model of hyperbolic geometry must differ from Euclidean geometry in some essential way (Exercise 15).

The Beltrami–Klein model is constructed by deforming the hyperbolic geometry of the unit disk. Naively speaking, each of the hyperbolic geodesics of the unit disk model is "straightened out" by being projected onto its chord as illustrated in Fig. 14.1. The difficulty, of course, lies in the fact that every point of the unit disk lies on many geodesics and hence it is not at all clear that all of the geodesics can be straightened out simultaneously. That this can happen is yet another of those amazing coincidences with which this subject abounds. Once this transformation is defined, it is used to endow the unit disk with a new way for measuring distances. The resulting geometry on the unit disk is the Beltrami–Klein model. We shall describe its distances, geodesics, angles, and reflections.

14.2 THE BELTRAMI–KLEIN MODEL

The following proposition lays down the groundwork for our required distortion of the hyperbolic geometry of the unit disk. Here and elsewhere in this chapter it will be assumed that every geodesic has both of its endpoints on the unit circle and if those endpoints are A and B then the geodesic will be denoted by $h(AB)$. The unit circle will be denoted by u and the inversion $I_{O,1}$ which leaves it pointwise fixed will be denoted by I.

Proposition 14.1. *The chords joining the endpoints of each of the hyperbolic geodesics containing a fixed point P of the unit disk are all concurrent.*

PROOF: If P happens to be the center O of the unit circle u, then the

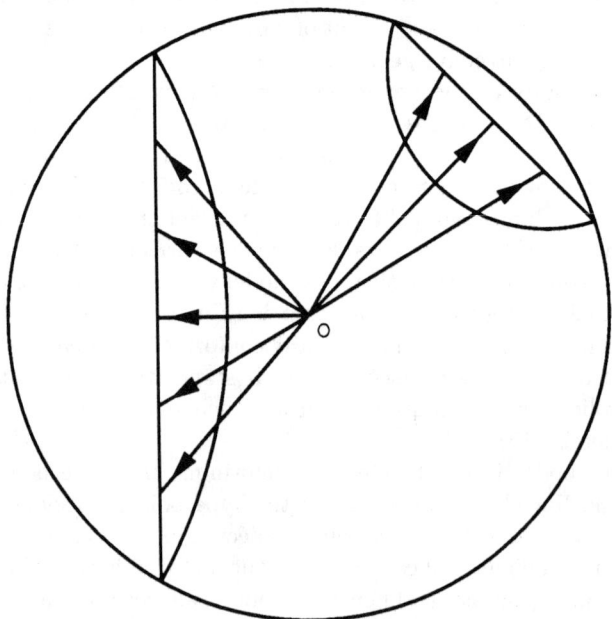

FIGURE 14.1.

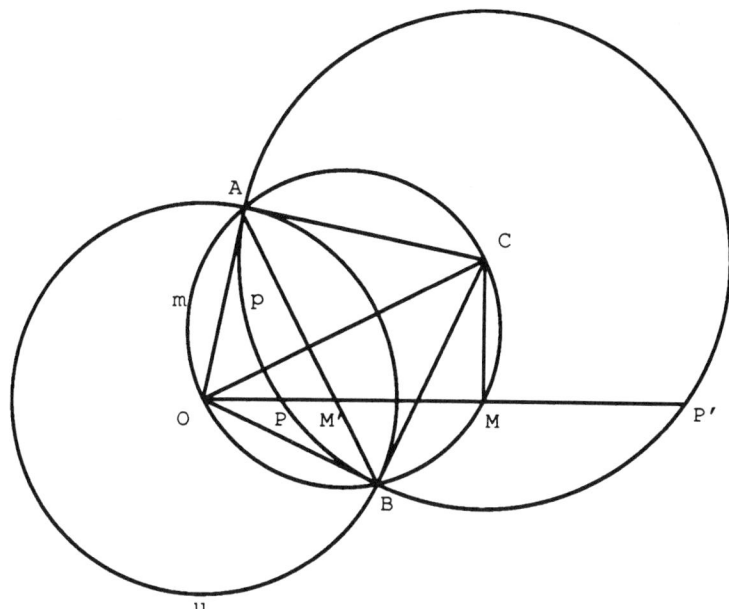

FIGURE 14.2.

conclusion is obvious since all these chords are diameters and they all contain O. We now assume that P is distinct from O. Let p be any circle orthogonal to u and let C be its Euclidean center (Fig. 14.2). Since p is orthogonal to u it follows from Proposition 3.5 that $I(p) = p$ and hence p also contains the point $P' = I(P)$. Hence, if M is the projection of the point C onto the straight line joining the points O, P, P', then M is the Euclidean midpoint of the segment PP', and so its position is independent of the positions of C and p, as long as p contains P and is orthogonal to u.

Let A and B be the endpoints of the geodesic p. We shall show that the segment AB necessarily contains the point $M' = I(M)$ (whose position depends only on that of the point P). Note that $I(AB)$ is a circle m that contains the points A, B, and O. Since

$$\angle OAC + \angle OBC = \frac{\pi}{2} + \frac{\pi}{2} = \pi,$$

it follows that the quadrilateral $OACB$ is cyclic and hence m also contains the point C. Since $\angle OBC$ is a right angle it follows that OC is a diameter of m and so, since $\angle OMC$ is also a right angle we now conclude that M is also on m. Consequently $I(M)$ is on $I(m) = AB$. Since the position of $I(M)$ depends only on that of P we are done.

q.e.d.

The above proposition makes the following definition of the transformation β of the unit disk onto itself unequivocal. If X is any point of the unit disk, let AB be any chord such that the hyperbolic geodesic $h(AB)$ contains X. Then $\beta(X)$ is the intersection of the radius OX with the chord AB. In the exceptional case where X is the origin O, we define $\beta(O) = O$. To obtain an analytical description of β we choose the chord AB to be perpendicular to the radius OX (Fig. 14.3), so that the points C and M of Fig. 14.2 coincide. Since M is the midpoint of X and $I(X) = 1/\bar{X}$ it follows that

$$M = \frac{1}{2}\left(X + \frac{1}{\bar{X}}\right) = \frac{|X|^2 + 1}{2\bar{X}} \tag{1}$$

It was shown in the course of the proof of Proposition 14.1 that $\beta(X) = I(M)$ and hence

$$\beta(X) = \frac{1}{\bar{M}} = \frac{2X}{1 + |X|^2} \tag{2}$$

For example, if $X = (1-i)/2$, then $\beta(X) = 2(1-i)/3$. In Exercise 11 the readers are asked to show that

$$\beta^{-1}(X) = \frac{1 - \sqrt{1 - |X|^2}}{|X|^2} X \tag{3}$$

The inverse point $\beta^{-1}(X)$ can also be easily located in a geometrical fashion. For any point X, let AB and CD be two chords containing it. Then $\beta^{-1}(X)$ is the intersection of the two hyperbolic geodesics $h(AB)$ and $h(CD)$. Alternately, $\beta^{-1}(X)$ is the intersection of $h(AB)$ with the radius OX.

The *Beltrami–Klein length* of any curve C in the unit disk is now defined as the hyperbolic length of the curve $\beta^{-1}(C)$ in the unit disk model. If AB is any chord of the unit circle, then, by definition, $\beta^{-1}(AB) = h(AB)$ and since $h(AB)$ is a geodesic of the unit disk model, it follows that AB is a geodesic of the Beltrami–Klein model. This may seem somewhat facile but, as was mentioned earlier, the brunt of the argument is actually contained in Proposition 14.1 which demonstrates that β simultaneously transforms all of the geodesics of the unit disk model into chords of the unit circle. In principle we now know how to compute the Beltrami–Klein distance between any two points X and Y. It equals the hyperbolic distance of the points $\beta^{-1}(X)$ and $\beta^{-1}(Y)$ in the unit disk model. This is an awkward procedure at best and the later Proposition 14.5 yields a more direct computation. We now go on to examine the effect that the transformation β has on the angles and the rigid motions of the unit disk model.

The actual angle determined by two straight lines of the Beltrami–Klein

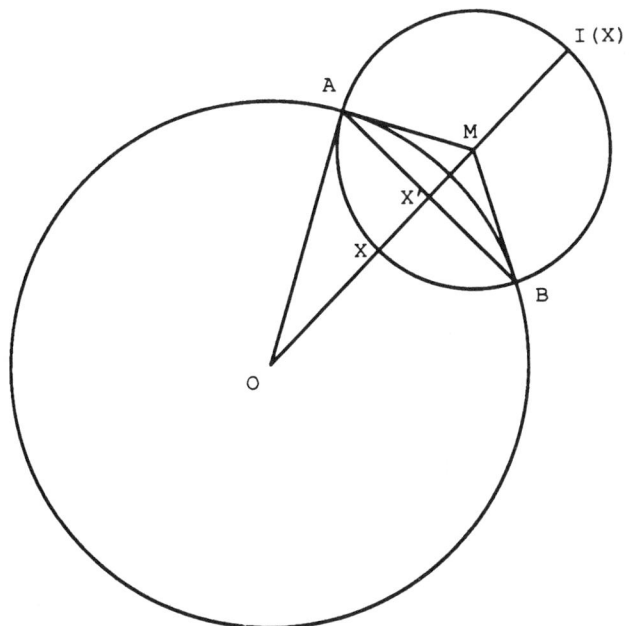

FIGURE 14.3.

model is of course identical with the Euclidean angle they determine, namely, it is the portion of the unit disk that lies between two of their rays. However, in order to obtain a geometry in which all right angles are equal (Euclid's Postulate 4) it is necessary to measure the angles of the Beltrami–Klein model in the same way that its distances are obtained, that is, by pulling back to the unit disk model. For example, let AB and CD be the horizontal and vertical chords of the unit disk that intersect in the point $X(.6,.6)$ of Fig. 14.4. If the corresponding hyperbolic geodesics $h(AB)$ and $h(CD)$ intersect in the point Y, then the Beltrami–Klein measure of $\angle AXD$ is, by definition, equal to the hyperbolic, and hence also Euclidean, measure of $\angle AYD$. To derive the latter, note that an argument similar to that used in the proof of Proposition 6.2 allows us to conclude that

$$\angle AYD = \pi - \angle NYM$$

where M and N are the respective Euclidean centers of $h(AB)$ and $h(CD)$. Since $M = I(X')$ in Fig. 14.3, the coordinates of M and N in Fig. 14.4 are $(0, 5/3)$ and $(5/3, 0)$. The coordinates of Y, by (3), are

$$\left(\frac{5 - \sqrt{7}}{6}, \frac{5 - \sqrt{7}}{6}\right).$$

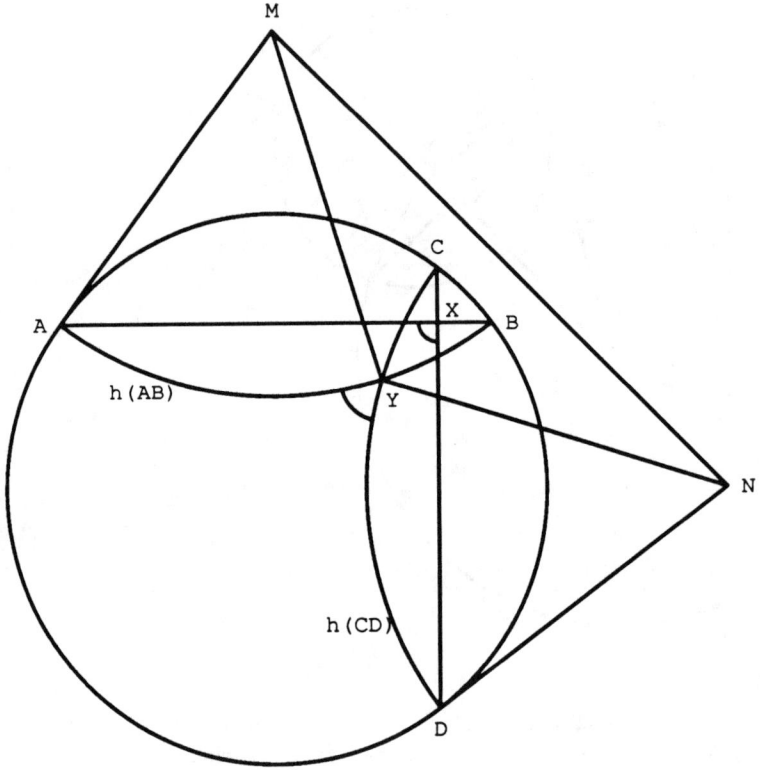

FIGURE 14.4.

The sides of $\triangle MYN$ are then seen to have lengths

$$\frac{5\sqrt{2}}{3}, \quad \frac{4}{3}, \quad \text{and} \quad \frac{4}{3}.$$

An application of the Euclidean Law of Cosines to $\triangle MYN$ then yields the final result that the Beltrami–Klein measure of $\angle AXD$ is

$$\angle AYD = \pi - \cos^{-1}\left(-\frac{9}{16}\right) = \cos^{-1}\left(\frac{9}{16}\right) \approx 56°.$$

It is clear that the foregoing example demonstrates how the Beltrami–Klein measure of any angle can be computed. However, we are still lacking in a method for easily recognizing and constructing right angles. Fortunately, such a method is available. First the reader is reminded that if A is any point outside the unit circle, then I_w denotes the hyperbolic reflection

$$\frac{w(-\bar{z})+1}{(-\bar{z})+\bar{w}}$$

of the unit disk model.

Proposition 14.2. *Let AB be a chord of the unit disk, and let M be the center of $h(AB)$. Then every chord of the unit disk whose extension passes through M is perpendicular to AB in the Beltrami–Klein sense.*

PROOF: Let I_M be the hyperbolic reflection of the unit disk model whose axis is the geodesic $h(AB)$ of Fig. 14.5. Then, since $h(AB)$ is orthogonal to the unit circle u, it follows from Proposition 3.5 that $I_M(u) = u$. If CD is any chord of u whose extension passes through M, then $I_M(CD) = CD$. Hence we may conclude that I_M interchanges C and D. But then I_M transforms $h(CD)$ into some hyperbolic geodesic through D and C. Since $h(CD)$ is the only hyperbolic geodesic through C and D, it follows that I_M fixes $h(CD)$ and so $h(CD)$ is orthogonal to $h(AB)$. By definition, the chords CD and AB are then perpendicular in the Beltrami–Klein sense.

q.e.d.

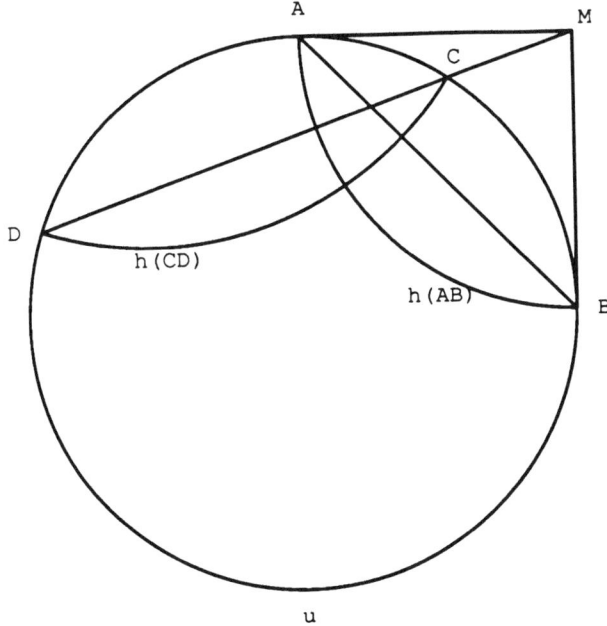

FIGURE 14.5.

We next turn to the reflections of the Beltrami–Klein model. By definition, these are transformations of the type $\beta \circ I_M \circ \beta^{-1}$ where I_M is any hyperbolic reflection of the unit disk model. This indirect definition is computationally effective. For example, let us consider the Beltrami–Klein reflection in the chord from 1 to i. The corresponding hyperbolic geodesic is centered at $1 + i$. As mentioned above, the corresponding hyperbolic reflection has the analytic expression

$$I_{1+i}(z) = \frac{-(1+i)\bar{z} + 1}{-\bar{z} + (1-i)}.$$

Applying expressions (2) and (3) above,

$$\beta \circ I_{1+i} \circ \beta^{-1}\left(\frac{2(1-i)}{3}\right) = \beta \circ I\left(\frac{1-i}{2}\right) = \beta\left(\frac{4+2i}{5}\right) = \frac{8+4i}{9}.$$

Thus the Beltrami–Klein reflection in the chord joining 1 to i carries the point $2(1-i)/3$ to the point $4(2+i)/9$.

Effective as this procedure is, it is lacking in visual directness. To remedy this defect we offer another perspective on the Beltrami–Klein reflections. Given any point M outside the unit disk, we define the *perspectivity* P_M to be the following transformation of the interior of the unit disk. If X is any point of the unit disk, let AB be any chord containing it (Fig. 14.6), let C be the other intersection of the straight line MA with the unit circle u, and let D be the other intersection of the straight line MB with u. Then $Y = P_M(X)$ is the intersection of the straight line MX with the chord CD. It is of course not at all clear that the transformation P_M is well defined, since the foregoing

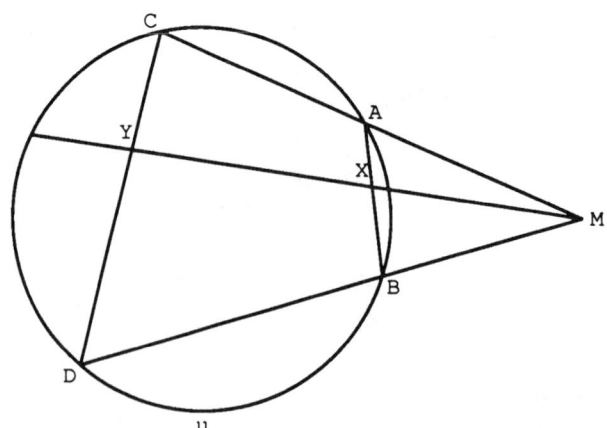

FIGURE 14.6.

construction of Y calls for an arbitrary choice of the chord AB. We shall see that $Y = P_M(X)$ is in fact independent of the choice of this chord.

Proposition 14.3. *If M is any point outside the unit disk then*

$$\beta \circ I_M \circ \beta^{-1} = P_M.$$

PROOF: Let X be any point in the interior of the unit disk and let the chords AB, CD be used to construct $Y = P_M(X)$ in Fig. 14.7. Let EF be the chord containing X and Y. Then, as observed above, $X' = \beta^{-1}(X)$ is the intersection of $h(EF)$ with $h(AB)$, and $Y' = \beta^{-1}(Y)$ is the intersection of $h(EF)$ with $h(CD)$.

Let S and T be the points of contact of the tangents from M to the unit circle u. It follows from Proposition 14.2 that the chords EF and ST are perpendicular in the Beltrami–Klein sense, and hence, by definition, that the geodesics $h(ST)$ and $h(EF)$ are orthogonal. Consequently, the hyperbolic reflection I_M fixes the geodesic $h(EF)$. The same hyperbolic reflection also maps A onto C and B onto D, and so it transforms the geodesic $h(AB)$ onto the geodesic $h(CD)$. Consequently I_M transforms the intersection X' to the intersection Y'. Hence

$$Y = \beta(Y') = \beta(I_M(X')) = \beta(I_M(\beta^{-1}(X))) = \beta \circ I_M \circ \beta^{-1}(X).$$

q.e.d.

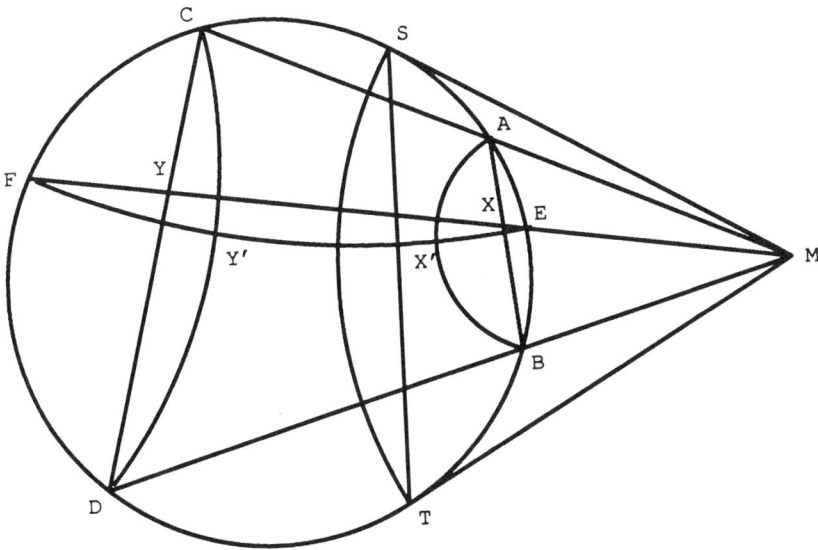

FIGURE 14.7.

The readers can easily verify directly that in the example preceding this proposition the points $1+i$, $2(1-i)/3$, and $4(2+i)/9$ are indeed collinear.

Now that we have a sufficient supply of Beltrami–Klein rigid motions at our disposal, we can derive an expression for the Beltrami–Klein distance that is very similar to the cross ratio expressions of Propositions 13.7 and 13.18. First, however, it is necessary to derive a very fundamental property of the cross ratio, one that is strongly related to the roots of this notion in projective geometry. This lemma refers to the cross ratio of four collinear points. As was seen in Proposition 13.3 the cross ratio of four such collinear points is necessarily real. Moreover, if the four points z_1, z_2, z_3, z_4 appear on the line in that very order, then both of the ratios

$$\frac{z_1 - z_3}{z_1 - z_2} \quad \text{and} \quad \frac{z_4 - z_3}{z_4 - z_2}$$

are positive and so is their ratio (z_1, z_2, z_3, z_4). Thus $\ln(z_1, z_2, z_3, z_4)$ is well defined.

Lemma 14.4. *Let m and n be two Euclidean straight lines, and let P be a point on neither line. If A, B, C, D, are points of m and A', B', C', D' are the intersections of the straight lines PA, PB, PC, PD with n, then*

$$(A, B, C, D) = (A', B', C', D').$$

PROOF: Since the points A, B, C, D are collinear, it follows that

$$(A, B, C, D) = \frac{\dfrac{A-C}{A-B}}{\dfrac{D-C}{D-B}} = \frac{\dfrac{AC}{AB}}{\dfrac{DC}{DB}}$$

where the last expression is the straightforward ratio of Euclidean lengths of Euclidean segments. A similar expression holds for the cross ratio (A', B', C', D'). However, by several applications of the Euclidean Law of Sines to Fig. 14.8,

$$AC = \text{Sin}(\angle APC)PC/\text{Sin}(\angle CAP)$$
$$AB = \text{Sin}(\angle APB)PB/\text{Sin}(\angle BAP)$$
$$DC = \text{Sin}(\angle CPD)PC/\text{Sin}(\angle PDC)$$
$$DB = \text{Sin}(\angle BPD)PB/\text{Sin}(\angle PDB).$$

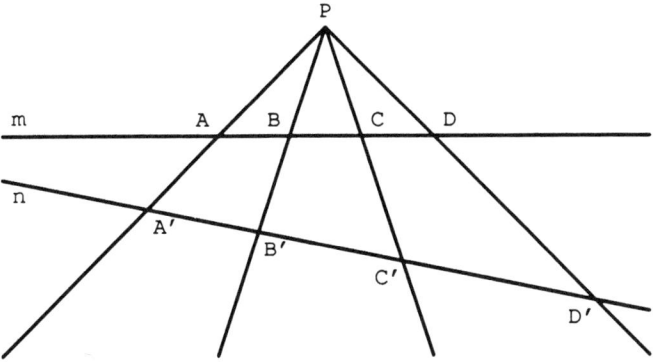

FIGURE 14.8.

Since $\angle CAP = \angle BAP$ and $\angle PDC = \angle PDB$ it follows that

$$(A, B, C, D) = \frac{\dfrac{\operatorname{Sin}(\angle APC)}{\operatorname{Sin}(\angle APB)}}{\dfrac{\operatorname{Sin}(\angle CPD)}{\operatorname{Sin}(\angle BPD)}} \qquad (4)$$

In as much as the angles appearing in (4) all have their common vertex at P it is clear that the identical trigonometric expression holds for the cross ratio (A', B', C', D'). Hence

$$(A, B, C, D) = (A', B', C', D').$$

q.e.d.

Theorem 14.5. *Let X, Y be any two points of the unit disk, and let A, B be the endpoints of the chord containing them. If these points appear in the order A, X, Y, B, then the Beltrami–Klein distance of the points X and Y is*

$$\left|\frac{1}{2}\ln(A, X, Y, B)\right|.$$

PROOF: Let M be the intersection of the lines joining A and B to -1 and 1 respectively (Fig. 14.9). Let $X'(a, 0)$ and $Y'(b, 0)$ be the respective x-intercepts of the lines MX and MY. Since $X' = P_M(X)$ and $Y' = P_M(Y)$ it follows that the Beltrami–Klein distance of the points X and Y equals that of the points X' and Y'. Moreover, it follows from Lemma 14.4 that

$$(A, X, Y, B) = (-1, X', Y', 1).$$

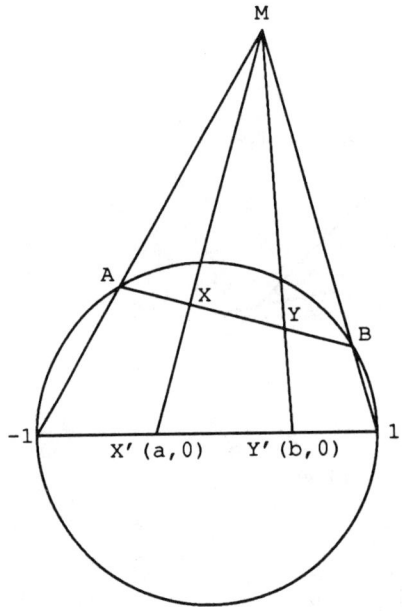

FIGURE 14.9.

Thus, it suffices to prove the theorem for the point X' and Y'. Now, the Beltrami–Klein distance of the point X' and Y' equals the hyperbolic unit disk distance between the points $\beta^{-1}(X') = (a',0)$ and $\beta^{-1}(Y') = (b',0)$ where, by (3),

$$a' = \frac{1-\sqrt{1-a^2}}{a} \quad \text{and} \quad b' = \frac{1-\sqrt{1-b^2}}{b}.$$

By Proposition 13.17 this distance equals equal the absolute value of

$$2\int_{a'}^{b'} \frac{dx}{1-x^2} = \ln\frac{\frac{1+b'}{1-b'}}{\frac{1+a'}{1-a'}} = \ln\frac{\frac{\sqrt{1+b}}{\sqrt{1-b}}}{\frac{\sqrt{1+a}}{\sqrt{1-a}}}$$

$$= \frac{1}{2}\ln\frac{\frac{-1-b}{-1-a}}{\frac{1-b}{1-a}} = \frac{1}{2}(-1,b,a,1) = \frac{1}{2}(-1,X',Y',1) = \frac{1}{2}(A,X,Y,B).$$

q.e.d.

It should be noted that there is a Riemann metric that describes the length of an arbitrary curve in the Beltrami–Klein model. The reader is referred to Exercises 12–14 for the details.

14.3 EXERCISES

1. Which chords are perpendicular to the x-axis in the Beltrami–Klein sense?
2. Let AB and CD be chords of the unit disk. Prove that if the extension of AB contains the center of $h(CD)$, then the extension of CD contains the center of $h(AB)$.
3. Find $P_{2i}(\frac{1}{2})$.
4. Determine the Beltrami–Klein distance of the points $\frac{1}{2}$ and $\frac{i}{2}$.
5. What is the Cartesian equation of the Beltrami–Klein perpendicular bisector to the line segment joining $\frac{i}{2}$ and $\frac{1}{4}$?
6. Construct a Beltrami–Klein equilateral triangle each of whose angles measures $45°$.
7. Determine the Beltrami–Klein measure of the angles between the x-axis and the straight lines joining the following pairs of points:
   a) i and 1, b) i and $\frac{1}{2}$, c) i and 0.
8. Construct a Beltrami–Klein right angled pentagon.
9. Determine the Beltrami–Klein area of the right triangle whose vertices are 0, $\frac{1}{2}$, and $\frac{i}{2}$.
10. Determine the Beltrami–Klein area of the Euclidean square whose vertices are $(0,0), (a,0), (0,a), (a,a)$.
11. Prove that
$$\beta^{-1}(X) = \frac{1 - \sqrt{1 - |X|^2}}{|X|^2} X.$$
12. Show that the Riemann metric
$$\frac{(1-y^2)\,dx^2 + 2xy\,dx\,dy + (1-x^2)\,dy^2}{(1-x^2-y^2)^2}$$
agrees with the Beltrami-Klein distance along the chords of the unit circle.
13. Show that the Riemann metric
$$\frac{(1-y^2)\,dx^2 + 2xy\,dx\,dy + (1-x^2)\,dy^2}{(1-x^2-y^2)^2}$$
agrees with the Beltrami–Klein length for all the curves in the unit disk.
14. Use the previous exercise to determine the Beltrami–Klein length of the parabola $y = x^2$ for $0 \leq x \leq \frac{1}{2}$.
15. Prove that there is no conformal model of hyperbolic geometry whose geodesics are Euclidean straight lines. (Hint: ponder the question of what the area of any triangle in such a model could be.)
16. Prove that there is no half-plane analog of the Beltrami–Klein model. In other words, show that if any geometry on the half-plane has Euclidean straight lines as

its geodesics, then it is not hyperbolic geometry. (One way to do this is to show that the total area of such a geometry is necessarily finite.)
17. Prove that every circle of the Beltrami–Klein model is necessarily a Euclidean ellipse.

15
A Brief History of Non-Euclidean Geometry

15.1 HISTORY

Euclid wrote his books circa 300 B.C. Since Euclid avoids employing Postulate 5 in the proofs of his first twenty eight propositions, historians have been led to speculate that Euclid himself had qualms about it. While this seems likely, there is no hard evidence of such concerns for the first seven centuries following the publication of *The Elements*. Proclus (410–485) wrote a commentary on Euclid's Book I that contains the first documented wranglings with this postulate. He mentions and criticizes a proof by Ptolemy (2nd Century A.D.) of the dependence of the parallel postulate on the others. Having done this he proceeds to give his own faulty proof:

> "I say that, *if any straight line cuts one of two parallels, it will cut the other also.*
>
> For let AB, CD be parallel, and let EFG cut AB [Fig. 15.1]; I say that it will cut CD also.
>
> For, since BF, FG are two straight lines from one point F, they have, when produced indefinitely, a distance greater than any magnitude, so that it will also be greater than the interval between the parallels. Whenever, therefore, they are at a distance from one another greater than the distance between the parallels, FG will cut CD."

The readers will no doubt have noticed that the above argument "proves" what we now know as Playfair's Postulate. Proclus then goes on to prove that Playfair's Postulate does indeed imply Euclid's Postulate 5. The flaw in Proclus's proof is its assumption that the distance between points on intersecting lines grows indefinitely as the points recede from the intersection of the lines. A little thought on the readers' part will convince them that this assumption is every bit as strong as the conclusion it attempts to prove, if not stronger.

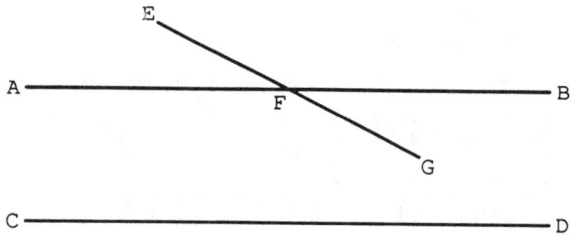

FIGURE 15.1.

Proclus's approach is typical of that of his successors in that he attempts to establish Euclid's Postulate 5 as a theorem. In other words, he believes that Postulate 5 *depends* on the other postulates in the sense that it can be deduced from them by a sufficiently long chain of reasoning. Repeated attempts to prove this dependence were made by Arab mathematicians during the middle ages, and by several European mathematicians during the Renaissance. None of these turned to be rigorous and the reader is referred to B.A. Rosenfeld's book for a fairly detailed discussion of these attempts and their flaws. In the final analysis, it could always be shown that each such proof contained a hidden assumption that was just as strong as the parallel postulate.

It is generally agreed that the modern history of non-Euclidean geometry begins with the book *Euclides ab omni naevo vindicatus* written by Girolamo Saccheri (1667–1733). In this book Saccheri attempted to prove the aforesaid dependence of Euclid's Postulate 5 by a *reductio ad absurdum* argument. He assumed that the statement of this postulate was false and went on to derive a contradiction, thus proving, in his mind, that Postulate 5 did indeed depend on the other psotulates. However, his derivation of a contradiction was flawed; his reasoning contained an error and there really was no contradiction where he thought he saw one. Fortunately, he was meticulous enough a mathematician to produce a fair amount of valid mathematics before he arrived at the supposed contradiction. In view of the subsequent events, the conclusions he derived from the first four postulates and from the negation of the fifth one (and in which negation he had no faith) became the cornerstone of non-Euclidean geometry.

In 1763 G.S. Klugel, after examining many "demonstrations" of Postulate 5 expressed doubts as to its demonstrability. This negative point of view was not to be endorsed and vindicated for more than half a century. The mathematicians of the 18'th century by and large still believed that the fifth postulate was indeed a consequence of the first four. J. H. Lambert (1728–1777) wrote a tract *Theory of Parallels* which was published posthumously. In this tract he carried on the valid portions of Saccheri's work much further and derived many more propositions. Just like his predecessor he assumed the falsity of Postulate 5 and proceeded from there. He derived the formulas for (the areas of triangles in both hyperbolic and elliptic geometry and noted their similarity to the analogous formula on the surface of the sphere. This analogy led him to comment that "I am almost inclined to draw the conclusion that the third hypothesis [hyperbolic geometry] arises with an imaginary spherical surface." In light of the subsequent develoments, this was indeed a deep insight. Nevertheless, Lambert did go on negate his imaginative leap and to "demonstrate" the validity of the Parallel Postulate by obtaining an erroneous contradiction.

A.M. Legendre (1752–1833) repeated Saccheri's and Lambert's attempts to prove Postulate 5 by contradiction. He too made an error in deriving his contradiction and he too produced much valuable mathematics before he arrived at this error. Some of his interesting results were reproduced in Chapter 10.

A comment might be in order on why such excellent mathematicians as Lambert and Legendre committed their errors. Ideally, geometry is a purely abstract science, dealing only with concepts rather than physical entities. It is clear even from Euclid's first definition, "A point is that which has no part" that he did not have a physical dot in mind. This attempt at distancing from the real world was carried to an extreme by the 19'th century geometer Steiner who reportedly taught his classes in the dark in order to prevent his students from employing their observations of the universe about them as a tool in their geometrical arguments. All such attempts at divorcing the physical intuition from their geometrical reasoning notwithstanding, these geometers, being creatures of flesh and blood, always fell short of complete success. It is indeed doubtful that such pure mental manipulations are within the grasp of the human mind. Be that as it may, after assuming the falsity of the Parallel Postulate each of these mathematicians arrived at a conclusion that so strongly contradicted their observations of their **physical** universe that they could not help but conclude that they had arrived at a **logical** inconsistency. Lambert, for example, expected the locus of all points equidistant from a straight line also to be a straight line. Legendre, on the other hand, tacitly assumed that all triangles have the same angle sum. We now know, of course, that neither of these statements is valid in the non-Euclidean context.

C.F. Gauss (1777–1855) was probably the first mathematician to entertain serious doubts about the demonstrability of Postulate 5 and to conceive of a valid non-Euclidean geometry. Unfortunately, his work on the subject is not well documented. In 1829 N.I. Lobachevsky (1793–1856) published the article *On the principles of geometry* in which he assumed that given a line m and a point P not on m, there exist more than one line through P that are parallel to m, and went on to develop hyperbolic geometry in a synthetic manner. Moreover, he advanced an argument, albeit somewhat faulty, proving that this hyperbolic geometry is just as logically consistent as its Euclidean predecessor. This argument was reproduced in Section 11.3 above. Thus, for the first time we see a mathematician propounding non-Euclidean geometry for its own sake rather than as a tool for proving its own invalidity. Lobachevsky even went further to demonstrate the utility of his creation. He showed that a certain new integral could be evaluated by a geometrical argument in the non-Euclidean context. Thus, non-Euclidean geometry could be used to solve problems that arose in the Euclidean domain. J. Bolyai (1802–1860) discovered the same geometry simultaneously and independently and published his observations as an article titled "*Supplement containing the absolutely true science of space, independent of the truth or falsity of Euclid's axiom XI [5]*". This article appeared as an appendix to a survey of attempts to prove the parallel postulate written by his father, F. Bolyai, and published in 1832.

The first complete proof of the logical consistency of the hyperbolic plane was produced in 1868 by E. Beltrami (1835–1900). In the first of the two papers he published that year Beltrami pointed out that the trigonometry of the geodesics of the pseudosphere, a surface of Euclidean geometry that Minding had investigated as far back as 1840, was identical with the trigonometry of the hyperbolic plane. Consequently any self-contradiction that might arise in hyperbolic geometry would of necessity also constitute a self-contradiction of Euclidean geometry. In other words, Beltrami proved that hyperbolic geometry was at least as consistent as Euclidean geometry.

The above paragraph glosses over a limitation that somewhat mars this contribution of Minding and Beltrami. The pseudosphere happens to be a bounded surface, namely one with an edge. The hyperbolic plane, on the other hand, extends indefinitely in all directions. Consequently, Minding's pseudosphere can be used to model only *part* of the hyperbolic plane inside Euclidean space. Hilbert showed in 1901 that this difficulty is in fact quite serious. He proved that there is *no* surface in Euclidean space whose geometry represents that of the entire hyperbolic plane. However, in his second 1868 article Beltrami completely vindicated Bolyai and Lobachevsky by constructing each of our three models of hyperbolic geometry as a Riemannian geometry. In other words, he explicitly formulated the Riemann metrics that define the upper half-plane, the unit disk model, and the Beltrami–Klein model. He found the geodesics of these metrics and pointed out that their geometry was identical with the non-Euclidean geometry of Bolyai and Lobachevsky. Beltrami actually worked in spaces of arbitrarily high dimensions, and he does say that J. Liouville (1800–1882) had previously mentioned the two dimensional Poincaré metric for the upper half-plane in his editorial Note IV to G. Monge's (1746–1818) book on differential geometry. It is also a fact that Riemann had considered the n-dimensional version of the unit disk model as early as 1854, but there is no hard evidence that either Liouville or Riemann were aware of the relationship of their metrics to hyperbolic geometry. In 1871 F. Klein pointed out that one of Beltrami's models could be constructed within the framework of projective geometry, using tools that A. Cayley developed in

1859 to construct a unified treatment of both the Euclidean geometry of the plane and spherical geometry. Since then this model has become known as either the Beltrami–Klein model or the projective model for hyperbolic geometry. It was in the same article of Klein's that the non-Euclidean geometry of Bolyai and Lobachevsky was dubbed *hyperbolic geometry*.

All of these models only prove the relative consistency of hyperbolic geometry and they all evade the problem of finding a proof of the absolute consistency of either Euclidean or hyperbolic geometry. In 1931 Kurt Gödel (1907–1978) proved a theorem that implies, amongst many other consequences, that no such proof of absolute consistency of these geometries is possible. Thus, it would seem, Beltrami and Klein had indeed settled the consistency question as far as could be done. However, mathematicians have other, less formal and more utilitarian criteria for consistency by which theories are judged, and the Poincaré half-plane owes its acceptance to one of those.

At this point it is necessary to backtrack into the 18'th century. During this period calculus flourished at a tremendous rate. Integrals and differential equations received the attention of every good mathematician both because of their intrinsic interest and because of their applicability to other sciences. These mathematicians very quickly derived techniques that are sufficiently strong to allow for the evaluation of any integral of the form

$$\int \frac{P(x)}{Q(x)} dx \qquad (1)$$

where $P(x)$ and $Q(x)$ are any polynomials with real coefficients. The integration methods required for this are all included in every standard calculus text. The assertion that all integrals of type (1) can be evaluated assumes that $Q(x)$ can be factored into linear and quadratic polynomials, but tools for the implementation of such factorizations are readily available from the discipline of numerical analysis. The readers may recall that their calculus texts also included methods for cracking some integrals of the type

$$\int R(\sqrt{ax^2 + bx + c}) \, dx$$

where $R(x)$ is some rational function. The next integrand in order of difficulty would replace the above quadratic function by a cubic, e.g.,

$$\int \frac{dx}{\sqrt{ax^3 + bx^2 + cx + d}} \qquad (2)$$

These integrals received the name of elliptic integrals, since the evaluation of the arclength of the ellipse gave rise to some of them. They turned out to be much harder than their predecessors. The construction of some kind of a theory for handling these problems was one of the major tasks that the 18'th century mathematicians passed on to the 19'th century. This was accomplished by N. Abel (1802–1829) and C. Jacobi (1804–1851). Their solution was based on two revolutionary strategies. The first of these was to focus attention on the inverses of the elliptic integrals. In retrospect, this makes sense since, for example,

$$\int \frac{dx}{\sqrt{1-x^2}} = \sin^{-1} x + C \qquad (3)$$

and so it is the inverse of the above integral, namely $\sin x$, that is the well known function. The inverses of the elliptic integrals became known as elliptic functions.

The second strategy employed by Abel and Jacobi in their attack on the elliptic integrals called for the replacement of the real variable x by the complex variable z. This had the advantage of endowing the elliptic functions with simplifying periodicities that are lost when their domain is restricted to the **real line**. To understand this better the reader should recall that the Sine function, which is the "elliptic function" associated with the integral (3), has the well known periodicity

$$\sin(z + 2\pi) = \sin z.$$

Abel and Jacobi observed that the general elliptic functions associated with the integral (2) have even stronger periodicities when their domain is extended to the **complex plane**. Specifically, if $f(z)$ denotes such an elliptic function, then there exist two complex numbers t and w such that

$$\frac{t}{w} \quad \text{is not real}$$

and

$$f(z) = f(z+t) = f(z+w) \quad \text{for all } z \qquad (4)$$

The existence of this double periodicity greatly facilitated the study of the elliptic functions.

It is clear that equations (4) are tantamount to saying that for any z,

$$f(z + mt + nw) = f(z) \quad \text{for all integers } m, n.$$

Thus the elliptic function $f(z)$ has the same value for all the points that belong to the grid consisting of all the points

$$\{z + mt + nw \mid m \text{ and } n \text{ are integers}\}.$$

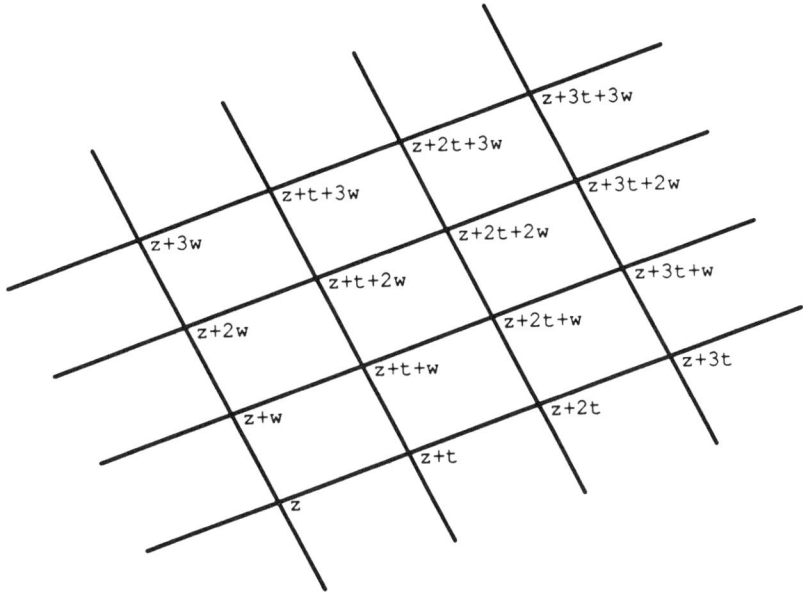

FIGURE 15.2.

For each z, these points describe a grid all of whose cells consist of copies of the parallelogram with vertices $z, z+t, z+t+w, z+w$ (Fig. 15.2). In a very definite sense, such a periodicity characterizes the elliptic functions. Loosely speaking, any differentiable function $f(z)$ that satisfies the periodicities of (4) is in fact an elliptic function.

Having disposed of these elliptic integrals, mathematicians went on to look at more complicated integrands, where the degree of the polynomial was arbitrary and the radical was replaced by more complicated radicals or combinations thereof. G.F.B. Riemann (1826–1866) proposed a very deep theory that allowed them to come to grips with such integrals. Just like the theory of elliptic functions Riemann's theory replaced the real variable x with its complex counterpart z. Riemann, however, did not focus on the inverses, perhaps because he could see no hope for finding helpful periodicities. After all, it was well known that all the two dimensional periodic patterns of the plane were essentially identical with those possessed by the elliptic functions, and any reasonable function which satisfied such a periodicity would of necessity be an elliptic function of the type disposed of by Abel and Jacobi.

The breakthrough was provided by H. Poincaré (1854–1912). At first he tried unsuccessfully to prove that the periodicities that would be required by the higher order analogs of the elliptic functions were too complicated to exist. When this attempt failed he reversed his direction and constructed a class of functions which did possess the required repetitive patterns, although the periodicities did not have a simple geometric interpretation. He called these *Fuchsian functions*, after H. Fuchs, the nineteenth century mathematician whose pioneering work in this area drew Poincaré's attention to it. We are fortunate enough to have Poincaré's own words (Science and Method) to describe the next step.

> "Just at this time I left Caen, where I was then living, to go on a geological excursion under the auspices of the school of mines. The changes of travel made me forget my mathematical work. Having reached Coutances, we entered an omnibus to go some place or other. At the moment when I put my foot on the step the idea came to me, without anything in my former thoughts seeming to have paved the way for it, that the transformations I had used to define the Fuchsian functions were identical with those of non-Euclidean geometry. I did not verify the idea; I should not have had time, as, upon taking my seat in the omnibus, I went on with my conversation already commenced, but I felt a perfect certainty. On my return to Caen, for conscience' sake I verified the results at my leisure."

Thus Poincaré realized that non-Euclidean geometry allowed for periodicities that were impossible in the Euclidean plane and that these periodicities could be used to get a firmer grip on the integrals considered by Riemann. For example, Fig. 6.10 has hyperbolic periodicities that are clearly impossible in the Euclidean plane. In order to facilitate his investigations of the periodicities permitted in the hyperbolic plane he defined the Riemann metric

$$\frac{dx^2 + dy^2}{y^2}$$

on the upper half plane, a device which, as we have seen, does indeed transform this domain into a model of hyperbolic geometry. Poincaré classified the resulting symmetries and named them *Fuchsian Groups*. These groups, as well as the Poincaré half-plane are an integral part of modern mathematics and are the subject of much current research. And so non-Euclidean geometry, which was conceived by Saccheri as merely a device for proving its own non existence, has become an essential part of the solution of problems which arose from Euclidean geometry.

15.2 EXERCISES

1. Find out the technical details of the contributions of the following mathematicians to non-Euclidean geometry.
   a)   Al-Nirizi

b) Nasir-Eddin
c) Omar Khayyam
d) Gerolamo Saccheri
e) Friedrich Ludwig Wachter
f) Bernhard Friedrich Thibaut
g) Karl Friedrich Gauss
h) Ferdinand Karl Schweikart
i) Franz Adolf Taurinus
j) Arthur Cayley
k) Sophus Lie

16

Spheres and Horospheres

16.1 INTRODUCTION

In Chapters 4–8 we constructed a model of hyperbolic geometry within the context of the Euclidean plane. It is more than likely that the reader has by now speculated about the converse problem of constructing a replica of the Euclidean plane inside hyperbolic geometry. This is the subject matter of this chapter.

The Poincaré model of the hyperbolic plane was constructed by imposing a new Riemann metric (the Poincaré metric) on the upper half-plane. Like all Riemann metrics, this metric presupposes a Cartesian coordinate system, which, in its turn, relies heavily on the uniqueness of parallels as specified in Playfair's Postulate. There is no obvious way in which this approach can be adapted to the converse problem. The hyperbolic plane has no known analog to a Cartesian coordinate system. In fact, as far the author has been able to ascertain, there is no known way of constructing Euclidean geometry inside the hyperbolic *plane*. That no such construction has been discovered may of course be due to the fact that no such construction might exist. However, the impossibility of such a construction has not yet been demonstrated and so the question is, as mathematicians like to say, open. The author believes that the reason this question remains unresolved is not so much due to its inherent difficulty as to the fact that a closely related question has been solved and the solution is so elegant and satisfying that mathematicians have lost interest in the original problem.

It so happens that the hyperbolic geometry that Bolyai and Lobachevsky constructed is actually three dimensional rather than two dimensional. Having developed their spatial tools, they then narrowed their attention to the hyperbolic plane and its trigonometry. This curious order of events is ignored by most expositors of the subject, but is actually crucial. What gave Bolyai and

258 THE POINCARÉ HALF-PLANE

Lobachevsky faith in the consistency of the axioms of the hyperbolic plane was the demonstrable consistency of its trigonometry. As we saw earlier, the formulas of hyperbolic trigonometry can be obtained from those of spherical trigonometry by multiplying all lengths by $\sqrt{-1}$, and so, it was claimed, any inconsistency in the hyperbolic plane would necessarily entail an inconsistency in the Euclidean geometry of the sphere. However, to derive the formulas of hyperbolic trigonometry both Bolyai and Lobachevsky found it necessary to first develop the absolute geometry of three dimensional space. In doing this they discovered two very surprising facts:

> *The geometry of the surfaces of hyperbolic spheres is identical with the geometry of the surfaces of Euclidean spheres;*
>
> *Inside hyperbolic space there are surfaces, called horospheres, whose geometry is indistinguishable from the geometry of the Euclidean plane.*

It is the author's belief that this observation about the horosphere's geometry satisfied later mathematicians' curiosity about the possibility of constructing a copy of Euclidean geometry inside the hyperbolic plane.

This chapter is dedicated to the precise formulation and the proof of the above two facts within the context of a three dimensional analog of the Poincaré metric. The readers should keep in mind, however, that Bolyai and Lobachevsky proved them synthetically, i.e., as theorems of absolute geometry.

16.2 HYPERBOLIC SPACE AND ITS RIGID MOTIONS

Since it is our goal in this chapter to compare and contrast a variety of geometries, it behooves us to define the notion of a geometry. For the purposes of this book, a *geometry* consists of a portion of space together with a specification of how the lengths of the curves that reside in this portion are to be computed. Examples of such geometries are Euclidean geometry, the Poincaré upper half-plane, the unit disk model, any geometry defined by a Riemann metric, spherical geometry, and the Euclidean geometry of any surface in R^3. We now go on to define and examine a new such geometrical creation of Poincaré's.

Hyperbolic space was modelled in upper half-space (to be defined momentarily) by Poincaré in a manner that is entirely analogous to his construction of the upper half-plane. Thus, the same distinction that differentiates between the hyperbolic plane and the upper half-plane also holds for hyperbolic space and the upper half-space. The latter is merely a model for the former. Nevertheless, we shall adhere to the convention set in the concluding paragraphs of Chapter 4 according to which the elements of the

SPHERES AND HOROSPHERES

model will still be described as being hyperbolic. Thus, for example, we shall use the terms hyperbolic lines and hyperbolic spheres when referring to certain subsets of the half-space model. This terminology is very convenient and should cause the reader no undue difficulties.

Assume that a Cartesian coordinate system has been imposed on Euclidean space. Let H^3 denote the upper half-space of Euclidean space, namely the set of all points (x, y, z) such that z is positive. Given any curve $\gamma(t)$, $a \leq t \leq b$, in H^3, its *hyperbolic length* is defined as the integral

$$\int_a^b \frac{\sqrt{dx^2 + dy^2 + dz^2}}{z} dt. \tag{1}$$

Example 16.1. Let us compute the hyperbolic length of the Euclidean line segment parametrized as $\gamma(t) = (3t - 1, t + 2, 2t + 1)$, $1 \leq t \leq 3$. This is, of course, the line segment joining the point $(2, 3, 3)$ to the point $(8, 5, 7)$. The parametric equations of the underlying line are

$$x = 3t - 1, \quad y = t + 2, \quad z = 2t + 1,$$

and so

$$dx = 3\,dt, \quad dy = dt, \quad dz = 2\,dt.$$

Substitution into (1) then yields

$$\int_1^3 \frac{\sqrt{9\,dt^2 + dt^2 + 4\,dt^2}}{2t + 1} = \int_1^3 \sqrt{\frac{14}{2t+1}}\,dt = \frac{\sqrt{14}}{2} \ln \frac{7}{3}$$

as the hyperbolic length of the given segment.

Example 16.2. The vertical Euclidean line segment joining the points (a, b, c_1) and (a, b, c_2) is parametrized as

$$x = a, \quad y = b, \quad z = t, \quad c_1 \leq t \leq c_2.$$

Hence, along this line,

$$dx = 0, \quad dy = 0, \quad dz = dt.$$

Accordingly, the hyperbolic length of this line is

$$\int_{c_1}^{c_2} \frac{\sqrt{0^2 + 0^2 + dt^2}}{t} = \int_{c_1}^{c_2} \frac{dt}{t} = \ln \frac{c_2}{c_1}.$$

Example 16.3. Next let us compute the hyperbolic length of the curve $\gamma(t) = (t, t^2, t^3)$ from $t = 1$ to $t = 2$. Along this curve

$$dx = dt, \quad dy = 2t\,dt, \quad dz = 3t^2\,dt.$$

Hence the hyperbolic length of this curve is given by the definite integral

$$\int_1^2 \sqrt{\frac{1 + 4t^2 + 9t^4}{t^3}}\,dt \approx 2.35.$$

The reader must by now be fully aware that the knowledge of even only some of the rigid motions of a geometry constitutes an invaluable tool in its investigation. This holds for H^3 as well. A **hyperbolic rigid motion** of the upper half-space H^3 is a transformation of H^3 into itself that preserves the lengths of curves. The complete classification of the rigid motions of H^3 lies beyond the scope of this book. Instead we will content ourselves with pointing out that some classes of well known Euclidean rigid motions of the Euclidean space R^3 also induce hyperbolic rigid motions of H^3.

Proposition 16.4. *Let a, b, α be fixed real numbers. Then the transformations*

$$\rho(x, y, z) = (x\cos\alpha + y\sin\alpha, -x\sin\alpha + y\cos\alpha, z)$$
$$\tau(x, y, z) = (x + a, y + b, z)$$

are hyperbolic rigid motions of H^3.

PROOF: We first consider the transformation $\rho(x, y, z)$. Let

$$\gamma(t) = [x(t), y(t), z(t)] \quad a \le t \le b$$

be a curve in H^3 and let its image under the transformation ρ be

$$\gamma^*(t) = [u(t), v(t), w(t)] \quad a \le t \le b,$$

where

$$u(t) = x(t)\cos\alpha + y(t)\sin\alpha$$
$$v(t) = -x(t)\sin\alpha + y(t)\cos\alpha$$
$$w(t) = z(t).$$

In order to show that ρ is indeed a rigid motion, we must verify that

$$\int_{\gamma^*} \frac{\sqrt{du^2 + dv^2 + dw^2}}{w} = \int_\gamma \frac{\sqrt{dx^2 + dy^2 + dz^2}}{z}.$$

This, however, is straightforward since

$$du = \frac{\partial u}{\partial x} dx + \frac{\partial u}{\partial y} dy = \cos\alpha\, dx + \sin\alpha\, dy$$

$$dv = \frac{\partial v}{\partial x} dx + \frac{\partial v}{\partial y} dy = -\sin\alpha\, dx + \cos\alpha\, dy$$

$$dw = dz,$$

and so

$$\text{Hyperbolic length of } \gamma^* = \int_{\gamma^*} \frac{\sqrt{du^2 + dv^2 + dw^2}}{w}$$

$$= \int_{\gamma} \frac{\sqrt{(\cos^2\alpha + \sin^2\alpha)\,dx^2 + (\sin^2\alpha + \cos^2\alpha)\,dy^2 + dz^2}}{z}$$

$$= \int_{\gamma} \frac{\sqrt{dx^2 + dy^2 + dz^2}}{z} = \text{hyperbolic length of } \gamma.$$

Thus the transformation ρ is indeed a hyperbolic rigid motion. The proof that τ is also a rigid motion is left to Exercise 1.

q.e.d.

The transformation τ of the above proposition is of course a Euclidean translation whose action is parallel to the xy plane. On the other hand, ρ is a Euclidean rotation of Euclidean space by the countercloskwise angle α with the z axis functioning as the axis of rotation, and τ is the horizontal translation that maps the point $(0, 0, z)$ to the point (a, b, z) for any z. Consequently, the composition $\tau \circ \rho \circ \tau^{-1}$ fixes every point (a, b, z) and so it must be the Euclidean rotation by the counterclockwise angle α whose axis is the vertical line containing the point $(a, b, 0)$. Since ρ and τ are also hyperbolic rigid motions, it follows that this last rotation about an arbitrary vertical axis is also a hyperbolic rigid motion. We formalize these facts as the following corollary.

Corollary 16.5. *a) Every Euclidean rotation whose axis is perpendicular to the xy plane induces a hyperbolic rigid motion of H^3.*
b) Every Euclidean translation whose action is parallel to the xy plane induces a hyperbolic rigid motion of H^3.

□

Some aspects of 3-dimensional hyperbolic geometry are already well known to us. For example, note that on the horizontal plane $z = 1$ we have $dz = 0$ for every curve, and hence the hyperbolic length of γ given by the integral of (1) agrees with its Euclidean length

$$\int_\gamma \sqrt{dx^2 + dy^2}.$$

Thus the hyperbolic geometry of the plane $z = 1$ is identical with its Euclidean geometry. Similarly, on the vertical plane $x = 0$, we have $dx = 0$ for every curve, and so the hyperbolic length of any curve γ on this plane is

$$\int_\gamma \frac{\sqrt{dy^2 + dz^2}}{z}$$

which, after an appropriate rewriting of letters, is clearly identical with the 2-dimensional Poincaré metric of the upper half-plane. In other words, the restriction of the 3-dimensional Poincaré metric on H^3 to the half plane $x = 0$, $y > 0$ results in a 2-dimensional Poincaré metric on the same half-plane. In view of Propostions 16.4 and 16.5 the same holds for the restriction of hyperbolic geometry to any plane that is perpendicular to the xy plane.

16.3 HYPERBOLIC GEODESICS

This section contains no surprises. Exactly those curves turn out to be geodesics of H^3 that the reader expects to be geodesics. A circle in R^3 is said to be *vertical* if its plane is perpendicular to the xy plane. We now state three propositions that are the three dimensional analogs of Propositions 4.1, 4.3, and Theorem 4.4. The reader is referred to Fig. 16.1 for an illustration of Proposition 16.6.

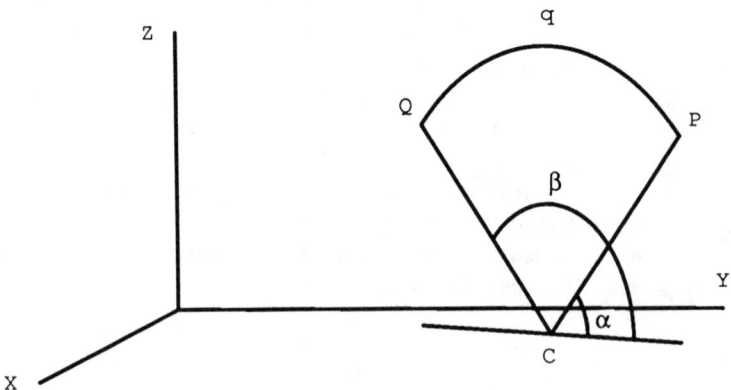

FIGURE 16.1.

Proposition 16.6. *Let q be a vertical circle in H^3 with center $C(a,b,0)$ and radius r. If P and Q are points of q such that the radii CP and CQ make angles $\alpha \leq \beta$ respectively, with the projection of q onto the xy plane, then the*

$$\text{hyperbolic length of arc} \quad PQ = \ln \frac{\csc \beta - \cot \beta}{\csc \alpha - \cot \alpha}.$$

PROOF: As was observed above, the three dimensional Poincaré metric, when restricted to the upper yz plane, defines on it a geometry that is identical with the Poincaré upper half-plane. Hence, by Proposition 4.4, the proposition holds when the circle is contained in the yz plane. The comments following Proposition 16.5 yield the validity of the proposition for any plane perpendicular to the xy plane.

q.e.d.

The following proposition was proved in Example 16.2.

Proposition 16.7. *The hyperbolic length of the Euclidean line segment joining the points (a, b, z_1) and (a, b, z_2) is*

$$\ln \frac{z_2}{z_1}.$$

□

The above two propositions yield the geodesics of H^3.

Theorem 16.8. *The geodesic segments of the Poincaré upper half-space are either*
a) arcs of vertical Euclidean semicircles that are centered on the xy plane,
or
b) segments of Euclidean straight lines that are perpendicular to the xy plane.

PROOF: Let $P(x, y, z)$ and $Q(x, y, z)$ be any two points of H^3 that are joined by a curve γ. Two cases must be considered.

CASE 1: The line segment PQ is not perpendicular to the xy plane. Let Π be the plane containing the points P, Q that is perpendicular to the xy plane, and let C be the intersection of the perpendicular bisector of the Euclidean segment PQ with the xy plane. There clearly exist a rotation and a translation of the types described in Proposition 16.4 whose composition transforms the plane Π into the yz plane and the point C into the origin (Fig. 16.2). Since these transformations are both hyperbolic and Euclidean rigid motions, they transform hyperbolic geodesics into hyperbolic geodesics and vertical

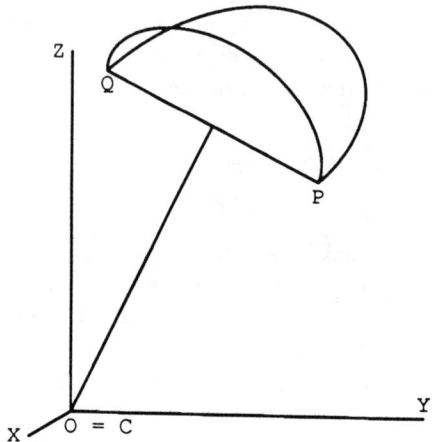

FIGURE 16.2.

Euclidean semicircles into vertical Euclidean semicircles. We may now assume without loss of generality that the plane Π is the yz plane and that the point C is indeed the origin.

Let $\gamma(t) = [x(t), y(t), z(t)]$ be a parametrized curve joining the points P and Q. If we employ the same spherical coordinate system that was used in Chapter 11, then

$$x(t) = r(t) \sin u(t) \cos v(t)$$
$$y(t) = r(t) \sin u(t) \sin v(t)$$
$$z(t) = r(t) \cos u(t).$$

Consequently, using somewhat abbreviated notation,

$$dx = (r' \sin u \cos v + r u' \cos u \cos v - r \sin u\, v' \sin v)\, dt$$
$$dy = (r' \sin u \sin v + r u' \cos u \sin v + r \sin u\, v' \cos v)\, dt$$
$$dz = (r' \cos u - r u' \sin u)\, dt$$

and so

$$dx^2 + dy^2 + dz^2 = [(r' \sin u \cos v + r u' \cos u \cos v - r \sin u\, v' \sin v)^2$$
$$+ (r' \sin u \sin v + r u' \cos u \sin v + r \sin u\, v' \cos v)^2$$
$$+ (r' \cos u - r u' \sin u)^2]\, dt^2$$

$$= [(r')^2 \sin^2 u (\cos^2 v + \sin^2 v) + r^2(u')^2 \cos^2 u (\cos^2 v + \sin^2 v)$$
$$+ r^2(v')^2 \sin^2 u (\sin^2 v + \cos^2 v)$$
$$+ 2rr'u' \sin u \cos u (\cos^2 v + \sin^2 v)$$
$$+ (r')^2 \cos^2 u - 2rr'u' \sin u \cos u + r^2(u')^2 \sin^2 u] dt^2$$
$$= [(r')^2 + r^2(u')^2 + r^2(v')^2 \sin^2 u] dt^2.$$

Hence the hyperbolic length of the curve γ is

$$\int_{t_1}^{t_2} \frac{\sqrt{(r')^2 + r^2(u')^2 + r^2(v')^2 \sin^2 u}}{r \cos u} dt \qquad (2)$$

where t_1, t_2 are the values of the parameter t that correspond to the points P, Q respectively. We shall show that if γ is a geodesic then it must necessarily be contained in the yz plane. Since the geometry of this plane is identical with that of the Poincaré half-plane, we will then be justified in concluding that γ is indeed the arc of a vertical semicircle centered at the origin.

Let δ be the curve obtained by setting $v = \frac{\pi}{2}$ in the description of γ. Then δ is a curve joining P and Q, lying completely in the yz plane, with parametrization

$$\delta(t) = [x(t), y(t), z(t)]$$

where

$$x(t) = 0$$
$$y(t) = r(t) \sin u(t)$$
$$z(t) = r(t) \cos u(t)$$

and $r(t)$, $u(t)$ are the same functions that were used to describe γ above.

The length of δ is obtained by substituting $v' = 0$ in expression (2) and so equals

$$\int_{t_1}^{t_2} \frac{\sqrt{(r')^2 + r^2(u')^2}}{r \cos u} dt \qquad (3)$$

Elementary properties of definite integrals guarantee that the integral of (2) is greater than or equal to that of (3), with equality holding if and only if v' is identically 0, i.e., if and only if v is constant. However, since the points P and Q

are both in the yz plane it follows that both the initial and terminal values of v are $\pi/2$. Hence the length of γ is always greater than or equal that of its projection δ with equality holding if and only if $\gamma = \delta$. Thus, if γ is a geodesic it must be completely contained in the yz plane. As was noted above, this guarantees that γ is indeed the arc of a vertical semicircle.

The treatment of the second case where the Euclidean line segment joining P and Q is vertical to the xy plane is relegated to Exercise 3.

q.e.d.

16.4 THE STEREOGRAPHIC PROJECTION OF SPHERES

A *hyperbolic sphere* of radius R is the locus of all the points of H^3 that have hyperbolic distance R from some fixed point C. We already saw in Chapter 5 that every hyperbolic circle of the upper half-plane is also a Euclidean circle. Similarly, every hyperbolic sphere of H^3 is also a Euclidean sphere. To see this note that every hyperbolic sphere can be obtained in the following manner. Take a vertical circle and rotate it 180° about the vertical line containing the center of the circle. Since every hyperbolic circle is also a Euclidean circle, and since this rotation is both a Euclidean and a hyperbolic rigid motion, it follows that the surface traced out by the circle is both a Euclidean and a hyperbolic sphere. The following proposition follows immediately from Proposition 5.2.

Proposition 16.9. *If a sphere has Euclidean center (a,b,c) and a Euclidean radius r, then it has the hyperbolic center (A,B,C) and the hyperbolic radius R, where*

$$A = a, \quad B = b, \quad C = \sqrt{c^2 - r^2}, \quad R = \frac{1}{2}\ln\frac{c+r}{c-r}$$

and

$$a = A, \quad b = B, \quad c = C\cosh(R), \quad r = C\sinh(R).$$

□

While every hyperbolic sphere in H^3 is also a Euclidean sphere (and vice versa), the hyperbolic and Euclidean metrics of H^3 define different geometries on each sphere. Consider, for example, the sphere with Euclidean radius 1 and Euclidean center $(0,0,2)$. The Euclidean length of any meridian is, of course, π. On the other hand, by the considerations following Proposition 5.3, the hyperbolic length of this same meridian is

$$\frac{1}{2}\frac{2\pi}{\sqrt{2^2 - 1^2}} = \frac{\pi}{\sqrt{3}}.$$

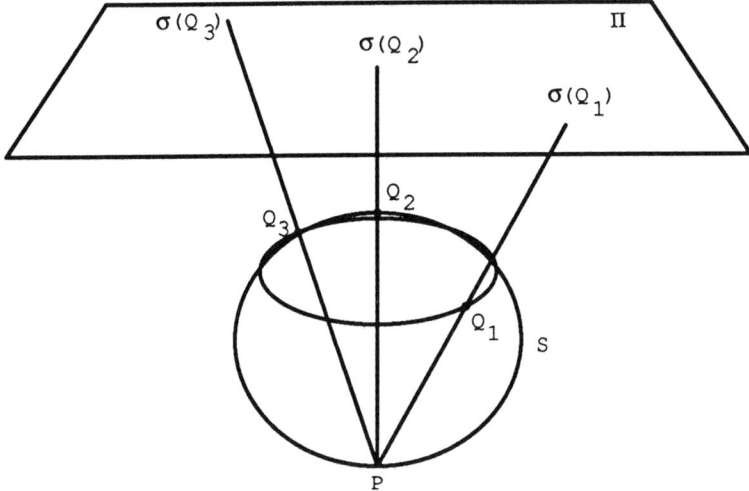

FIGURE 16.3.

These two different lengths mean that the hyperbolic and Euclidean metrics of H^3 induce different geometries on the same sphere, a not surprising observation. What is very surprising, however, is that given any sphere in H^3 there is another sphere such that the **hyperbolic** geometry of the first sphere is indistinguishable from the **Euclidean** geometry of the second. Loosely speaking, we shall prove this by "flattening" out every sphere onto some plane and then demonstrating that the image of the Euclidean geometry of one sphere is indistinguishable from the image of the hyperbolic geometry of the other sphere.

We begin by defining the aforementioned "flattening" of spheres. Let S be any sphere in R^3, let P be its lowest point (its south pole), and let Π be any plane that is parallel to the xy plane and does not contain P. If Q is any point of the sphere S that is distinct from P, then the *stereographic projection* Q' of Q onto the plane Π is the point of intersection of the line PQ with the plane Π (Fig. 16.3). The transformation $\sigma(Q) = Q'$ is called the *stereographic projection* of S onto Π. Note that, strictly speaking, the domain of σ is not S, rather it is S with its south pole P removed. Nevertheless, just as we think of an inversion as a transformation of the entire plane, we shall treat the stereographic projection as though it is defined on the whole sphere. This has the advantage of simplifying the notation and terminology. When a distinction is absolutly necessary, we shall refer to the sphere minus its south pole as the *punctured sphere*. The following lemma expresses the stereographic projection in terms of the coordinates of Q.

Lemma 16.10. *Let S be a sphere with Euclidean center (0,0,c) and Euclidean radius r. Then the stereographic projection of S onto the plane $z = d$ maps the point (x,y,z) of the sphere onto the point (u,v,d) where*

$$x = \frac{2ur(d-c+r)}{u^2 + v^2 + (d-c+r)^2}$$

$$y = \frac{2vr(d-c+r)}{u^2 + v^2 + (d-c+r)^2}$$

$$z = \frac{(u^2 + v^2)(c-r) + (d-c+r)^2(c+r)}{u^2 + v^2 + (d-c+r)^2}.$$

PROOF: The verification of this lemma is straightforward but somewhat lengthy. Let (u, v, d) be an arbitrary point of the target plane Π. It suffices to show that the point (x, y, z), whose coordinates are given in the statement of the lemma, is collinear with the points (u, v, d) and $(0, 0, c - r)$ (the south pole of the given sphere), and that this point actually lies on the given sphere. This means that one needs to verify that the coordinates (x, y, z) of Q satsify the following two relations

$$\frac{x-0}{u-0} = \frac{y-0}{v-0} = \frac{z-(c-r)}{d-(c-r)} \tag{4}$$

$$x^2 + y^2 + (z-c)^2 = r^2. \tag{5}$$

The details have been relegated to Exercises 4 and 5.

q.e.d.

We next formalize the notion of indistinguishability between geometries. Let S and T be two geometries and let f be a function from S to T. The function S is said to be an *isometry* if it satisfies the following two conditions:

a) *f* has an inverse,
b) if γ is any finite curve on S, then the length of γ on S equals the length of $f(\gamma)$ on T.

The functions U and V of Chapter 13 that map the upper half-plane and the unit disk model onto each other are examples of such isometries. Every rigid motion is an isometry of its underlying geometry with itself.

Proposition 16.11. *Let S be any sphere with Euclidean center (a,b,c) and Euclidean radius r, and let Π be the plane $z = d$. If $D = d - c + r$, then the stereographic projection σ of S onto Π establishes an isometry between the*

Euclidean geometry of S and the geometry defined by the Riemann metric

$$\frac{4D^2r^2(dx^2+dy^2)}{[x^2+y^2+D^2]^2} \tag{6}$$

on the plane Π. Moreover, if S^+ is the portion of the punctured sphere S that lies in the upper half-space H^3, then σ establishes an isometry between the hyperbolic geometry of S^+ and the geometry defined by the Riemann metric

$$\frac{4D^2r^2(dx^2+dy^2)}{[(x^2+y^2)(c-r)+D^2(c+r)]^2} \tag{7}$$

on the portion $\sigma(S^+)$ of the plane Π.

Before proceeding with the proof of this proposition let us clarify the meaning of the second part. Since the Poincaré metric is only defined in the upper half-space H^3 it is necessary, when discussing the hyperbolic geometry of surfaces, to restrict attention to those parts of the surfaces that lie in H^3. Thus if the sphere S is centered on the xy plane, the portion S^+ is its upper hemisphere. If the sphere S has radius R and is centered at $(0, 0, R)$ then it is tangent to the xy plane and S^+ is a punctured sphere. Of course, if the sphere S is completely contained in H^3 then $S^+ = S$.

PROOF of Proposition 16.11: To show that the Euclidean geometry of S is isometric to the Riemann metric of (6) on the plane Π, we must show that for any curve γ on S

$$\int_\gamma \sqrt{dx^2+dy^2+dz^2} = \int_{\sigma(\gamma)} \frac{2Dr\sqrt{dx^2+dy^2}}{x^2+y^2+D^2} \tag{8}$$

Since by Proposition 16.4 the horizontal Euclidean translations are both Euclidean and hyperbolic isometries we may assume that the spheres in question are centered at a point on the z axis. Thus we may assume that the sphere has Euclidean center $(0, 0, c)$, Euclidean radius r, and that it is being projected onto the plane $z = d$. By Lemma 16.10 the stereographic projection maps the point (x, y, z) of the sphere S onto the point (u, v, d) where

$$x = \frac{2uDr}{u^2+v^2+D^2}$$

$$y = \frac{2vDr}{u^2+v^2+D^2}$$

and so a straightforward calculation yields

$$\frac{\partial x}{\partial u} = \frac{2Dr(v^2 - u^2 + D^2)}{(u^2 + v^2 + D^2)^2} \quad \frac{\partial x}{\partial v} = \frac{-4uvDr}{(u^2 + v^2 + D^2)^2}$$

$$\frac{\partial y}{\partial u} = \frac{-4uvDr}{(u^2 + v^2 + D^2)^2} \quad \frac{\partial y}{\partial v} = \frac{2Dr(u^2 - v^2 + D^2)}{(u^2 + v^2 + D^2)^2}.$$

Moreover, it follows from equation (4) above that

$$z = \frac{Dx}{u} + c - r = \frac{Dy}{v} + c - r$$

and hence

$$\frac{\partial z}{\partial u} = \frac{-4uD^2 r}{(u^2 + v^2 + D^2)^2}$$

$$\frac{\partial z}{\partial v} = \frac{-4vD^2 r}{(u^2 + v^2 + D^2)^2}.$$

Consequently,

$$dx^2 + dy^2 + dz^2 = \left(\frac{\partial x}{\partial u}du + \frac{\partial x}{\partial v}dv\right)^2 + \left(\frac{\partial y}{\partial u}du + \frac{\partial y}{\partial v}dv\right)^2 + \left(\frac{\partial z}{\partial u}du + \frac{\partial z}{\partial v}dv\right)^2$$

$$= \left[\left(\frac{\partial x}{\partial u}\right)^2 + \left(\frac{\partial y}{\partial u}\right)^2 + \left(\frac{\partial z}{\partial u}\right)^2\right]du^2 + \left[\left(\frac{\partial x}{\partial v}\right)^2 + \left(\frac{\partial y}{\partial v}\right)^2 + \left(\frac{\partial z}{\partial v}\right)^2\right]dv^2$$

$$+ 2\left[\frac{\partial x}{\partial u}\frac{\partial x}{\partial v} + \frac{\partial y}{\partial u}\frac{\partial y}{\partial v} + \frac{\partial z}{\partial u}\frac{\partial z}{\partial v}\right]dudv$$

$$= \frac{4D^2 r^2[(v^2 - u^2 + D^2)^2 + 4u^2 v^2 + 4u^2 D^2]}{(u^2 + v^2 + D^2)^4}du^2$$

$$+ \frac{4D^2 r^2[4u^2 v^2 + (u^2 - v^2 + D^2)^2 + 4v^2 D^2]}{(u^2 + v^2 + D^2)^4}dv^2$$

$$+ \frac{16uvD^2 r^2[-(v^2 - u^2 + D^2) - (u^2 - v^2 + D^2) + 2D^2]}{(u^2 + v^2 + D^2)^4}dudv$$

$$= \frac{4D^2 r^2 (v^2 + u^2 + D^2)^2}{(u^2 + v^2 + D^2)^4}du^2 + \frac{4D^2 r^2 (u^2 + v^2 + D^2)^2}{(u^2 + v^2 + D^2)^4}dv^2$$

$$= \frac{4D^2 r^2 (du^2 + dv^2)}{(u^2 + v^2 + D^2)^2}.$$

This verifies both equation (8) and the proposition's assertion regarding the Euclidean geometry of S. Turning to the hyperbolic geometry of S^+, it must be shown that for any curve γ on S^+

$$\int_\gamma \frac{\sqrt{dx^2 + dy^2 + dz^2}}{z} = \int_{\sigma(\gamma)} \frac{2Dr\sqrt{dx^2 + dy^2}}{(x^2 + y^2)(c - r) + D^2(c + r)} \tag{9}$$

However, using the above applications of the chain rule, we get

$$\frac{dx^2 + dy^2 + dz^2}{z^2} = \frac{\dfrac{4D^2 r^2 (du^2 + dv^2)}{(u^2 + v^2 + D^2)^2}}{\left[\dfrac{(u^2 + v^2)(c - r) + D^2(c + r)}{u^2 + v^2 + D^2}\right]^2}$$

$$= \frac{4D^2 r^2 (du^2 + dv^2)}{[(u^2 + v^2)(c - r) + D^2(c + r)]^2},$$

which verifies both equation (9) and the proposition's assertion regarding the hyperbolic geometry of S^+.

q.e.d.

16.5 THE GEOMETRY OF SPHERES AND HOROSPHERES

We are now in position to prove the two theorems announced in the introductory section of this chapter.

Theorem 16.12. *The hyperbolic geometry of any sphere S in H^3 is isometric with the Euclidean geometry of some sphere S^*.*

PROOF: By Proposition 16.4 we may assume that the given sphere S has the point $(0, 0, c)$ as its Euclidean center. Assume further that its Euclidean radius is r. It will be shown that the hyperbolic geometry of S is isometric to the Euclidean geometry of the sphere S^* with Euclidean center $(0, 0, c^*)$ and Euclidean radius r^* where

$$c^* = c - r\sqrt{\frac{c + r}{c - r}} + \frac{r}{\sqrt{c^2 - r^2}}$$

and

$$r^* = \frac{r}{\sqrt{c^2 - r^2}}.$$

This will accomplished by showing that the stereographic projection of S onto the plane $z = c$ establishes an isometry of the hyperbolic geometry of S with the same geometry (metric) as does the stereographic projection of the Euclidean geometry of the sphere S^* onto that same plane. The verification of this coincidence involves only a straightforward application of Proposition 16.11. According to this proposition, the stereographic projection of S onto the plane $z = c$ establishes an isometry between the hyperbolic metric on S and the metric

$$\frac{4r^4(dx^2 + dy^2)}{[(x^2 + y^2)(c - r) + r^2(c + r)]^2} \tag{10}$$

on the plane $z = c$, since in this case $D = c - c + r = r$. On the other hand, the stereographic projection of the sphere S^* establishes an isometry of its Euclidean metric with the following metric on the plane $z = c$:

$$\frac{4D^{*2}r^{*2}(dx^2 + dy^2)}{[x^2 + y^2 + D^{*2}]^2} \tag{11}$$

where

$$D^* = c - c^* + r^*$$
$$= c - \left[c - r\sqrt{\frac{c+r}{c-r}} + \frac{r}{\sqrt{c^2 - r^2}}\right] + \frac{r}{\sqrt{c^2 - r^2}}$$
$$= r\sqrt{\frac{c+r}{c-r}}.$$

If the values of c^*, r^*, D^* are now substituted into (11) we have

$$\frac{4D^{*2}r^{*2}(dx^2 + dy^2)}{[x^2 + y^2 + D^{*2}]^2} = \frac{\dfrac{4r^4}{(c-r)^2}(dx^2 + dy^2)}{\left[x^2 + y^2 + \dfrac{r^2(c+r)}{c-r}\right]^2}$$

$$= \frac{4r^4(dx^2 + dy^2)}{[(x^2 + y^2)(c-r) + r^2(c+r)]^2}$$

which is identical with (10), the stereographic projection of the hyperbolic geometry of S.

q.e.d.

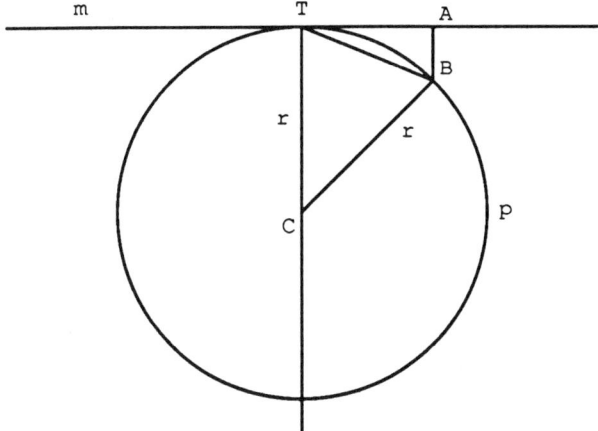

FIGURE 16.4.

The astute reader will have noticed that the above proof suffers from a minor deficiency. The projections of the spheres S and S^* onto the plane $z = c$ are not defined at their respective south poles. Consequently the alleged isometry between S and S^* is in fact only an isometry when these poles are omitted. While it is clear that this is in reality merely a fine point, mathematical rigor requires that it be patched up. This can be accomplished either by an involved discussion of limits and the continuity of distances, or else by the method indicated in Exercise 10.

In Exercise 7 the reader is asked to demonstrate that the hyperbolic geometry of the sphere with hyperbolic radius R is isometric with the Euclidean geometry of the sphere with Euclidean radius $\sinh(R)$.

A *horosphere* is either a horizontal plane of H^3 or the surface of a Euclidean sphere that is tangent to the xy plane and, except for the point of contact, is completely contained in the upper half-space. This definition relies on the identification of hyperbolic 3-space with H^3, and it should be of interest to the reader to see how this surface is defined within the context of *synthetic* hyperbolic geometry. For this purpose, let us momentarily digress into Euclidean geometry. Let p be a circle of radius r and center C that is tangent to the line m at T (Fig. 16.4). Let A be an arbitrary point of the line m and let B be a point on the circle p such that AB is perpendicular to the tangent line m. It is clear that such a B must exist for sufficiently large r. Now $\triangle CTB$ is isosceles and hence

$$\angle BTC = \frac{\pi}{2} - \frac{1}{2} \angle TCB \tag{12}$$

Assume now that the line m and the points T and A are held fixed while the point C recedes to infinity along the line TC (which remains perpendicular to m). Thus the radius r becomes indefinitely large and B, remaining on the circle p, ascends towards the fixed point A. A slight modification of the argument that was employed in the proof of Theorem 1.1 to show that $\angle DD_nA$ of Figure 1.11 converges to 0 implies that $\angle TCB$ of Fig. 16.4 also converges to 0 as r becomes indefinitely large and C recedes to infinity. Consequently, $\angle BTC$ converges to $\pi/2$ and so B actually converges to A as r diverges to infinity. In other words, the limiting position of the circle p is the tangent line m.

Equation (12) relies on the fact that the sum of the angles of every Euclidean triangle is p, which is not the case for hyperbolic triangles. In fact, in the hyperbolic plane this equation must be replaced by the inequality

$$\angle BTC < \frac{\pi}{2} - \frac{1}{2}\angle TCB$$

and hence there is no reason to conclude that in the hyperbolic plane $\angle BTC$ converges to a right angle as the radius of p diverges to infinity, nor that the circle p has the tangent line m as its limiting position. In fact, it doesn't. To see this, let us examine this process in the context of the Poincaré upper half-plane (Fig. 16.5).

Three positions of the point C on its way to "infinity" are denoted by C_1, C_2, C_3, and the corresponding circles are denoted by p_1, p_2, p_3. Note that as C recedes to hyperbolic infinity we, the Euclidean observers, see it converging to the point C_∞, and that the limiting position of the circles p is the Euclidean

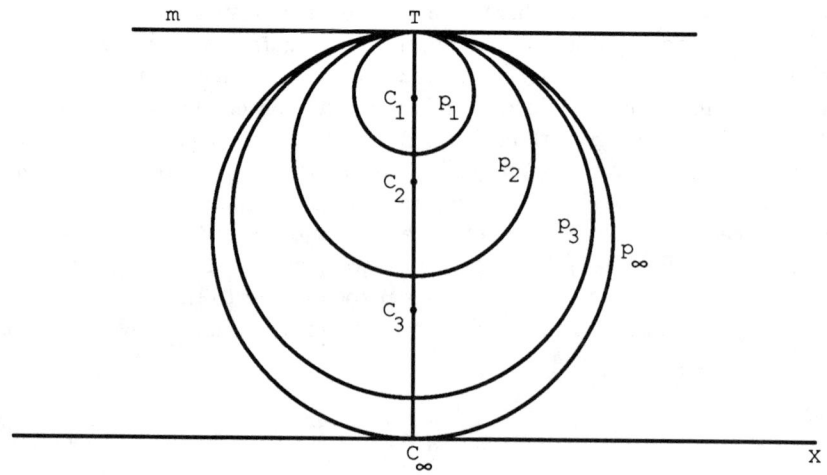

FIGURE 16.5.

circle p_∞ which is not a hyperbolic straight line. In fact, the inversion centered at C_∞ with radius TC_∞ is a hyperbolic rigid transformation which maps the curve p_∞ onto the Euclidean tangent line m. Thus, to the inhabitants of the Poincaré upper half-space p_∞ and m have indistinguishable geometric properties.

Euclidean circles of the plane that are tangent to the x-axis but are otherwise completely contained in the upper half-plane are nowadays called *horocycles*. Lobachevsky refered to them as oricycles whereas Bolyai, at the suggestion of Gauss, called them paracycles. They are crucial to the investigations of the deeper properties of the hyperbolic plane and its higher dimensional analogs. The reader is reminded that they were already encountered in Chapter 9 as the flow lines of certain hyperbolic translations. It is clear that just as the horocycle can be thought of as the limiting position of hyperbolic circles whose center is receding to infinity, so the horosphere can be thought of as the limiting position of hyperbolic spheres whose center is receding to infinity. It is also clear that every horosphere can be obtained by rotating a horocycle about an appropriate axis.

Theorem 16.13. *Every horosphere is isometric to the Euclidean plane.*

PROOF: A typical horosphere S has Euclidean center (a, b, c) and Euclidean radius $r = c$. Consequently, taking $d = c$, so that the target plane P of Proposition 16.11 is the equatorial plane of S, the second part of this same proposition asserts that the hyperbolic geometry of the horosphere is isometric with the geometry defined by the Euclidean metric

$$dx^2 + dy^2$$

on the plane Π.

q.e.d.

16.6 EXERCISES

1. Show that the transformation τ of Proposition 16.4 is a hyperbolic rigid motion.
2. Find the hyperbolic midpoint of the Euclidean line segment joining $(1, 2, 3)$ to $(1, 2, 8)$.
3. Prove part b of Theorem 16.8.
4. Verify relation (4).
5. Verify relation (5).
6. Explain carefully why the plane with metric $dx^2 + dy^2$ is isometric to the plane with metric $a\,dx^2 + b\,dy^2$ for any positive real numers a, b.
7. Show that the hyperbolic geometry of a sphere with hyperbolic radius R is isometric with the Euclidean geometry of a sphere of Euclidean radius $\sinh(R)$.

8. If D is a solid that is completely contained in the upper half-space, its hyperbolic volume is defined as

$$\iiint_D \frac{dxdydz}{z^3}.$$

Show that the hyperbolic volume of a sphere with hyperbolic radius R is

$$2\pi[\sinh(R)\cosh(R) - R].$$

9. Let C be any point of Euclidean 3-space R^3 and let k be any positive real number. The inversion $J_{C,k}$ is the transformation of R^3 which maps any point P to a point P' such that C, P, P' are collinear, $CP \cdot CP' = k^2$, and C is outside the line segment PP'. Prove that if C is on the xy plane then $J_{C,k}$ induces a hyperbolic rigid motion of H^3.

10. Construct a direct isometry between the spheres S and S^* of Theorem 16.12.

Appendix
Proofs of Some of Euclid's Propositions

This appendix consists of the statements and proofs of some of the propositions of Books I and III of *Euclid's Elements* that are mentioned in Chapter 1. Whenever the proof given here is not based on Euclid's, its author is explicitly mentioned. All the other proofs differ from those of Euclid only in minor details.

Proposition 5. *In isosceles triangles the angles at the base are equal to one another; and, if the equal straight lines be produced further, the angles under the base will be equal to one another.*

PROOF (Pappus): In $\triangle ABC$ let $AB = AC$. Then, by Proposition 4 (alias SAS),

$$\triangle ABC \cong \triangle ACB.$$

Consequently, $\angle ABC = \angle ACB$.

q.e.d.

Proposition 6. *If in a triangle two angles be equal to one another, the sides which subtend the equal angles will also be equal to one another.*

PROOF: Let ABC be a triangle in which $\angle ABC = \angle ACB$ (Fig. A.1). It is required to show that AB and AC are equal. If that is not the case, suppose AB is the greater. Let D be a point in the interior of AB such that $BD = AC$. It follows from SAS that

$$\triangle DBC \cong \triangle ACB,$$

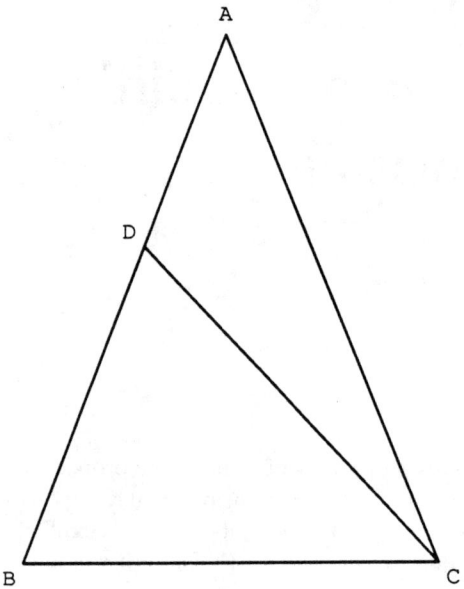

FIGURE A.1.

which contradicts Common Notion 5 (that the whole is greater than the part).

q.e.d.

Proposition 7. *Given two straight lines constructed on a straight line (from its extremities) and meeting in a point, there cannot be constructed on the same straight line (from its extremities) and on the same side of it, two other straight lines meeting in another point and equal to the former two respectively, namely, each to that which has the same extremity.*

PROOF: Suppose this proposition is false. Then there exists a quadrilateral $ABDC$ (Fig. A.2) in which

$$AC = AD \quad \text{and} \quad BC = BD.$$

But then, since by Proposition 5 the angles at the base of an isosceles triangle are equal, we have the following contradiction:

$$\angle CDB > \angle CDA = \angle DCA > \angle DCB = \angle CDB.$$

q.e.d.

PROOFS OF SOME OF EUCLID'S PROPOSITIONS   **279**

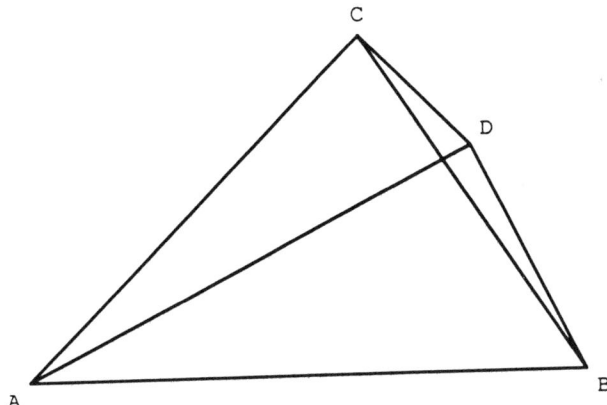

FIGURE A.2.

Proposition 8. *If two triangles have two sides equal to two sides respectively, and have also the base equal to the base, they will also have the angles equal which are contained by the equal straight lines.*

PROOF: Apply one of the triangles to the other along one pair of equal edges so that both triangles lie on the same side of their now common edge. If the proposition were false this configuration would yield a counterexmaple to Proposition 7.

q.e.d.

Proposition 9. *To bisect a given rectilinial angle.*

PROOF: Let $\angle BAC$ be the angle which is to be bisected (Fig. A.3). On the sides AB and AC chose points D and E such that $AD = AE$. Let $\triangle DEF$ be equilateral. Then, by Proposition 8 (alias *SSS*),

$$\triangle ADF \cong \triangle AEF$$

and hence $\angle DAF = \angle EAF$. Thus the line AF bisects the given angle.

q.e.d.

Proposition 10. *To bisect a given finite straight line.*

PROOF (Apollonius): Let AB be the line segment to be bisected (Fig. A.4). Let C and D be the intersections of the two circles with radii AB and centers A and B respectively, and let E be the intersection of CD with AB.

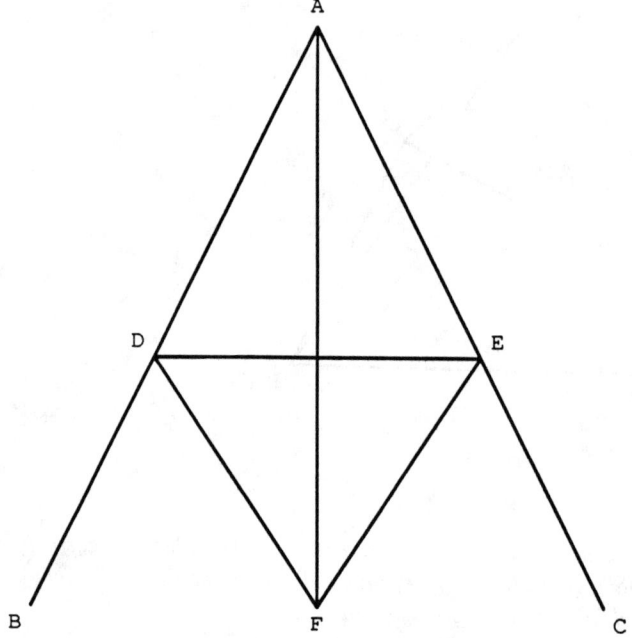

FIGURE A.3.

It follows from SSS that

$$\triangle ACD \cong \triangle BCD.$$

Hence we may conclude that $\angle ACE = \angle BCE$, and so, by *SAS*,

$$\triangle ACE \cong \triangle BCE.$$

Consequently E is the midpoint of AB.

q.e.d.

Proposition 11. *To draw a straight line at right angles to a given straight line from a given point on it.*

PROOF: Let C be the given point on the given straight line AB (Fig. A.5). Let D and E be two distinct points on AB such that $CD = CE$, and let $\triangle DEF$ be equilateral. An *SSS* argument shows that

$$\triangle FDC \cong \triangle FEC.$$

PROOFS OF SOME OF EUCLID'S PROPOSITIONS   **281**

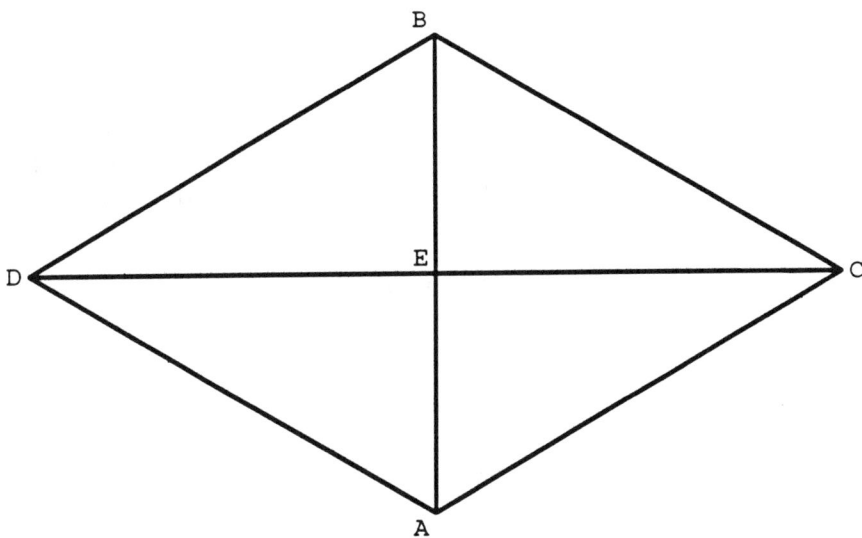

FIGURE A.4.

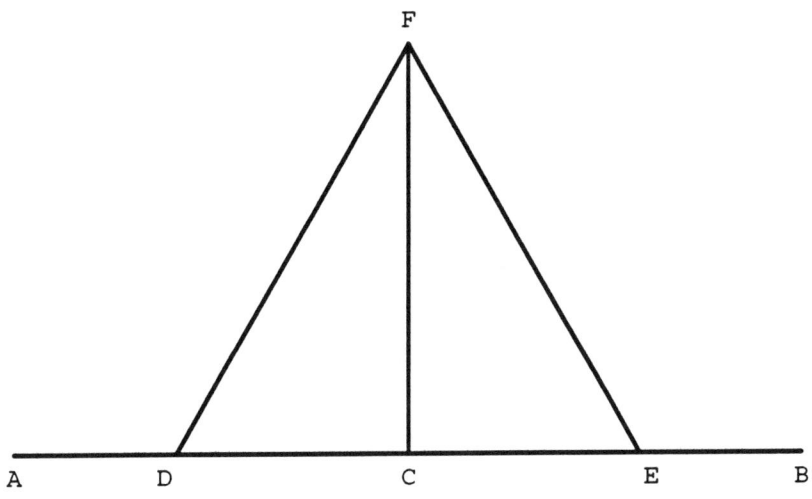

FIGURE A.5.

Hence ∠FCD = ∠FCE and each is a right angle.

q.e.d.

Proposition 12. *To a given infinite straight line, from a given point which is not on it, to draw a perpendicular straight line.*

PROOF: Let AB be the given (infinite) straight line, and C the given point which is not on it (Fig. A.6). Let D be any point on the other side of AB, and suppose that the circle with radius CD centered at C intersects the line AB at the points G and E. Finally, let H be the midpoint of GE. An *SSS* argument establishes that

$$\triangle GHC \cong \triangle EHC$$

and consequently

$$\angle GHC = \angle EHC$$

and both are right angles. Thus CH is the required perpendicular.

q.e.d.

Euclid is a little hampered by the fact that his definition of angles excludes the "straight angle", namely the angle of measure 180°. This makes his proofs of the following three propositions seem somewhat pedantic.

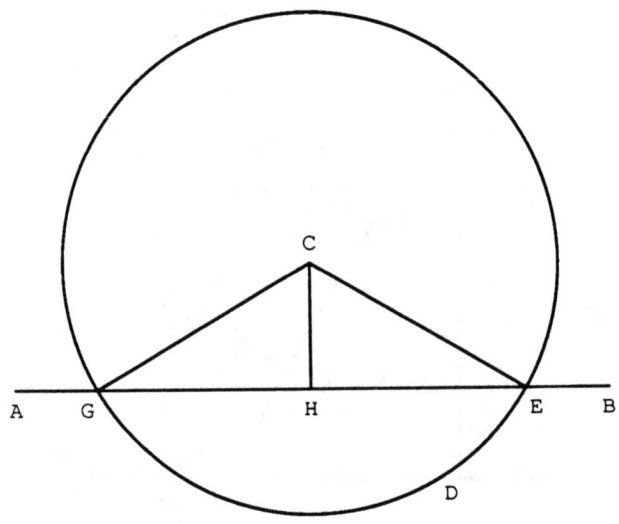

FIGURE A.6.

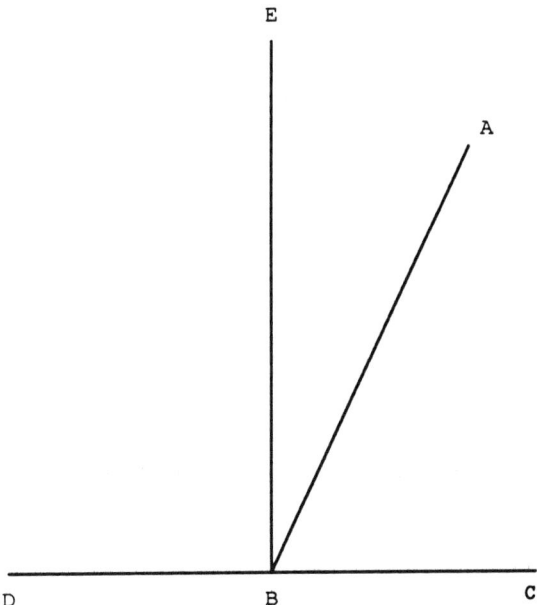

FIGURE A.7.

Proposition 13. *If a straight line set up on a straight line make angles, it will make either two right angles or angles [whose sum is] equal to two right angles.*

PROOF: Let any straight line AB set up on the straight line CD make the angles CBA and ABD (Fig. A.7). If these two angles are equal, then they are both right angles and we are done. If not, let BE be perpendicular to CD. Then

$$\angle CBA + \angle ABD = \angle CBA + \angle ABE + \angle EBD = \angle CBE + \angle EBD$$

which last two are in fact two right angles.

q.e.d.

Proposition 14. *If with any straight line, and at a point on it, two straight lines not lying on the same side make the adjacent angles to two right angles, the two straight lines will be in a straight line with one another.*

PROOF: Let AB be the given straight line, and at the point B on it, let the two straight lines BC and BD (lying on opposite sides of AB) form the angles ABC and ABD whose sum equals two right angles (Fig. A.8). It is required to show that the lines BC and BD are in the same straight line.

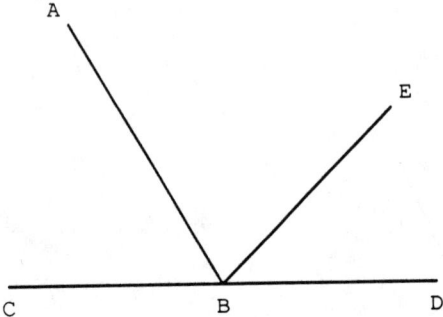

FIGURE A.8.

Suppose such is not the case and that in fact BE is the extension of BC. Then, by Proposition 13, the sum of the angles ABC and ABE is also two right angles and hence

$$\angle ABD = \angle ABE.$$

This contradicts Common Notion 5 and so we are done.

q.e.d.

Proposition 15. *If two straight lines cut one another, they make the vertical angles equal to one another.*

PROOF: Let the straight lines AB and CD intersect at the point E (Fig. A.9). We shall show that the angles AEC and DEB are equal. By Proposition 13

$$\angle DEA + \angle AEC = \text{two right angles}$$

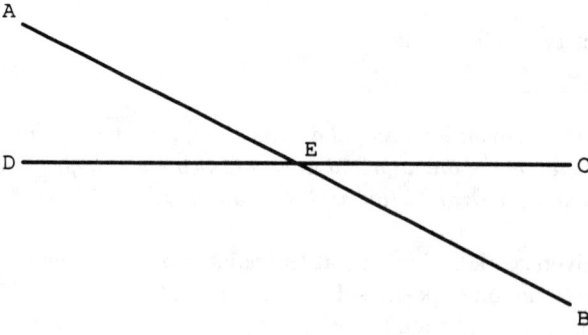

FIGURE A.9.

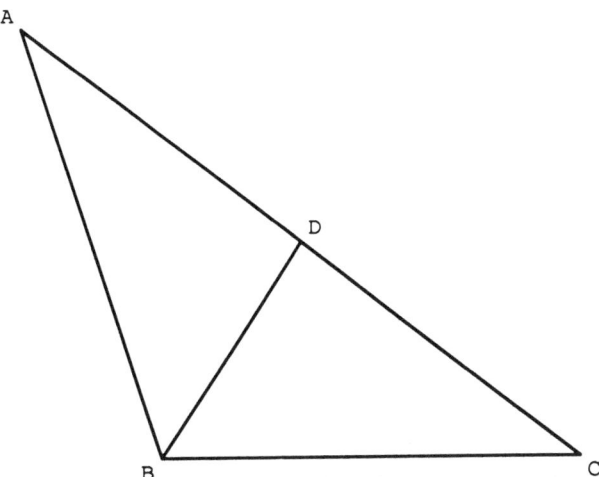

FIGURE A.10.

and

$$\angle DEA + \angle DEB = \text{two right angles}.$$

Hence,

$$\angle AEC = \angle DEB.$$

q.e.d.

Proposition 18. *In any triangle the greater side subtends the greater angle.*

PROOF: Let ABC be a triangle having the side AC greater than the side AB (Fig. A.10). In the interior of AC chose a point D such that $AB = AD$. Successive applications of Common Notion 5, Proposition 5, and Proposition 16 yield the conclusion

$$\angle CBA > \angle DBA = \angle BDA > \angle BCA.$$

q.e.d.

Proposition 19. *In any triangle the greater angle is subtended by the greater side.*

PROOF: Let ABC be a triangle in which $\angle ABC > \angle ACB$. By Propositions 5 and 18, if side AC were equal to, or less than, side AB, then we would have respectively $\angle ABC = \angle ACB$ or $\angle ABC < \angle ACB$ both of which contradict the given information. Hence side AC must be greater than side AB.

q.e.d.

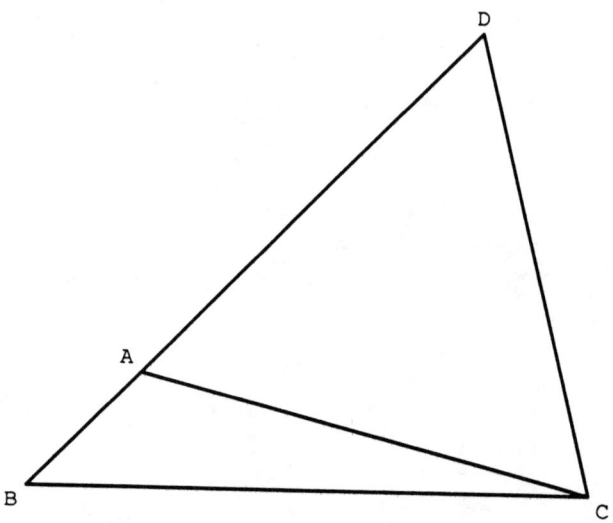

FIGURE A.11.

Proposition 20. *In any triangle two sides taken together in any manner are greater than the remaining one.*

PROOF: Let ABC be a triangle (Fig. A.11). It will be shown that $BA + AC > BC$.

Extend BA to a point D such that $AD = AC$, so that $\triangle ADC$ is isosceles and hence $\angle ADC = \angle ACD$. Hence

$$\angle BDC = \angle DCA < \angle DCB,$$

and so

$$BC < BD = BA + AD = BA + AC.$$

q.e.d.

Proposition 22. *Out of three straight lines which are equal to three given straight lines, to construct a triangle: thus it is necessary that two of the straight lines, taken in any manner, should be greater than the remaining one.*

PROOF: Let a, b, c be three straight lines such that the sum of any two exceeds the third. If FG is any line segment of length b draw two circles with centers F, G and radii a, c respectively. Suppose these circles intersect each other at the point K. Then $\triangle FGK$ is the required triangle.

q.e.d.

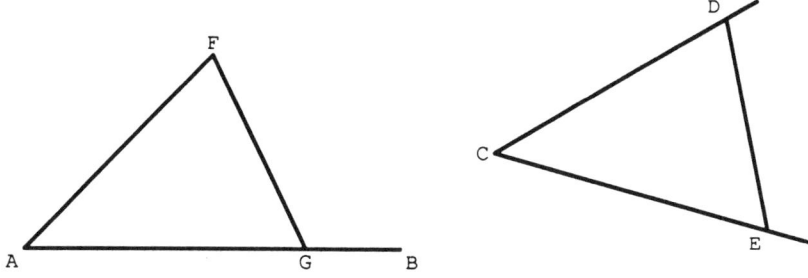

FIGURE A.12.

Proposition 23. *On a given straight line and at a point on it to construct a rectilinial angle equal to a given rectilineal angle.*

PROOF: Let AB be the given straight line, A a point on it, and $\angle DCE$ the given angle (Fig. A.12).

On the sides CD and CE respectively, let D and E be arbitrary points. By the previous proposition it is possible to constuct a $\triangle AGF$ such that G is on the line AB and

$$AG = CE, \quad AF = CD, \quad \text{and} \quad FG = DE.$$

Then $\triangle FAG \cong \triangle DCE$ and so $\angle FAG = \angle DCE$.

q.e.d.

Rather than prove Proposition 26 in its full generality we confine ourselves to its better known half, alias *ASA*. The lesser known portion, which goes under the name of *SAA* is of course an immediate corollary of *ASA* and the later theorem which states that the sum of the angles of every triangle is equal to two right angles. That it is indeed independent of the Parallel Postulate is not a very well known fact. Euclid's proof of these two congruence theorems is very lengthy because he is apparently avoiding the use of rigid motions (applications, in his terminology). In the interest of brevity we prove only *ASA* and do make use of the notion of application since Euclid already did so in his proof of Proposition 4.

Proposition 26′. *If two triangles have the two angles equal to two angles respectively, and the sides adjoining these angles are also equal, then they will also have the remaining sides equal to the remaining sides and the remaining angle to the remaining angle.*

PROOF (Al-Nirizi): Let ABC and DEF (Fig. A.13) be two triangles in which

$$\angle ABC = \angle DEF, \quad \angle ACB = \angle DFE, \quad \text{and} \quad BC = EF.$$

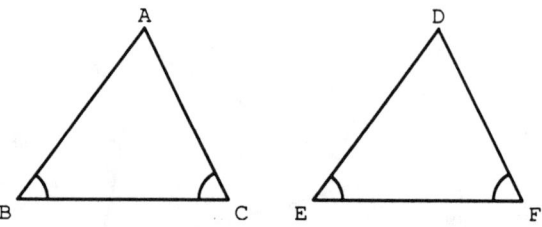

FIGURE A.13.

Apply △ABC to △DEF so that B and C coincide with E and F respectively. Because of the given equalities of angles, this can be done so that the straight lines BA and CA coincide with the straight lines ED and FD. It then follows from Postulate 1 that A must coincide with D. Hence

$$\angle BAC = \angle EDF, \quad BA = ED, \quad \text{and} \quad CA = FD.$$

q.e.d.

Proposition 27. *If a straight line falling on two straight lines make the alternate angles equal to one another, the straight lines will be parallel to one another.*

PROOF: Let the straight line EF falling on the two straight lines AB and CD make the alternate angles ∠AEF and ∠EFD equal to one another (Fig. A.14). Suppose the lines AB and CD are not parallel but intersect in a point G. Then ∠AEF is an exterior angle and ∠EFD is an opposite interior angle relative to △EFG. By Proposition 16 the former angle is greater than the latter, which contradicts the proposition's hypothesis.

q.e.d.

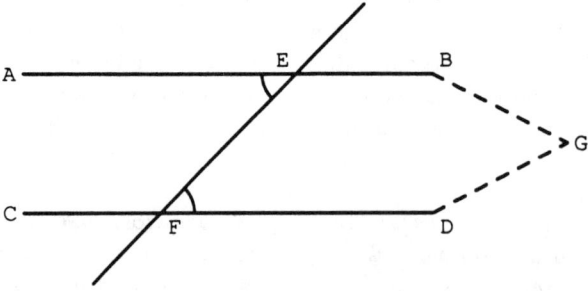

FIGURE A.14.

Proposition 28. *If a straight line falling on two straight lines make the exterior angle equal to the interior and opposite angle on the same side, or* the interior angles on the same side equal to two right angles, the straight lines will be parallel to one another.

PROOF: The validity of this proposition is an immediate corollary of Propositions 13, 15, and 27.

□

FROM BOOK III

Proposition 18. *If a straight line touch a circle, and a straight line be joined from the centre to the point of contact, the straight line so joined will be perpendicular to the tangent.*

PROOF: Let P be any point of a circle with center C, and let PT be perpendicular to the radius CP. In $\triangle CPT$, PT, being a hypotenuse, is greater that the radius CP. Hence the point T lies outside the given circle. It follows that, except for P, every point of the line PT lies outside the given circle. This means that PT is tangent to the given circle and so the line tangent to the given circle at P is indeed perpendicular to the radius CP.

q.e.d.

Proposition 20. *In a circle, the angle at the center is double of the angle at the circumference, when the angles have the same circumference as base.*

PROOF: Let C be the center of a circle, AmB an arc, and P a point of the circle that is not on this arc (Fig. A.15). It is required to prove that

$$\angle ACB = 2\angle APB.$$

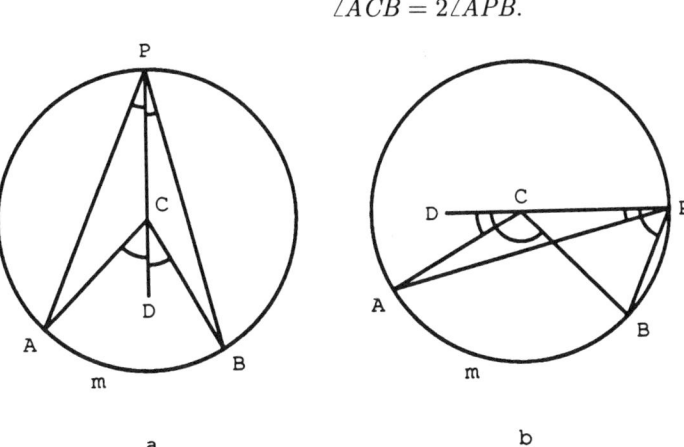

FIGURE A.15.

Draw the straight line PCD. Since ∠ACD is exterior to the isosceles △ACP, it follows that

$$\angle ACD = 2\angle APC.$$

Similarly

$$\angle BCD = 2\angle BPC.$$

When these two equations are added in Fig. A.15a and subtracted in Fig. A.15b we obtain the required equation.

q.e.d.

Proposition 22. *The opposite angles of quadrilaterals in circles are equal to two right angles.*

PROOF: Let ABCD be a quadrilateral whose vertices are on the circumference of a circle centered at O (Fig. A.16). Draw the radii OB and OD. If α is the central angle at O that corresponds to the arc BAD, and if γ is the central angle

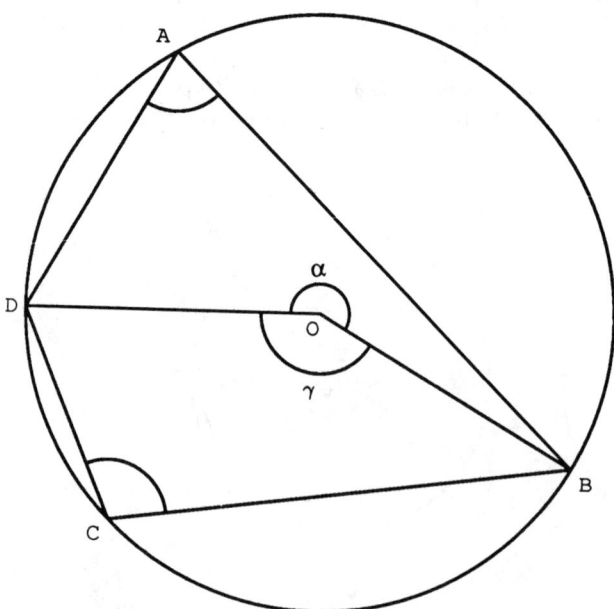

FIGURE A.16.

at O that corresponds to the arc DCB, then, by Proposition 20 above,

$$\angle DAB = \frac{\gamma}{2} \quad \text{and} \quad \angle BCD = \frac{\alpha}{2}.$$

Consequently,

$$\angle DAB + \angle BCD = \frac{\gamma + \alpha}{2} = \frac{2\pi}{2} = \pi.$$

q.e.d.

Proposition 31. *In a circle, the angle in a semicircle is right.*

PROOF: Let AB be a diameter of a circle centered at O, and let P be a point on its circumference. By Proposition 20 above,

$$\angle APB = \frac{1}{2} \angle AOB = \frac{1}{2}\pi.$$

q.e.d.

Proposition 32. *If a straight line touch a circle, and from the point of contact there be drawn across, in the circle, a straight line cutting the circle, the angles which it makes with the tangent will be equal to the angles in the alternate segments of the circle.*

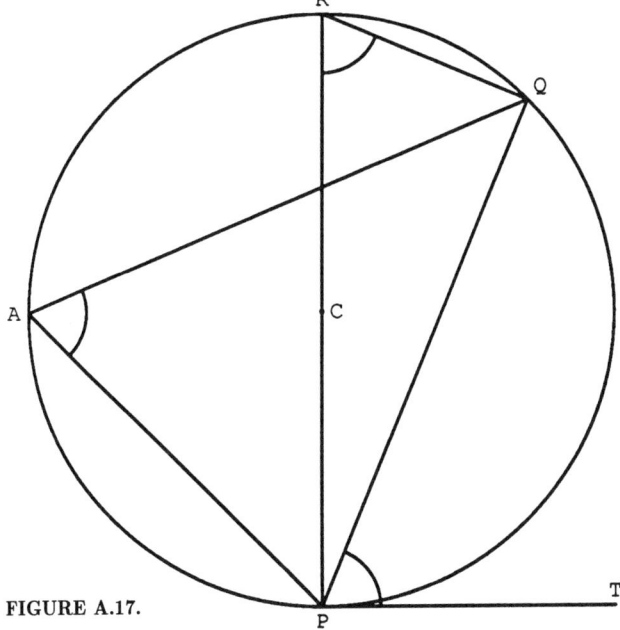

FIGURE A.17.

PROOF: In a given circle centered at C, let PQ be a chord, let PT be a tangent, and let A be a point on the arc not enclosed by the ray PT and the chord PQ (Fig. A.17). It is required to prove that

$$\angle PAQ = \angle TPQ.$$

Let PR be a diameter. By Proposition 20 above,

$$\angle PAQ = \frac{1}{2} \angle PCQ = \angle PRQ.$$

However, by Propositions 18 and 31, $\angle TPR$ and $\angle RQP$ are both right angles and hence, bearing in mind that the sum of the angle of any triangle is two right angles,

$$\angle PAQ = \angle PRQ = \frac{\pi}{2} - \angle QPR = \angle TPQ.$$

q.e.d.

Bibliography

Ahlfors, L. V., *Complex Analysis*, New York: McGraw Hill, 1953.
Bachmann, F., *Aufbau der Geometrie aus dem Spiegelungsbegrieff*, Berlin: Springer-Verlag, 1973
Baker, C. W., *Introduction to Topology*, Wm. C. Brown Publishers, Dubuque: 1991.
Beltrami, E., Saggio di interpretazione della geometria non-euclidea, *Gior. Mat.* 6 (1868), 284–312.
——— , Teoria fondamentale degli spazii di curvatura costante, *Annali di Mat.* ser II 2 (1868), 232–255.
Beardon, A., *The Geometry of Discrete Groups*, Springer-Verlag: New York 1983.
Bolyai, J., Appendix, *The Theory of Space, with Introduction, Comments, and Addenda* (F. Karteszi, editor), Amsterdam: North-Holland, 1987.
Bonola, R., *Non-Euclidean Geometry, a Critical and Historical Study of its Developments*, New York: Dover, 1955.
Buseman, H., and Kelly, P. J., *Projective Geometry and Projective Metrics*, New York: Academic Press, 1953.
Caratheodory C., *Theory of Functions of a Complex Variable*, 2 vols., New York: Chelsea 1954.
Cayley, A., Sixth memoir upon quintics, *Phil. Trans.* 149(1859) 61–91.
Chinn, W. G., and Steenrod, N. E., *First Concepts of Topology*, Random House, New York: 1966.
Coxeter, H. S. M., *Non-Euclidean Geometry*, 5th ed., Toronto: University of Toronto Press, 1968.
Eves, H., *A Survey of Geometry*, 2 vols., Boston: Allyn and Bacon, 1966.
Fenchel, W., *Elementary Geometry in Hyperbolic Space*, Walter de Gruyter: Berlin, 1989.
Gauss, K. F., *General Investigations of Curved Surfaces* (A. Hiltebeitel and J. Morehead, translators), New York: Raven Press, 1965.
Greenberg, M. J., *Euclidean and Non-Euclidean Geometries, Development and History*, 2d ed., San Francisco: Freeman, 1980.
Heath T.L., *The Thirteen Books of Euclid's Elements*, 2d. ed., 3 vols., New York: Dover 1956.
Hilbert, D., *The Foundations of Geometry*, Chicago: The Open Court Publishing Company, 1902.

Hilbert, D., *Grundlagen Der Geometrie*, 7th ed., B. G. Teubner, 1930.
Kells, L.M., *Plane and Spherical Trigonometry*, 2nd ed., New York: McGraw Hill 1940.
Klein, F., Ueber die sogennante Nicht-Euklidische Geometrie, Math. Ann. 4(1871), 573–625.
Lehner, J., *A Short Course in Automorphic Functions*, New York: Holt, Rinehart and Winston, 1966.
Liebman, H., *Nichteuklidische Geometrie*, J. G. Goschen: Berlin 1923.
Lobachevsky, N., *Geometric Researches on the Theory of Parallels* (George B. Halstead, translator), Chicago: Open Court Publishing Co., 1914.
Lyndon, R.C., *Groups and Geometry*, New York: Cambridge University Press 1985.
Magnus, W., *NonEuclidean Tesselations and their Groups*, New York: Academic Press 1974.
Martin, G.E., *Transformation Geometry, an Introduction to Symmetry*, New York: Springer-Verlag, 1982.
Maskit, B., *On Poincaré's Theorem for Fundamental Polygons*, Advances in Math. 7(1971), 219–230.
Massey, W. S., *Algebraic Topology: An Introduction*, New York: Harcourt Brace & World, 1967.
Millman, R. S., and Parker, G. D., *Elements of Differential Geometry*, Englewood Cliffs: Prentice Hall, 1977.
Milnor, J., Hyperbolic geometry: The first 150 years, *Bull. Amer. Math. Society*, 6(1982), 9–24.
Moise, E. E., *Elementary Geometry from an Advanced Standpoint*, Addison Wesley, 1974.
Monge, G., *Application de L'Analyse a la Geometrie*, (J. Liouville, editor), 5th edition, Paris: Bachelier 1850.
O'Neill, B., *Elementary Differential Geometry*, New York: Academic Press, 1966.
Pasch, Moritz, *Vorlesungen uber Neuere Geometrie*, Leipzig: Teubner, 1882.
Pedoe, D., *A Course of Geometry*, New York: Cambridge University Press 1970.
Poincaré, H., *Theory of Fuchsian Groups, Papers on Fuchsians Functions* (John Stillwell, translator), New York: Springer-Verlag, 1985.
Poincaré, H., *Science and Method* (Francis Maitland, translator), New York: Dover, 1952.
Proclus, *Commentary on the first book of Euclid* (Glenn R. Morrow, translator), Princeton: Princeton University Press, 1970.
Riemann, B., Ueber die Hypothesen welche der Geometrie zu Grunde Liegen, *Abh. K. G. Wiss. Gottingen* 13 (from his Inaugural Address of 1854).
Rosenfeld, B.A., *A History of Non-Euclidean Geometry, Evolution of the Concept of a Geometric Space* (Abe Shenitzer, translator), New York: Springer-Verlag, 1988.
Schattschneider, D., *The Open Book: The Notebooks & Periodic Drawings of M. C. Escher: The Notebooks & Symmetry Drawings of M. C. Escher*, San Francisco: Freeman, 1990.
Schwarz, H.A., Uber diejenigen Falle, in Welchem die Gaussische hypergeometrische Reiher eine algebraische Function ihres vierten Elementes darstellt, *J. f. reine und angew. Math.* 75(1872), 292–335.
Schwerdtfeger, H., *Geometry of Complex Numbers*, New York: Dover, 1979.
Smogorzhevsky, S.A., *Lobachevskian Geometry* (V. Kisin, translator), Moscow: Mir Publishers, 1976.
Wallace, E.C. and West, S., *Roads to Geometry*, Englewood Cliffs: Prentice Hall, 1992.

Index

A

AAA hyperbolic congruence, 103
Abel, N., 252
Absolute geometry, 20, 28, 48, 49, 88, 161–166
Almagest, 167
Altitude, 32, 129
Angle, 4
    Beltrami–Klein, 237
    bisection, 15
    exterior, 16
    hyperbolic, 70, 71
    oriented, 38
    rectilineal, 4
    right, 5, 8, 15, 83
    spherical, 171
    vertical, 16
Antipode, 175
Application, 9, 14
Arclength, 190
Area
    Beltrami–Klein, 245
    element, 193
    Euclidean, 8
    Euclidean triangle, 22
    hyperbolic, 110–118
    parallelogram, 9, 22
    Riemann metric, 203
    spherical, 175
    surface, 193
    triangle, 22
Argument, 132
ASA congruence, 18–19
Axiom of Archimedes, 27, 31

B

Beltrami, E., 250
Beltrami–Klein
    length, 236
    model, 233–246
    perpendicular, 239
    reflection, 240
    rigid motion, 240–242
Bisection, 15, 70
Bolyai, F., 250
Bolyai, J., 119, 167, 250, 257, 275
Bowed geodesic, 69

C

Cayley, A., 250
Ceva, 32, 129, 181
Chasles, M., 35
Cocyclic points, 210
Common notions, 9–10, 111–112
Complex numbers, 131–160, 207–232, 252–253
Composition, 36
Conjugate, 134
Coxeter, H.S.M., 229
Cross product, 172
Cross ratio, 209–213, 227, 242–243
Curvature, 183–205

D

Defect, 114, 162
Definitions (Euclid's), 3
Democritus, 2
Developable, 186
Dot product, 172

E

Elements, The, 2, 247, 277
Elliptic function, 252
Elliptic geometry, 179–181
Elliptic integral, 252
Escher, M., 229
Euclid, 2, 247
Euclidean
    congruence, 23, 26
    flow diagram, 143
Eudoxus, 2
Euler, L., 16, 168, 176

F

Fix, 36
Fixed point, 36
Fixed set, 36, 152
Flow line, 144
Foundations of Geometry, The, 23
Fuchsian functions, 254
Fuchsian groups, 254

G

Gauss, C. F., 73, 183, 196, 200, 249, 275
Gauss map, 184
Gaussian curvature, 189, 196
Geodesic triangle, 188, 204
Geodesic ray, 95
Geodesic segment, 67, 169, 188
Geometry, 258
    absolute, 20, 27–33, 48–49, 90, 161–166
    differential, 183–205
    Euclidean, 3
    hyperbolic, 88, 251
    neutral, 20
    Riemannian, 73–78, 198–205, 225–226
    spherical, 167–181
Girard, A., 176
Glide–reflection, 44
Godel, K., 251
Great circle, 169
Group, 35

H

Half-plane, 64
Half-space, 258
Harriot, T., 176
Hilbert, D., 1, 23, 35, 250
Hilbert's axioms, 22–27
Horocycle, 275
Horosphere, 258, 273, 275

Hyperbolic
    congruence, 103
    flow diagram, 145–155, 213–221
    geodesic, 69, 262
    geometry, 88, 251
    Law of Cosines, 125
    Law of Sines, 125
    length, 64, 259
    reflection, 89
    regular polygon, 105–106
    right triangle, 98, 122, 124
    rigid motion, 71–73, 88–90, 136–160, 213–225, 260–261, 276
    rotation, 89, 222–223, 260–261
    space, 258
    straight line, 69
    tesselation, 104–107, 228–230
    Theorem of Pythagoras, 122–124
    translation, 146, 222
    triangle, 93–103, 114–116, 125–128
    trigonometry, 119–129
    volume, 276

I

Identity, 36
Infinity, 4
Inner product, 172
Inverse, 38, 47, 143
Inversion, 53, 62, 71, 276
Involution, 39
Isometry, 268

J

Jacobi, C., 252

K

Klein, F., 35, 233, 250–251
Klugel, G. S., 248

L

Lagrange, J. L., 173, 183
Lambert, J. H., 248–249
Law of Cosines, 125
Law of Sines, 125
Legendre, A. M., 249
Line, 3–4, 23
Liouville, J., 250
Lobachevsky, N., 167, 174, 190, 249, 257, 275

M

Maple, 50, 62, 78, 108

Mathematica xiii, 50, 62, 78, 108
Measure, 8
Median, 32, 129, 181
Menelaus, 32, 129, 181
Metric
    Beltrami-Klein, 245
    Euclidean, 74
    Poincaré, 74, 259
    Polaris, 199
    Riemann, 74–76
    unit disk, 225
Minding, 189, 250
Modulus, 132
Moebius rigid motion, 146
Moebius transformation, 142, 208

O

On the Principles of Geometry, 249
Order, 24
Orthogonal, 58

P

Parallel lines, 6, 19–20, 25
Parallel Postulate, 25, 27, 166,
Pasch's axiom, 25, 31
Periodicity, 252
Perpendicular, 5, 15
Perpendicular bisector, 32, 91
Perspectivity, 240
Pieri, M., 35
Plane, 23
Playfair's Postulate, 25, 28, 84, 247
Poincaré upper half-plane, 63, 88
Poincaré, H., 253–254
Point, 3, 23
Polaris, 198–199
Postulates, 1–4 7–8, 79–84
Postulate, 5 8, 20, 84, 166
Proclus, 247
Projective model, 233
Proportion, 22
Pseudosphere, 190
Ptolemy, 167, 247
Punctured sphere, 267

R

Rectangle, 164
Rectilineal, 4–5
Reflection, 39, 135
Riemann, G. F. B., 73, 253
Riemannian curvature, 201

Right triangle, 21, 124
Rigid motion, 35–50, 132–136
Rotation, 38, 134

S

Saccheri, G., 248
SAS congruence, 13–14
Schwarz, H. A., 229
Similar triangles, 22
Sophists, 2
Sphere, 168–181, 258, 266–273
Spherical
    Law of Cosines, 173, 181
    Law of Sines, 173, 181
    right triangle, 175
    Theorem of Pythagoras, 175
    triangle, 172
    trigonometry, 167–181
SSS congruence, 15
Standard position, 93
Stereographic projection, 267
Straight line, 4
Straight geodesic, 69
Sum of angles
    absolute triangle, 16, 161–166
    Euclidean triangle, 21, 28
    hyperbolic triangle, 99
Surface, 4, 183–205
Symmetry, 50, 52

T

Tesselation, 108
Thales, 1
Theorem of Pythagoras, 21, 124
Topology, 76
Total curvature, 185
Transformation, 36
Translation, 37, 132

V

Vertical semicircle, 262

$\arg(z)$, 132
$\beta(X)$, 236
$\bar{z}$, 134
$(z_1,\ z_2,\ z_3,\ z_4)$, 209
$d(Q,\ P)$, 36
D_κ, 148
γ_{AB}, 44
$h(AB)$, 234

$h(P, Q)$, 69
$ha(R)$, 111
H^3, 259
Id, 36
I_w, 223
I_p, 80
$I_{C,k}$, 53
$j(P, Q)$, 214
$|z|$, 132

$M_{\alpha,\gamma}$, 151
$P_M(X)$, 240
$R_{C,\alpha}$, 38
ρ_m, 39
S^+, 269
$T_{\alpha,\beta,\gamma}$, 150
$U(z)$, 213
$V(z)$, 214